Lecture Notes in Mathematics

Volume 2269

This series reports on new developments in all areas of mathematics and their applications - quickly, informally and at a high level. Mathematical texts analysing new developments in modelling and numerical simulation are welcome. The type of material considered for publication includes:

1. Research monographs 2. Lectures on a new field or presentations of a new angle in a classical field 3. Summer schools and intensive courses on topics of current research.

Texts which are out of print but still in demand may also be considered if they fall within these categories. The timeliness of a manuscript is sometimes more important than its form, which may be preliminary or tentative.

More information about this series at http://www.springer.com/series/304

Ulrich Bunke • Alexander Engel

Homotopy Theory with Bornological Coarse Spaces

 Springer

Ulrich Bunke
Faculty of Mathematics
University of Regensburg
Regensburg, Germany

Alexander Engel
Faculty of Mathematics
University of Regensburg
Regensburg, Germany

ISSN 0075-8434   ISSN 1617-9692   (electronic)
Lecture Notes in Mathematics
ISBN 978-3-030-51334-4   ISBN 978-3-030-51335-1   (eBook)
https://doi.org/10.1007/978-3-030-51335-1

Mathematics Subject Classification: 19K99, 51F99

Contents

Chapter 1
Introduction

Coarse geometry was invented by J. Roe (see e.g. [Roe93a, Roe03]) in connection with applications to the index theory of Dirac-type operators on complete Riemannian manifolds. To capture the index of these operators K-theoretically or numerically J. Roe furthermore constructed first examples of coarse homology theories, namely coarse K-homology, coarse ordinary homology, and coarsifications of locally finite homology theories.

The symbol class of a Dirac-type operator is a locally finite K-homology class. The transition from the symbol class to the coarse index of the Dirac operator proceeds in two steps. The first step sends the symbol class to the coarse symbol class in the coarsification of the locally finite K-homology. This step removes the local information and builds a coarsely invariant object which is still of topological nature. The second step consists of an application of the coarse assembly map [Roe03, Sec. 5]. The resulting coarse index captures global analytic properties of the Dirac operator, in particular its invertibility. The K-theoretic coarse Baum–Connes conjecture [HR95] asks under which conditions this assembly map is an isomorphism. We added the adjective "K-theoretic", since we now understand this assembly map as a specialization of a more general construction [BEb].

Information about the K-theoretic coarse assembly map has many implications to index theory, geometry and topology. In good situations surjectivity implies that some coarse K-theory classes can be realized as indices of generalized Dirac operators. Because the analysis of Dirac operators is well-developed, this allows conclusions about properties of coarse K-theory classes, e.g. vanishing of delocalized traces. On the other hand, injectivity has applications to the positive scalar curvature question. Indeed, uniform positive scalar curvature implies that the spin Dirac operator is invertible so that its index vanishes. One can then conclude that its symbol class vanishes at least coarsely, and this has consequences for the topology of the manifold. Finally, if the K-theoretic coarse assembly map for the group G with the word metric is an isomorphism, then this implies via the

© The Editor(s) (if applicable) and The Author(s), under exclusive licence
to Springer Nature Switzerland AG 2020
U. Bunke, A. Engel, *Homotopy Theory with Bornological Coarse Spaces*,
Lecture Notes in Mathematics 2269, https://doi.org/10.1007/978-3-030-51335-1_1

descent principle the Novikov conjecture which has implications to the topology of manifolds.

The initial motivation for this book was to uncover the basic structure of the proof of the K-theoretic coarse Baum–Connes conjecture [Wri05]. Our main new insight is the following. The proof of the coarse Baum–Connes conjecture [Wri05] under the assumption of finite asymptotic dimension is not specific to K-theory. The argument actually shows that a natural transformation between two coarse homology theories (if suitably axiomatized) is an isomorphism on a space of finite asymptotic dimension if it is an isomorphism on discrete spaces. This is analogous to the fact that a natural transformation between homology theories satisfying the Eilenberg–Steenrod axioms (except the dimension axiom) induces an isomorphism of their values on CW-complexes, if it induces an isomorphism on the one-point space. This idea can be applied to the K-theoretic Baum–Connes conjecture by interpreting the K-theoretic coarse assembly map as a natural transformation between coarse homology theories in the sense above. This argument will only be completed in [BEb] while the present book provides the construction of the homology theory appearing as the target of the assembly map.

The first goal of the present book is to propose a set of axioms for coarse homology theories which make this idea work. The basic category containing the objects of interest is the category of bornological coarse spaces **BornCoarse** (Definition 2.11). A bornological coarse space is a set equipped with a coarse structure and a compatible bornology. A morphism between bornological coarse spaces is a map which is controlled and proper. The idea of this definition is to work with the minimal amount of structure. Metric spaces present bornological coarse spaces, but the datum of a metric is a much finer structure and not preserved by isomorphisms in **BornCoarse**. Moreover there are important examples of bornological coarse spaces which can not be presented by metric spaces, e.g. the spaces with the hybrid structure occurring in the argument of [Wri05] or continuously controlled coarse structures.

In order to formulate the basic axioms for a coarse homology theory technically, we will use the language of ∞-categories [Cis19, Lur09]. This theory is probably one of the most rapidly accepted new theories of mathematics and by now indispensible in modern homotopy theory. We have tried to minimize the required technical knowledge about ∞-categories by using it in a model independent way. A reader unfamiliar with this language should consider this book as an opportunity to learn the language of ∞-categories in action. But it is not the place for explaining many internal technical details of the language.

A coarse homology theory (Definition 4.22) is a functor

$$E : \textbf{BornCoarse} \to \textbf{C}$$

whose target is some stable ∞-category. This functor must be coarsely invariant, excisive and u-continuous, and it must annihilate flasque bornological coarse spaces like the ray $[0, \infty)$.

On the one hand the choice of these axioms is motivated by the application explained above. On the other hand they are natural from the point of view of a homotopy theory of bornological coarse spaces which we will explain in the following.

On the morphism sets of **BornCoarse** we have the equivalence relation called closeness (Definition 3.13). Morphisms in **BornCoarse** which are invertible up to closeness are called coarse equivalences (Definition 3.14). In coarse homotopy theory one studies coarse spaces up to coarse equivalence. Furthermore, one considers flasque bornological coarse spaces (Definition 3.21) as trivial. Coarse homology theories are defined such that they provide homotopy theoretic invariants of bornological coarse spaces which in view of the excision axiom are in addition local in a certain sense.

For different set-ups and axioms for coarse homology theories see [Mit01, Hei].

As our goal is to show results for arbitrary coarse homology theories, it is natural to prove them for a universal coarse homology theory (Definition 4.3)

$$\text{Yo}^s : \textbf{BornCoarse} \to \textbf{Sp}\mathcal{X},\qquad\qquad(1.1)$$

where **Sp**$\mathcal{X}$ is called the category of coarse motivic spectra. For X in **BornCoarse** we call $\text{Yo}^s(X)$ in **Sp**$\mathcal{X}$ the coarse motivic spectrum of X. Using modern homotopy theory the construction of this universal coarse homology theory is completely standard. It starts with the Yoneda embedding

$$\text{yo} : \textbf{BornCoarse} \to \textbf{PSh}(\textbf{BornCoarse}).$$

In order to explain the right-hand side note that for a ∞-category $\mathcal{C}$ we write **PSh**$(\mathcal{C}) := \textbf{Fun}(\mathcal{C}^{\text{op}}, \textbf{Spc})$ for the ∞-category of **Spc**-valued presheaves. The ∞-category **Spc** is the ∞-category of spaces which e.g. contains the mapping spaces between objects in arbitrary ∞-categories. In general we view (without explicitly mentioning) ordinary categories as ∞-categories using the nerve functor.

In order to construct (1.1) one then applies to **PSh**(**BornCoarse**) various localizations forcing the properties which are dual to the axioms for a coarse homology theory. This yields the unstable version (Definition 3.34)

$$\text{Yo}: \textbf{BornCoarse} \to \textbf{Spc}\mathcal{X},$$

where **Spc**$\mathcal{X}$ is the category of coarse motivic spaces. The stable version (1.1) is then obtained by applying a stabilization functor.

We now formulate one of the main results of the present paper. We consider the minimal cocomplete stable full subcategory

$$\textbf{Sp}\mathcal{X}\langle\mathcal{A}_{\text{disc}}\rangle \subseteq \textbf{Sp}\mathcal{X}$$

of **Sp**$\mathcal{X}$ containing coarse motivic spectra of the elements of the set $\mathcal{A}_{\text{disc}}$ of discrete bornological coarse spaces. A discrete bornological coarse space has the simplest possible coarse structure on its underlying set, but its bornology might be interesting.

Theorem 1.1 (Theorem 5.59) *If X is a bornological coarse space of weakly finite asymptotic dimension (Definition 5.58), then $\mathrm{Yo}^s(X) \in \mathbf{Sp}\mathcal{X}\langle\mathcal{A}_{\mathrm{disc}}\rangle$.*

By the universal property of Yo^s a coarse homology theory with values in $\mathbf{C}$ is the same as a colimit preserving functor $\mathbf{Sp}\mathcal{X} \to \mathbf{C}$. The following corollary is now immediate. Let $E \to F$ be a transformation between $\mathbf{C}$-valued coarse homology theories, and let X be in **BornCoarse**.

Corollary 1.2 (Corollary 6.43) *Assume:*

1. *The induced map $E(Y) \to F(Y)$ is an equivalence for all discrete Y in* **BornCoarse**.
2. *X has weakly finite asymptotic dimension.*

   Then $E(X) \to F(X)$ is an equivalence.

Under more restrictive assumptions we also have the following result which is closer to the classical uniqueness results in ordinary algebraic topology. It requires an additional additivity assumption on the coarse homology theory.

Theorem 1.3 (Theorem 6.44) *Assume:*

1. *E, F are additive.*
2. *The induced map $E(*) \to F(*)$ is an equivalence.*
3. *X has weakly finite asymptotic dimension.*
4. *X has the minimal compatible bornology.*

   Then $E(X) \to F(X)$ is an equivalence.

The proof of Theorem 1.1 follows ideas of [Wri05] and uses the coarsening space. This is a simplicial complex build from an anti-Čech system of coverings of the bornological coarse space in question (see Sect. 5.6.1). The condition of weakly finite asymptotic dimension translates to finite-dimensionality of the coarsening space. We equip this coarsening space with the hybrid coarse structure (Definition 5.10) which interpolates between the coarse structure of the original bornological coarse space and the local metric geometry of the simplicial complex. We then show that the coarsening space with the hybrid structure is flasque (Theorem 5.55). As a consequence its motivic coarse spectrum vanishes, and the motivic coarse spectrum of the original space is equivalent up to suspension to the germs-at-∞ (see below) of the coarsening space. Using the Decomposition Theorem and the Homotopy Theorem also described below we can then decompose these germs-at-∞ into motivic coarse spectra of a collection of cells, and finally of discrete spaces (see Sect. 5.5.2). In this argument we argue by induction by the dimension and therefore the finiteness of the latter is important. Similar techniques (with different words) are also used elsewhere, e.g. in the proof of the Farell–Jones conjecture in [BLR08].

The technical details involved in these arguments are tricky. One goal of Chap. 5 is to develop the background in great generality and in a ready-to-use way. After introducing hybrid structures we provide the following two main results which we can only formulate roughly at this place. We think of the hybrid structure as a coarse structure constructed from a uniform structure. The first is the Homotopy Theorem.

Theorem 1.4 (Theorem 5.26) *Uniform homotopy equivalences between uniform spaces induce equivalences of the coarse motivic spectra of the spaces equipped with the associated hybrid structures.*

Note that the hybrid structure on a uniform space X also depends on an exhaustion $\mathcal{Y}$ of the uniform space. We consider the coarse motivic spectrum $\mathrm{Cofib}(\mathrm{Yo}^s(\mathcal{Y}) \to \mathrm{Yo}^s(X))$ as the motive of the germ-at-∞ of X relative to the exhaustion $\mathcal{Y}$. The second result is the Decomposition Theorem.

Theorem 1.5 (Theorem 5.22) *The germ-at-∞ is excisive for coarsely and uniformly excisive decompositions.*

These results will be used in subsequent papers [BEKW20, BEb, BKWa].

The next goal of this book is to construct various examples of coarse homology theories satisfying our axioms. In particular we will discuss the classical examples of coarse ordinary homology, coarse K-homology, and coarsifications of locally finite homology theories. In all cases we must extend the classical definitions to functors defined on the whole category **BornCoarse** with values in an appropriate stable ∞-category (e.g. chain complexes $\mathbf{Ch}_\infty$ for ordinary coarse homology or spectra $\mathbf{Sp}$ for coarse K-homology).

Extending and lifting the construction of coarse ordinary homology is essentially a straightforward matter. The results of Sect. 6.3 can be summarized as follows:

Theorem 1.6 *There exists a coarse homology theory* $H\mathcal{X} : \mathbf{BornCoarse} \to \mathbf{Ch}_\infty$ *such that* $\pi_* H\mathcal{X} : \mathbf{BornCoarse} \to \mathbf{Ab}^{\mathbb{Z}\mathrm{gr}}$ *is the classical ordinary coarse homology.*

In Chap. 7 we provide a systematic study of locally finite homology theories and their coarsifications. The part on locally finite homology theories could be understood as a generalization of [WW95]. A locally finite homology theory (Definition 7.27) is a functor

$$F : \mathbf{TopBorn} \to \mathbf{C},$$

where **TopBorn** is the category of topological bornological spaces (i.e., topological spaces with an additional compatible bornology, see Definition 7.2), and $\mathbf{C}$ is a complete and cocomplete stable ∞-category. Besides the usual homotopy invariance and excision properties we require the local finiteness condition:

$$\lim_{B} F(X \setminus B) \simeq 0,$$

where the limit runs over the bounded subsets of X.

We show that every object C of $\mathbf{C}$ gives rise to a locally finite homology theory

$$(C \wedge \Sigma_+^{\infty,\text{top}})^{\text{lf}} : \mathbf{BornCoarse} \to \mathbf{C}.$$

Another example of completely different origin is the $\mathbf{Sp}$-valued analytic locally finite K-homology $K^{\text{an,lf}}$ which is constructed using Kasparov-KK-theory. We have $K^{\text{an,lf}}(*) \simeq KU$ and it is an interesting question to compare $K^{\text{an,lf}}(*)$ with its homotopy theoretic version $(KU \wedge \Sigma_+^{\infty,\text{top}})^{\text{lf}}$.

Motivated by this question we provide a general classification result for locally finite homology theories which we again only formulate in a rough way at this point.

Proposition 1.7 (Proposition 7.43) *The values of a locally finite homology theory on nice spaces are completely determined by the value of the theory on the point.*

Since locally finite simplicial complexes are nice we can conclude:

Corollary 1.8 (Corollary 7.67) $(KU \wedge \Sigma_+^{\infty,\text{top}})^{\text{lf}}$ *and* $K^{\text{an,lf}}$ *coincide on topological bornological spaces which are homotopy equivalent to locally finite simplicial complexes.*

We think that this result is folklore, but we could not locate a proof in the literature.

In the present book our main usage of locally finite homology theories is as an input for the coarsification machine. The main result of Sect. 7.2 can be formulated as follows.

Theorem 1.9 *To every locally finite homology theory F we can associate functorially a coarse homology theory QF such that $QF(*) \simeq F(*)$.*

The coarse homology QF is called the coarsification of F. The coarsification of the analytic locally finite homology theory $QK^{\text{an,lf}}$ is of particular importance since it is the domain of the K-theoretic coarse assembly map which we explain in Sect. 8.10.

The last example of a coarse homology theory considered in the present book is the coarse topological K-homology. In contrast to the construction of the coarse ordinary homology theory, lifting the classical definition of coarse K-homology from a group-valued functor defined on proper metric spaces to a spectrum-valued functor defined on $\mathbf{BornCoarse}$ is highly non-trivial. The whole Chap. 8 is devoted to this problem. Its short summary can be stated as follows (Theorems 8.79 and 8.88):

Theorem 1.10 *There exists a coarse homology theory $K\mathcal{X} : \mathbf{BornCoarse} \to \mathbf{Sp}$ such that $\pi_* K\mathcal{X}(X)$ is the classical coarse K-homology for proper metric spaces X.*

The construction of the functor $K\mathcal{X}$ proceeds in two steps. In the first one associates to a bornological coarse space X a Roe category $\mathbf{C}^*(X)$ (Definition 8.74). This is a C^*-category which generalizes the classical construction of a Roe algebra. The functor $K\mathcal{X}$ is then obtained from $\mathbf{C}^*$ by composing with a topological K-theory functor for C^*-categories (Definition 8.56). In Chap. 8 will also give a selfcontained presentation of the necessary background on C^*-categories and their K-theory.

In the following we indicate why the construction of $K\mathcal{X}$ required some new ideas. The classical construction presents $K\mathcal{X}_*$ in terms of the K-theory groups of Roe algebras. Note that the construction of a Roe algebra (Definition 8.34) involves the choice of an X-controlled Hilbert space and is therefore not strictly functorial in the space X. The idea behind using the Roe category is to ensure functoriality by working with the C^*-category of all possible choices. Unfortunately, the details are more complicated. For example, in order to get the correct K-theory the X-controlled Hilbert space used in the construction of the Roe algebra must be ample (Definition 8.2). But ample Hilbert spaces only exist under additional assumptions on X (Proposition 8.20).

The morphism spaces in $\mathbf{C}^*(X)$ are defined as closures of spaces of operators of controlled propagation. There is a slightly weaker condition called quasi-locality which leads to a slightly bigger C^*-category $\mathbf{C}^*_{\mathrm{ql}}(X)$. By composing the functor $\mathbf{C}^*_{\mathrm{ql}}$ with the topological K-theory functor for C^*-categories we get a another coarse K-theory functor $K\mathcal{X}_{\mathrm{ql}}$. The inclusions $\mathbf{C}^*(X) \to \mathbf{C}^*_{\mathrm{ql}}(X)$ induce a natural transformation of coarse homology theories $K\mathcal{X} \to K\mathcal{X}_{\mathrm{ql}}$ which is obviously an equivalence on discrete bornological coarse spaces. So Corollary 1.2 has the following application. Let X be in **BornCoarse**.

Corollary 1.11 *If X has weakly finite asymptotic dimension, then $K\mathcal{X}(X) \to K\mathcal{X}_{\mathrm{ql}}(X)$ is an equivalence.*

After the preprint version of this book appeared, Špakula–Tikuisis [ŠT19] have shown that Corollary 1.11 is true under the weaker condition that X has finite straight decomposition complexity. Later, Špakula–Zhang generalized the result further to exact spaces [ŠZ20].

In order to apply Corollary 1.2 or Theorem 1.3 to a coarse assembly map it is necessary to interpret this coarse assembly map as a natural transformation between coarse homology theories in the sense above. In the present book this goal is only reached partially and only in the case of coarse K-theory, see Sect. 8.10.

In the following we explain some details. A coarse homology theory $E :$ **BornCoarse** $\to \mathbf{C}$ gives rise to the object $E(*)$ in $\mathbf{C}$. As explained above the object $E(*)$ in $\mathbf{C}$ gives rise to a locally finite homology theory

$$(E(*) \wedge \Sigma_+^{\infty,\mathrm{top}})^{\mathrm{lf}} : \mathbf{TopBorn} \to \mathbf{C}.$$

We can now form its coarsification

$$Q(E(*) \wedge \Sigma_+^{\infty,\mathrm{top}})^{\mathrm{lf}} : \mathbf{BornCoarse} \to \mathbf{C}. \tag{1.2}$$

Our original intention was to construct the coarse assembly map as a natural transformation

$$Q(E(*) \wedge \Sigma_+^{\infty,\mathrm{top}})^{\mathrm{lf}} \to E$$

since this was suggested by the classical example of coarse K-homology which
we discuss in Sect. 8.10. The true nature of the coarse assembly map as a natural
transformation between coarse homology theories was only uncovered in the follow-
up paper [BEb]. The main new insight explained in that paper is that (1.2) is not the
correct domain of the coarse assembly map. The correct domain is described in
[BEb]. In that paper we will study these domains called local homology theories
and the coarse assembly map in a motivic way.

But fortunately, as also shown in [BEb], the correct domain of the coarse
assembly map for E and $Q(E(*) \wedge \Sigma_+^{\infty, \mathrm{top}})^{\mathrm{lf}}$ are equivalent on nice spaces like
finite-dimensional simplicial complexes with their natural path metric or complete
Riemannian manifolds. Using the framework developed in this book we will show
in [BEb] a variety of isomorphism results for the coarse assembly map in general,
most of which were previously only known for the K-theoretic version.

After the first version of this book was written it turned out that the setup
developed here can be applied to other interesting problems. As in the case of the
proof of the K-theoretic coarse Baum–Connes conjecture [Wri05] it turned out that
the proof of the Novikov conjecture for algebraic K-theory in [GTY12, RTY14]
for groups whose underlying metric space has finite-decomposition complexity is
actually not specific to algebraic K-theory. In the following we will give a very
rough overview of this development. For a group G we consider the orbit category
$G\mathbf{Orb}$ of transitive G-sets and equivariant maps. For a functor $M : G\mathbf{Orb} \to \mathbf{C}$
and a family of subgroups $\mathcal{F}$ we then consider the assembly map

$$\mathrm{Ass}_{M, \mathcal{F}} : \operatorname*{colim}_{G_\mathcal{F}\mathbf{Orb}} M \to M(*), \tag{1.3}$$

where $G_\mathcal{F}\mathbf{Orb}$ is the full subcategory of orbits with stabilizers in $\mathcal{F}$, and $*$ is the
one-point orbit. The basic question is now under which conditions on G, $\mathcal{F}$ and M
this assembly map is an equivalence or at least split injective.

If M is the equivariant topological K-theory $K^{\mathrm{top}, G}$, and $\mathcal{F}$ is the family of finite
subgroups, then this assembly map is the homotopy theoretic version[1] of the Baum–
Connes assembly map. If M is the algebraic K-theory $K^{\mathrm{alg}}\mathbf{A}^G$ associated to an
additive category $\mathbf{A}$ with G-action and $\mathcal{F}$ is the family of virtually cyclic subgroups,
then this assembly map is the one appearing in the Farell–Jones conjecture. We refer
to [DL98] for the construction of the examples of the functors M mentioned above.
The main new development based on this book is the observation, that the proof
from [GTY12, RTY14] for algebraic K-theory only depends on the fact that the
functor $K^{\mathrm{alg}}\mathbf{A}^G$ extends to an equivariant coarse homology theory with some very
natural additional properties. The precise formulation of this condition is condensed
in the notion of a CP-functor introduced in [BEKWa].

The whole set-up introduce in the present book has a natural G-equivariant
generalization which is developed in [BEKW20]. The proof of the split injectivity

---

[1] In [HP04] it is stated that it coincides with the analytic definitions using KK-theory, but some
details of the proof must still be worked out.

for CP-functors is based on the descent principle whose details are presented in [BEKWa]. In this paper we also compare the assembly map (1.3) with an equivariant versions of the coarse assembly map. In order to apply the descent principle one must know that this equivariant generalization of the coarse assembly map is an equivalence. In [BEKW19] we show that this is the case even on a motivic level under the geometric assumption that the group has finite decomposition complexity.

In [BEKW20] we axiomatized equivariant coarse homology theories in complete analogy to the non-equivariant case considered in the present book. One of the additional structures on a coarse homology theory needed in order to verify the CP-condition are transfers. These transfers are discussed in [BEKWb].

As explained above, in order to show injectivity of the assembly map for a given functor on the orbit category we must extend it to an equivariant coarse homology theory. This motivated the construction of further examples of equivariant coarse homology theories:

1. equivariant coarse ordinary homology [BEKW20].
2. equivariant coarse algebraic K-homology associated to an additive category with G-action [BEKW20].
3. equivariant coarse Waldhausen K-homology of spaces [BKWb].
4. equivariant coarse K-homology associated to a left-exact ∞-category with G-action [BCKW].
5. topological equivariant coarse K-homology associated to a C^*-category with G-action [BEe].

Eventually, for all of these cases we can show injectivity results for the corresponding assembly map.

The Farell–Jones conjecture asserts that the assembly map (1.3) for the case $M = K^{\mathrm{alg}}\mathbf{A}$ and the family of virtually cyclic subgroups is an equivalence. Again it turned out that the proofs (under additional geometric assumptions on the group) given e.g. in [BLR08, Bar16] are not specific to algebraic K-theory. In [BKWa] we will show that the method extends to the K-theory functor for left-exact ∞-categories with G-action. This case subsumes all previously known cases of the Farell–Jones conjecture and provides a variety of new examples. The argument again depends on an extension of the functor on the orbit category to an equivariant coarse homology theory satisfying our axiomatics. This extension actually differs from the extension used for the injectivity results.

At the end of this overview we mention some further developments based on the foundations provided by this book. In [BEc] we introduce axioms for coarse cohomology theories by dualizing the axioms for coarse homology theories. Examples of coarse cohomology theories can be constructed by dualizing coarse homology theories. As an application we settle a conjecture of Klein [Kle01, Conj. on Page 455] that the dualizing spectrum of a group is a quasi-isometry invariant. We further discuss the pairing between coarse K-homology and coarse K-cohomology.

In [Cap] equivariant coarse cyclic and Hochschild homology was defined. These homology theories are related by a trace map with equivariant coarse algebraic K-

homology. In [BC20] a universal equivariant coarse algebraic K-theory functor with values in the motives of [BGT13] was constructed.

Coarse fixed point spaces and localization results for equivariant coarse homology theories were studied in [BCb]. In [BCa] we showed that coarse algebraic K-theory of additive categories is actually a lax symmetric monoidal functor. A similar fact for left-exact ∞-categories is a crucial ingredient of [BKWa].

Aspects of index theory using (equivariant) coarse homology theories were considered in [BEd, Buna].

The present book lays the foundations for all these development. It provides a reference for the technical material in Chaps. 5, 6, 7, and 8 which will be used in the subsequent papers. These papers employ with references the basic notions and constructions provided in Chaps. 2, 3, and 4.

This book develops the basics of coarse geometry in the framework of the category **BornCoarse** from scratch. This part should be accessible to students on the Bachelor level and could serve as a condensed introduction to the basic notions and basic examples of the field.

As mentioned above the discussion of coarse homology theories and the motivic approach requires some basic experience with the language of ∞-categories, say on the level of one-semester user-oriented introduction. The combination of Chaps. 2, 3, and 4 is a comprehensive introduction to coarse homotopy theory using the modern language of ∞-categories. This part could also be seen as a demonstration of the motivic approach which is completely analogous to the one in algebraic geometry or differential cohomology theory. The discussion of locally finite homology theories and their coarsification using the language of ∞-categories is new. Adopting the level reached in the preceeding sections does not require any additional ∞-categorial prerequisites.

In order to understand the discussion of coarse K-homology one needs background knowledge about C^*-algebras and their K-theory. But we will explain the necessary background about C^*-categories in detail.

The book served as the basis for PhD-level courses on coarse homotopy theory, coarse K-homology and assembly maps.

Acknowledgments The authors were supported by the SFB 1085 "Higher Invariants" funded by the Deutsche Forschungsgemeinschaft DFG. The second named author was also supported by the Research Fellowship EN 1163/1-1 "Mapping Analysis to Homology" of the Deutsche Forschungsgemeinschaft DFG.

Parts of the material in the present paper were presented in the course "Coarse Geometry" held in the summer term 2016 at the University of Regensburg. We thank the participants for their valuable input.

We thank Clara Löh for her helpful comments on a first draft of this paper. Furthermore we thank Daniel Kasprowski and Christoph Winges for suggesting various improvements, and Rufus Willett for helpful discussions.

Part I
Motivic Coarse Spaces and Spectra

Chapter 2
Bornological Coarse Spaces

In this chapter we introduce from scratch the category **BornCoarse** of bornological coarse spaces. We further provide basic examples for objects and morphisms of **BornCoarse** and finally we discuss some categorical properties of **BornCoarse**.

Remark 2.1 In this remark we fix the set-theoretic size issues. We choose a sequence of three Grothendieck universes whose elements are called very small, small and large sets.

The objects of the categories of geometric objects like **Set**, **Top**, **BornCoarse** considered below belong to the very small universe, while these categories themselves are small.

If not said differently, categories or ∞-categories will assumed to be small, and cocompleteness or completeness refers to the existence of all colimits or limits indexed by very small categories.

By this convention **Spc** and **Sp** are the small ∞-categories of very small spaces and spectra.

But we will also consider the large ∞-categories $\mathbf{Spc}^{\mathrm{la}}$ and $\mathbf{Sp}^{\mathrm{la}}$ of small spaces or small spectra. Furthermore, the categories of motivic coarse spaces $\mathbf{Spc}\mathcal{X}$ and $\mathbf{Sp}\mathcal{X}$ constructed below are large. □

2.1 Basic Definitions

In this section sets are very small sets.

Let X be a set, and let $\mathcal{P}(X)$ denote the set of all subsets of X.

Definition 2.2 A bornology on X is a subset $\mathcal{B}$ of $\mathcal{P}(X)$ which is closed under taking finite unions and subsets, and such that $\bigcup_{B \in \mathcal{B}} B = X$.

The elements of $\mathcal{B}$ are called the bounded subsets of X.

© The Editor(s) (if applicable) and The Author(s), under exclusive licence
to Springer Nature Switzerland AG 2020
U. Bunke, A. Engel, *Homotopy Theory with Bornological Coarse Spaces*,
Lecture Notes in Mathematics 2269, https://doi.org/10.1007/978-3-030-51335-1_2

The main role of bounded subsets in the present paper is to declare certain regions of X to be small in order to define further finiteness conditions. Let D be a subset of a space X with a bornology $\mathcal{B}$.

Definition 2.3 D is locally finite (resp., locally countable) if for every B in $\mathcal{B}$ the intersection $B \cap D$ is finite (resp., countable).

For a subset U of $X \times X$ we define the inverse by

$$U^{-1} := \{(x, y) \in X \times X : (y, x) \in U\}.$$

If U' is a second subset of $X \times X$, then we define their composition by

$$U \circ U' := \{(x, y) \in X \times X : (\exists z \in X : (x, z) \in U \text{ and } (z, y) \in U')\}.$$

Let X be a set.

Definition 2.4 A coarse structure on X is a subset $\mathcal{C}$ of $\mathcal{P}(X \times X)$ which is closed under taking finite unions, composition, inverses and subsets, and which contains the diagonal of X.

Remark 2.5 Coarse structures are not always required to contain the diagonal, especially in early publications on this topic. In these days coarse structures containing the diagonal were called unital. See Higson–Pedersen–Roe [HPR96, End of Sec. 2] for an example of a naturally occurring non-unital coarse structure. We are using here the nowadays common convention to require the coarse structures to contain the diagonal and correspondingly we drop the adjective 'unital'.   □

The elements of $\mathcal{C}$ will be called entourages of X or controlled subsets.

For a subset U of $X \times X$ and a subset B of X we define the U-thickening of B by

$$U[B] := \{x \in X : (\exists b \in B : (x, b) \in U)\}.$$

If X is equipped with a coarse structure $\mathcal{C}$, then we call the subsets $U[B]$ for U in $\mathcal{C}$ the controlled thickenings of B.

Definition 2.6 A bornology and a coarse structure are said to be compatible if every controlled thickening of a bounded subset is again bounded.

Definition 2.7 A bornological coarse space is a triple $(X, \mathcal{C}, \mathcal{B})$ of a set X with a coarse structure $\mathcal{C}$ and a bornology $\mathcal{B}$ such that the bornology and coarse structure are compatible.

We now consider a map $f : X \to X'$ between sets equipped with bornological structures $\mathcal{B}$ and $\mathcal{B}'$, respectively.

Definition 2.8

1. We say that f is proper if for every B' in $\mathcal{B}'$ we have $f^{-1}(B') \in \mathcal{B}$.
2. The map f is bornological if for every B in $\mathcal{B}$ we have $f(B) \in \mathcal{B}'$.

Assume that $f : X \to X'$ is a map between sets equipped with coarse structures $\mathcal{C}$ and $\mathcal{C}'$, respectively.

Definition 2.9 We say that f is controlled if for every U in $\mathcal{C}$ we have $(f \times f)(U) \in \mathcal{C}'$.

We now consider two bornological coarse spaces X and X'.

Definition 2.10 A morphism between bornological coarse spaces $f : X \to X'$ is a map between the underlying sets $f : X \to X'$ which is controlled and proper.

Definition 2.11 We let **BornCoarse** denote the small category of very small bornological coarse spaces and morphisms.

2.2 Examples

Example 2.12 Let X be a set and consider a subset A of $\mathcal{P}(X \times X)$. Then we can consider the minimal coarse structure $\mathcal{C}\langle A \rangle$ containing A. We say that $\mathcal{C}\langle A \rangle$ is generated by A.

Similarly, let W be a subset of $\mathcal{P}(X)$. Then we can consider the minimal bornology $\mathcal{B}\langle W \rangle$ containing W. Again we say that $\mathcal{B}\langle W \rangle$ is generated by W.

If for every generating entourage U in A and every generating bounded subset B in W the subsets $U[B]$ and $U^{-1}[B]$ are contained in finite unions of members of W, then the bornology $\mathcal{B}\langle W \rangle$ and the coarse structure $\mathcal{C}\langle A \rangle$ are compatible. $\quad\square$

Example 2.13 Every set X has a minimal coarse structure $\mathcal{C}_{min} := \mathcal{C}\langle \emptyset \rangle$ generated by the empty set. This coarse structure consists of all subsets of the diagonal $\mathrm{diag}_X \subseteq X \times X$ and is compatible with every bornological structure on X. In particular, it is compatible with the minimal bornological structure $\mathcal{B}_{min}$ which consists of the finite subsets of X.

The maximal coarse structure on X is $\mathcal{C}_{max} := \mathcal{P}(X \times X)$. The only compatible bornological structure is then the maximal one $\mathcal{B}_{max} := \mathcal{P}(X)$. $\quad\square$

Definition 2.14 We call a bornological coarse space $(X, \mathcal{C}, \mathcal{B})$ discrete if $\mathcal{C} = \mathcal{C}_{min}$.

Note that a discrete bornological coarse space has a maximal entourage diag_X.
We consider a coarse space $(X, \mathcal{C})$, an entourage U in $\mathcal{C}$, and a subset B of X.

Definition 2.15 The subset B called is U-bounded if $B \times B \subseteq U$.

If $(X, \mathcal{C}, \mathcal{B})$ is a bornological coarse space, then the bornology $\mathcal{B}$ necessarily contains all subsets which are U-bounded for some U in $\mathcal{C}$.

Example 2.16 Let $(X, \mathcal{C})$ be a coarse space. Then there exists a minimal compatible bornology $\mathcal{B}$. It is the bornology consisting of finite unions of the subsets of X which are bounded for some entourage of X. We also say that $\mathcal{B}$ is generated by $\mathcal{C}$. □

Example 2.17 Let $(X', \mathcal{C}', \mathcal{B}')$ be a bornological coarse space, and let $f : X \to X'$ be a map of sets. We define the induced bornological coarse structure by

$$f^*\mathcal{C} := \mathcal{C}\langle\{(f \times f)^{-1}(U') : U' \in \mathcal{C}'\}\rangle$$

and

$$f^*\mathcal{B} := \mathcal{B}\langle\{f^{-1}(B') : B' \in \mathcal{B}'\}\rangle \,.$$

Then

$$f : (X, f^*\mathcal{C}', f^*\mathcal{B}') \to (X', \mathcal{C}', \mathcal{B}') \,.$$

is a morphism of bornological coarse spaces. In particular, if f is the inclusion of a subset, then we say that this morphism is the inclusion of a bornological coarse subspace. □

Example 2.18 Let (X, d) be a metric space. In the present book we will allow infinite distances. Such metrics are sometimes called quasi-metrics. For every positive real number r we define the entourage

$$U_r := \{(x, y) \in X \times X : d(x, y) < r\}. \tag{2.1}$$

The collection of these entourages generates the coarse structure

$$\mathcal{C}_d := \mathcal{C}\langle\{U_r : r \in (0, \infty)\}\rangle \tag{2.2}$$

which will be called be the coarse structure induced by the metric.

We further define the bornology of metrically bounded sets

$$\mathcal{B}_d := \mathcal{B}\langle\{B_d(r, x) : r \in (0, \infty) \text{ and } x \in X\}\rangle \,,$$

where $B_d(r, x)$ denotes the metric ball of radius r centered at x.

We say that $(X, \mathcal{C}_d, \mathcal{B}_d)$ is the underlying bornological coarse space of the metric space (X, d). We will denote this bornological coarse space often by X_d. □

Recall that a path metric space is a metric space where the distance between any two points is given by the infimum of the lengths of all curves connecting these two points.

Let (X, d) be a path metric space, and let $\mathcal{C}_d$ be the associated coarse structure (2.2).

Lemma 2.19 *There exists an entourage U in $\mathcal{C}_d$ such that $\mathcal{C}_d = \mathcal{C}\langle U \rangle$.*

Proof Recall that the coarse structure $\mathcal{C}_d$ is generated by the set of entourages U_r, see (2.1), for all r in $(0, \infty)$. For a path metric space one can check that $\mathcal{C}_d = \mathcal{C}\langle\{U_r\}\rangle$ for any r in $(0, \infty)$. $\square$

Example 2.20 Let X be a set, and let U be an entourage on X. Then the coarse structure $\mathcal{C}\langle U\rangle$ is induced by a metric as in Example 2.18. In fact, we have $\mathcal{C}\langle U\rangle = \mathcal{C}_d$ for the quasi-metric

$$d(x, y) := \inf\{n \in \mathbb{N} \mid (x, y) \in V^n\},$$

where $V := (U \circ U^{-1}) \cup \mathrm{diag}_X$, and where we set $V^0 := \mathrm{diag}_X$.

   In general one can check that a coarse structure can be presented by a metric if and only if it is countably generated. $\square$

Example 2.21 Let Γ be a group. We equip Γ with the bornology $\mathcal{B}_{min}$ consisting of all finite subsets. For every finite subset B we can consider the Γ-invariant entourage $\Gamma(B \times B)$. The family of these entourages generates the canonical coarse structure

$$\mathcal{C}_{can} := \mathcal{C}\langle\{\Gamma(B \times B) : B \in \mathcal{B}_{min}\}\rangle.$$

Then $(\Gamma, \mathcal{C}_{can}, \mathcal{B}_{min})$ is a bornological coarse space. The group Γ acts via left multiplication on this bornological coarse space by automorphisms. This example is actually an example of a Γ-bornological coarse space which we will define now.
                                              $\square$

Definition 2.22 A Γ-bornological coarse space is a bornological coarse space $(X, \mathcal{C}, \mathcal{B})$ with an action of Γ by automorphisms such that the set $\mathcal{C}^\Gamma$ of Γ-invariant entourages is cofinal in $\mathcal{C}$.

   The foundations of the theory of Γ-bornological coarse spaces will be thoroughly developed in [BEKW20].

2.3   Categorical Properties of BornCoarse

The families of objects considered in this subsection are always indexed by very small sets.

Lemma 2.23 *The category* **BornCoarse** *has all non-empty products.*

Proof Let $(X_i, \mathcal{C}_i, \mathcal{B}_i)_{i \in I}$ be a family of bornological coarse spaces. On the set $X := \prod_{i \in I} X_i$ we define the coarse structure $\mathcal{C}$ generated by the entourages $\prod_{i \in I} U_i$ for all elements $(U_i)_{i \in I}$ in $\prod_{i \in I} \mathcal{C}_i$, and the bornology $\mathcal{B}$ generated by the subsets $B_j \times \prod_{i \in I \setminus \{j\}} X_i$ for all j in I and B_j in $\mathcal{B}_j$. The projections from $(X, \mathcal{C}, \mathcal{B})$ to the factors are morphisms. This bornological coarse space together with the projections represents the product of the family $(X_i, \mathcal{C}_i, \mathcal{B}_i)_{i \in I}$. $\square$

Note that **BornCoarse** does not have a final object and therefore does not admit the empty product.

Example 2.24 We will often consider the following variant $X \ltimes X'$ of a product of two bornological coarse spaces $(X, \mathcal{C}, \mathcal{B})$ and $(X', \mathcal{C}', \mathcal{B}')$. In detail, the underlying set of this bornological coarse space is $X \times X'$, its coarse structure is $\mathcal{C} \times \mathcal{C}'$, but its bornology differs from the cartesian product and is generated by the subsets $X \times B'$ for all bounded subsets B' of X'. The projection to the second factor $X \ltimes X' \to X'$ is a morphism, but the projection to the first factor is in general not. On the other hand, for a point $x \in X$ the map $X' \to X \ltimes X'$ given by $x' \mapsto (x, x')$ is a morphism. In general, the cartesian product does not have this property.   □

Lemma 2.25 *The category* **BornCoarse** *has all coproducts.*

Proof Let $(X_i, \mathcal{C}_i, \mathcal{B}_i)_{i \in I}$ be a family of bornological coarse spaces. We define the bornological coarse space $(X, \mathcal{C}, \mathcal{B})$ by

$$X := \bigsqcup_{i \in I} X_i, \quad \mathcal{C} := \mathcal{C}\langle \bigcup_{i \in I} \mathcal{C}_i \rangle, \quad \mathcal{B} := \{B \subseteq X \mid (\forall i \in I : B \cap X_i \in \mathcal{B}_i)\}.$$

Here we secretly identify subsets of X_i or of $X_i \times X_i$ with the corresponding subset of X or of $X \times X$, respectively. The obvious embeddings $X_i \to X$ are morphisms. The bornological coarse space $(X, \mathcal{C}, \mathcal{B})$ together with these embeddings represents the coproduct $\bigsqcup_{i \in I}(X_i, \mathcal{C}_i, \mathcal{B}_i)$ of the family in **BornCoarse**.   □

Remark 2.26 It turns out that the category **BornCoarse** has all non-empty very small limits and many more (but not all) very small colimits. We refer to [BEKW20, Sec. 2.2] for more details. Even better, one can modify the definition of a bornology slightly in order to obtain a complete and cocomplete category $\widetilde{\textbf{BornCoarse}}$ of generalized bornological coarse spaces without changing the homotopy theory developed below [Hei].

Let $(X_i, \mathcal{C}_i, \mathcal{B}_i)_{i \in I}$ be a family of bornological coarse spaces.

Definition 2.27 The free union $\bigsqcup_{i \in I}^{\text{free}}(X_i, \mathcal{C}_i, \mathcal{B}_i)$ of the family is the following bornological coarse space:

1. The underlying set of the free union is the disjoint union $\bigsqcup_{i \in I} X_i$.
2. The coarse structure of the free union is generated by the entourages $\bigcup_{i \in I} U_i$ for all families $(U_i)_{i \in I}$ with U_i in $\mathcal{C}_i$.
3. The bornology of the free union is given by $\mathcal{B}\langle \bigcup_{i \in I} \mathcal{B}_i \rangle$.

Remark 2.28 The free union should not be confused with the coproduct. The coarse structure of the free union is bigger, while the bornology is smaller. The free union plays a role in the discussion of additivity of coarse homology theories.   □

Definition 2.29 We define the mixed union $\bigsqcup_{i \in I}^{\text{mixed}}(X_i, \mathcal{C}_i, \mathcal{B}_i)$ such that the underlying coarse space is that of the coproduct and the bornology is the one from the free union.

Note that the identity morphism on the underlying set gives morphisms in **BornCoarse**

$$\coprod_{i \in I} X_i \xrightarrow{\text{mixed}} \coprod_{i \in I} X_i \xrightarrow{\text{free}} \coprod_{i \in I} X_i . \qquad (2.3)$$

Let $(X, \mathcal{C})$ be a set with a coarse structure. Then

$$\mathcal{R}_{\mathcal{C}} := \bigcup_{U \in \mathcal{C}} U \subseteq X \times X$$

is an equivalence relation on X.

Definition 2.30

1. We call the equivalences classes of X with respect to the equivalence relation $\mathcal{R}_{\mathcal{C}}$ the coarse components of X.
2. We will use the notation $\pi_0^{\text{coarse}}(X, \mathcal{C})$ for the set of coarse components of $(X, \mathcal{C})$.
3. We call $(X, \mathcal{C})$ coarsely connected if $\pi_0^{\text{coarse}}(X, \mathcal{C})$ has a single element.

We can apply the construction to bornological coarse spaces. If $(X_i)_{i \in I}$ is a family of bornological coarse spaces, then we have a bijection

$$\pi_0^{\text{coarse}}\left(\coprod_{i \in I} X_i\right) \cong \coprod_{i \in I} \pi_0^{\text{coarse}}(X_i) .$$

If a bornological coarse space has finitely many coarse components then it is the coproduct of its coarse components with the induced structures. But in the case of infinitely many components this is not true in general.

Example 2.31 Let $(X, \mathcal{C}, \mathcal{B})$ be a bornological coarse space. For an entourage U of X we consider the bornological coarse space

$$X_U := (X, \mathcal{C}\langle\{U\}\rangle, \mathcal{B}) ,$$

i.e., we keep the bornological structure but we consider the coarse structure generated by a single entourage. In **BornCoarse** we have an isomorphism

$$X \cong \operatorname*{colim}_{U \in \mathcal{C}} X_U$$

induced by the natural morphisms $X_U \to X$. This observation often allows us to reduce arguments to the case of bornological coarse spaces whose coarse structure is generated by a single entourage. $\qquad \square$

Example 2.32 Let $(X, \mathcal{C}, \mathcal{B})$ and $(X', \mathcal{C}', \mathcal{B}')$ be bornological coarse spaces. We will often consider the bornological coarse space

$$(X, \mathcal{C}, \mathcal{B}) \otimes (X', \mathcal{C}', \mathcal{B}') := (X \times X', \mathcal{C} \times \mathcal{C}', \mathcal{B} \times \mathcal{B}') \,.$$

We use the notation $\otimes$ in order to distinguish this product from the cartesian product. In fact, the bornology

$$\mathcal{B} \times \mathcal{B}' = \mathcal{B}\langle \{B \times B' : B \in \mathcal{B} \text{ and } B' \in \mathcal{B}'\} \rangle$$

is the one of the cartesian product in the category of bornological spaces and bornological[1] maps.

The product

$$- \otimes - : \mathbf{BornCoarse} \times \mathbf{BornCoarse} \to \mathbf{BornCoarse}$$

is a symmetric monoidal structure on **BornCoarse** with tensor unit $*$. Note that it is not the cartesian product in **BornCoarse**. The latter has no tensor unit since **BornCoarse** does not have a final object.   □

---

[1] Instead of proper.

Chapter 3
Motivic Coarse Spaces

In this chapter we introduce the ∞-category of motivic coarse spaces $\mathbf{Spc}\mathcal{X}$. Our goal is to do homotopy theory with bornological coarse spaces. To this end we first complete the category of bornological coarse spaces formally and then implement the desired geometric properties through localizations.

In the following $\mathbf{Spc}^{\mathrm{la}}$ is the large presentable ∞-category of small spaces. This category can be characterized as the universal large presentable ∞-category generated by a final object. We start with the large presentable ∞-category of space-valued presheaves

$$\mathbf{PSh}(\mathbf{BornCoarse}) := \mathbf{Fun}(\mathbf{BornCoarse}^{\mathrm{op}}, \mathbf{Spc}^{\mathrm{la}}) \,.$$

We construct the ∞-category of motivic coarse spaces $\mathbf{Spc}\mathcal{X}$ by a sequence of localizations (descent, coarse equivalences, vanishing on flasque bornological coarse spaces, u-continuity) of $\mathbf{PSh}(\mathbf{BornCoarse})$ at various sets of morphisms. Our goal is to encode well-known invariance and descent properties of coarse homology theories in a motivic way. The result of this construction is the functor

$$\mathrm{Yo} \colon \mathbf{BornCoarse} \to \mathbf{Spc}\mathcal{X}$$

(Definition 3.34) which sends a bornological coarse space to its motivic coarse space.

Technically one could work with model categories and Bousfield localizations or, as we prefer, with ∞-categories (Remark 3.9). A reference for the general theory of localizations is Lurie [Lur09, Sec. 5.5.4].

Our approach is parallel to constructions in $\mathbb{A}^1$-homotopy theory [Hoy17] or differential cohomology theory [BG, BNV16].

In the final Sect. 3.5 we show some further properties of Yo which could be considered as first steps into the field of coarse homotopy theory.

© The Editor(s) (if applicable) and The Author(s), under exclusive licence to Springer Nature Switzerland AG 2020
U. Bunke, A. Engel, *Homotopy Theory with Bornological Coarse Spaces*,
Lecture Notes in Mathematics 2269, https://doi.org/10.1007/978-3-030-51335-1_3

3.1 Descent

We introduce a Grothendieck topology τ_χ on **BornCoarse** and the associated ∞-category of sheafs. This Grothendieck topology will encode coarse excision.

Remark 3.1 The naive idea to define a Grothendieck topology on **BornCoarse** would be to use covering families given by coarsely excisive pairs [HRY93], see Definition 3.40. The problem with this approach is that in general a pullback does not preserve coarsely excisive pairs, see Example 3.42. Our way out will be to replace excisive pairs by complementary pairs of a subset and a big family, see Example 3.8.   □

Let X be a bornological coarse space.

Definition 3.2 A big family $\mathcal{Y}$ on X is a filtered family of subsets $(Y_i)_{i \in I}$ of X such that for every i in I and entourage U of X there exists j in I such that $U[Y_i] \subseteq Y_j$.

In this definition the set I is a filtered, partially ordered set and the map $I \to \mathcal{P}(X), i \mapsto Y_i$ is order-preserving where $\mathcal{P}(X)$ is equipped with the partial order given by the subset relation.

Example 3.3 Let $\mathcal{C}$ denote the coarse structure of X. If A is a subset of X, then we can consider the big family in X

$$\{A\} := (U[A])_{U \in \mathcal{C}} \tag{3.1}$$

generated by A. Here we use that the set of entourages $\mathcal{C}$ is a filtered partially ordered set with respect to the inclusion relation.   □

Example 3.4 If $(X, \mathcal{C}, \mathcal{B})$ is a bornological coarse space, its family $\mathcal{B}$ of bounded subsets is big. Indeed, the condition stated in Definition 3.2 is exactly the compatibility condition between the bornology and the coarse structure required in Definition 2.7. In this example we consider $\mathcal{B}$ as a filtered partially ordered set with the inclusion relation.   □

Let X be a bornological coarse space.

Definition 3.5 A complementary pair $(Z, \mathcal{Y})$ is a pair consisting of a subset Z of X and a big family $\mathcal{Y} = (Y_i)_{i \in I}$ such that there exists an index i in I with $Z \cup Y_i = X$.

Remark 3.6 Implicitly the definition implies that if a big family $(Y_i)_{i \in I}$ is a part of a complementary pair, then the index set I is not empty. But if the other component Z is the whole space X, then it could be the family $(\emptyset)$.   □

Example 3.7 Let $(Z, \mathcal{Y})$ be a complementary pair on a bornological coarse space X. Then Z has an induced bornological coarse structure. For its construction we apply the construction given in Example 2.17 to the embedding $Z \to X$. We define the family

$$Z \cap \mathcal{Y} := (Z \cap Y_i)_{i \in I}$$

of subsets of Z. This is then a big family on Z. ☐

Example 3.8 Let $(Z, \mathcal{Y})$ be a complementary pair on a bornological coarse space X and $f : X' \to X$ be a morphism. Then $f^{-1}\mathcal{Y} := (f^{-1}(Y_i))_{i \in I}$ is a big family on X' since f is controlled. Furthermore, $(f^{-1}(Z), f^{-1}\mathcal{Y})$ is a complementary pair on X'. ☐

Remark 3.9 In the present paper we work with ∞-categories (or more precisely with $(\infty, 1)$-categories) in a model-independent way. For concreteness one could think of quasi-categories as introduced by Joyal [Joy08] and worked out in detail by Lurie [Lur09, Lur17].

Implicitly, we will consider an ordinary category as an ∞-category using nerves. In order to construct a model for $\mathbf{Spc}^{\mathrm{la}}$ one could start with the large ordinary category $\mathbf{sSet}^{\mathrm{la}}$ of small simplicial sets. We let W denote the set of weak homotopy equivalences in $\mathbf{sSet}^{\mathrm{la}}$. Then

$$\mathbf{Spc}^{\mathrm{la}} \simeq \mathbf{sSet}^{\mathrm{la}}[W^{-1}]. \tag{3.2}$$

The ∞-category of spaces is also characterized by a universal property: it is the universal large presentable ∞-category generated by a final object (usually denoted by $*$).

If $\mathbf{C}$ is some small ∞-category, then we can consider the large category of space-valued presheaves

$$\mathbf{PSh}(\mathbf{C}) := \mathbf{Fun}(\mathbf{C}^{\mathrm{op}}, \mathbf{Spc}^{\mathrm{la}}).$$

We have the Yoneda embedding

$$\mathrm{yo} : \mathbf{C} \to \mathbf{PSh}(\mathbf{C}).$$

It presents the category of presheaves on $\mathbf{C}$ as the free cocompletion of $\mathbf{C}$. We will often use the identification (following from Lurie [Lur09, Thm. 5.1.5.6])[1]

$$\mathbf{PSh}(\mathbf{C}) \simeq \mathbf{Fun}^{\mathrm{lim}}(\mathbf{PSh}(\mathbf{C})^{\mathrm{op}}, \mathbf{Spc}^{\mathrm{la}}),$$

---

[1] Lurie's formula is the marked equivalence in the chain

$$\mathbf{PSh}(\mathbf{C}) = \mathbf{Fun}(\mathbf{C}^{\mathrm{op}}, \mathbf{Spc}^{\mathrm{la}}) \simeq \mathbf{Fun}(\mathbf{C}, \mathbf{Spc}^{la,\mathrm{op}})^{\mathrm{op}} \overset{!}{\simeq} \mathbf{Fun}^{\mathrm{colim}}(\mathbf{PSh}(\mathbf{C}), \mathbf{Spc}^{la,\mathrm{op}})^{\mathrm{op}}$$

$$\simeq \mathbf{Fun}^{\mathrm{lim}}(\mathbf{PSh}(\mathbf{C})^{\mathrm{op}}, \mathbf{Spc}^{\mathrm{la}}).$$

where the superscript lim stands for small limit-preserving functors. In particular, if E and F are presheaves, then the evaluation $E(F)$ in $\mathbf{Spc}^{\text{la}}$ is defined and satisfies

$$E(F) \simeq \lim_{(\text{yo}(X)\to F)\in \mathbf{C}/F} E(X)\,, \qquad (3.3)$$

where $\mathbf{C}/F$ denotes the slice category. Note that for X in $\mathbf{C}$ we have the equivalence $E(\text{yo}(X)) \simeq E(X)$. $\qquad\qquad\square$

For a big family $\mathcal{Y} = (Y_i)_{i\in I}$ we define

$$\text{yo}(\mathcal{Y}) := \operatorname*{colim}_{i\in I} \text{yo}(Y_i) \in \mathbf{PSh}(\mathbf{BornCoarse})\,.$$

Note that we take the colimit after applying the Yoneda embedding. If we would apply these operations in the opposite order, then in general we would get a different result. In order to simplify the notation, for a presheaf E we will often abbreviate $E(\text{yo}(\mathcal{Y}))$ by $E(\mathcal{Y})$. By definition we have the equivalence $E(\mathcal{Y}) \simeq \lim_{i\in I} E(Y_i)$.

Let E be in $\mathbf{PSh}(\mathbf{BornCoarse})$.

Definition 3.10 We say that E satisfies descent for complementary pairs if $E(\emptyset) \simeq *$ and if for every complementary pair $(Z, \mathcal{Y})$ on a bornological coarse space X the square

$$
\begin{array}{ccc}
E(X) & \longrightarrow & E(Z) \\
\downarrow & & \downarrow \\
E(\mathcal{Y}) & \longrightarrow & E(Z \cap \mathcal{Y})
\end{array}
\qquad (3.4)
$$

is cartesian in $\mathbf{Spc}$.

Lemma 3.11 *There is a Grothendieck topology τ_χ on $\mathbf{BornCoarse}$ such that the τ_χ-sheaves are exactly the presheaves which satisfy descent for complementary pairs.*

Proof We consider the largest Grothendieck topology τ_χ on $\mathbf{BornCoarse}$ for which all the presheaves satisfying descent for complementary pairs are sheaves. The main non-trivial point is now to observe that the condition of descent for a complementary pair $(Z, \mathcal{Y})$ on a bornological coarse space X is equivalent to the condition of descent for the sieve generated by the covering family of X consisting of Z and the subsets Y_i for all i in I. By the definition of the Grothendieck topology τ_χ this sieve belongs to τ_χ. Therefore a τ_χ-sheaf satisfies descent for complementary pairs. For more details we refer to [Hei]. $\qquad\square$

We let

$$\mathbf{Sh}(\mathbf{BornCoarse}) \subseteq \mathbf{PSh}(\mathbf{BornCoarse})$$

denote the full subcategory of τ_χ-sheaves. We have the usual sheafification adjunction

$$L : \mathbf{PSh}(\mathbf{BornCoarse}) \leftrightarrows \mathbf{Sh}(\mathbf{BornCoarse}) : \mathrm{incl}.$$

Lemma 3.12 *The Grothendieck topology τ_χ on* **BornCoarse** *is subcanonical.*

Proof We must show that for every X' in **BornCoarse** the representable presheaf $\mathrm{yo}(X')$ is a sheaf. Note that $\mathrm{yo}(X')$ comes from the set-valued presheaf $y(X')$ by postcomposition with the functor

$$\iota : \mathbf{Set}^{\mathrm{la}} \to \mathbf{sSet}^{\mathrm{la}} \to \mathbf{Spc}^{\mathrm{la}},$$

where

$$y : \mathbf{BornCoarse} \to \mathbf{PSh}_{\mathbf{Set}^{\mathrm{la}}}(\mathbf{BornCoarse})$$

is the usual one-categorical Yoneda embedding from bornological coarse spaces to set-valued presheaves on **BornCoarse**. Since ι preserves limits it suffices to show descent for the set-valued sheaf $y(X')$ in $\mathbf{PSh}_{\mathbf{Set}^{\mathrm{la}}}(\mathbf{BornCoarse})$. Let $(Z, \mathcal{Y})$ be a complementary pair on a bornological coarse space X. Then we must show that the natural map

$$y(X')(X) \to y(X')(Z) \times_{y(X')(Z \cap \mathcal{Y})} y(X')(\mathcal{Y})$$

is a bijection. The condition that there exists an i_0 in I such that $Z \cup Y_{i_0} = X$ implies injectivity. We consider now an element on the right-hand side. It consists of a morphism $f : Z \to X'$ and a compatible family of morphisms $g_i : Y_i \to X'$ for all i in I whose restrictions to $Z \cap Y_i$ coincide with the restriction of f. We therefore get a well-defined map of sets $h : X \to X'$ restricting to f and the g_i, respectively. In order to finish the argument it suffices to show that h is a morphism in **BornCoarse**. If B' is a bounded subset of X', then the equality

$$h^{-1}(B') = \left(g^{-1}(B') \cap Z\right) \cup \left(f_{i_0}^{-1}(B') \cap Y_{i_0}\right)$$

shows that $f^{-1}(B)$ is bounded. Therefore h is proper. We now show that h is controlled. Let U be an entourage of X. Then we have a non-disjoint decomposition

$$U = \left(U \cap (Z \times Z)\right) \cup \left(U \cap (Y_{i_0} \times Y_{i_0})\right) \cup \left(U \cap (Z \times Y_{i_0})\right) \cup \left(U \cap (Y_{i_0} \times Z)\right).$$

We see that

$$(h \times h)(U \cap (Z \times Z)) = (f \times f)(U \cap (Z \times Z))$$

and

$$(h \times h)(U \cap (Y_{i_0} \times Y_{i_0})) = (g_{i_0} \times g_{i_0})(U \cap (Y_{i_0} \times Y_{i_0}))$$

are controlled. For the remaining two pieces we argue as follows. Since the family $\mathcal{Y}$ is big there exists i in I such that $U \cap (Y_{i_0} \times Z) \subseteq U \cap (Y_i \times Y_i)$. Then

$$(h \times h)(U \cap (Y_{i_0} \times Z)) \subseteq (g_i \times g_i)(U \cap (Y_i \times Y_i))$$

is controlled. For the remaining last piece we argue similarly. We finally conclude that $(h \times h)(U)$ is controlled. $\square$

3.2 Coarse Equivalences

Coarse geometry is the study of bornological coarse spaces up to equivalence. In order to define the notion of equivalence we first introduce the relation of closeness between morphisms.

Let X, X' be bornological coarse spaces, and let $f_0, f_1 \colon X \to X'$ be a pair of morphisms.

Definition 3.13 We say that the morphisms f_0 and f_1 are close to each other if $(f_0 \times f_1)(\mathrm{diag}_X)$ is an entourage of X'.

Since the coarse structure on X' contains the diagonal and is closed under symmetry and composition the relation of being close to each other is an equivalence relation on the set of morphisms from X to X'.

Let $f : X \to X'$ be a morphism in **BornCoarse**.

Definition 3.14 We say f is an equivalence if there exists a morphism $g : X' \to X$ in **BornCoarse** such that $f \circ g$ and $g \circ f$ are close to the respective identities.

Example 3.15 Let X be a bornological coarse space and Y be a subset of X. If U is an entourage of X containing the diagonal, then the inclusion $Y \to U[Y]$ is an equivalence in **BornCoarse** if we equip the subsets with the induced structures. In order to define an inverse $g : U[Y] \to Y$ we simply choose for every point x in $U[Y]$ a point $g(y)$ in Y such that $(x, g(x))$ in U. $\square$

Example 3.16 Let (X, d) and (X', d') be metric spaces and $f \colon X \to X'$ be a map between the underlying sets. Then f is a quasi-isometry if there exists C, D, E in $(0, \infty)$ such that for all x, y in X we have

$$C^{-1} d'(f(x), f(y)) - D \leq d(x, y) \leq C d'(f(x), f(y)) + D$$

and for every x' in X' there exists x in X such that $d'(f(x), x') \leq E$.

If f is a quasi-isometry, then the map $f : X_d \rightarrow X'_d$ (see Example 2.18) is an equivalence.   □

Example 3.17 Invariance under equivalences is a basic requirement for useful invariants of bornological coarse spaces. Typical examples of such invariants with values in {false, true} are local and global finiteness conditions or conditions on the generation of the coarse or bornological structure. Here is a list of examples of such invariants for a bornological coarse space X.

1. X is locally countable (or locally finite), see Definition 8.15.
2. X has the minimal compatible bornology, see Example 2.16.
3. X has locally bounded geometry, see Definition 8.138.
4. X has bounded geometry, see Definition 7.77.
5. The bornology of X is countably generated, see Example 2.12.
6. The coarse structure of X has one (or finitely many, or countably many) generators, see Example 2.12.

The set $\pi_0^{\mathrm{coarse}}(X)$ of coarse components of X is an example of a **Set**-valued invariant, see Definition 2.30.   □

Our next localization implements the idea that a sheaf E in **Sh(BornCoarse)** should send equivalences in **BornCoarse** to equivalences in **Spc**^{la}. To this end we consider the coarse bornological space $\{0, 1\}$ with the maximal bornological and coarse structures. We now observe that two morphisms $f_0, f_1 : X \rightarrow X'$ are close to each other if and only if we can combine them to a single morphism $\{0, 1\} \otimes X \rightarrow X'$, see Example 2.32 for $\otimes$. For every bornological coarse space X we have a projection $\{0, 1\} \otimes X \rightarrow X$.

We now implement the invariance under coarse equivalences into **Sh(BornCoarse)** using the interval object $\{0, 1\}$ similarly as in the approach to $\mathbb{A}^1$-homotopy theory by Morel and Voevodsky [MV99, Sec. 2.3].

Let E be in **Sh(BornCoarse)**.

Definition 3.18 We call E coarsely invariant if for every X in **BornCoarse** the projection induces an equivalence

$$E(X) \rightarrow E(\{0, 1\} \otimes X) .$$

Lemma 3.19 *The coarsely invariant sheaves form a full localizing subcategory of* **Sh(BornCoarse)**.

Proof The subcategory of coarsely invariant sheaves is the full subcategory of objects which are local [Lur09, Def. 5.5.4.1] with respect to the small set of projections $\mathrm{yo}(\{0, 1\} \otimes X) \rightarrow \mathrm{yo}(X)$ for all X in **BornCoarse**. The latter are morphisms of sheaves by Lemma 3.12. We then apply [Lur09, Prop. 5.5.4.15] to finish the proof.   □

We denote the subcategory of coarsely invariant sheaves by $\mathbf{Sh}^{\{0,1\}}(\mathbf{BornCoarse})$. We have a corresponding adjunction

$$H^{\{0,1\}} : \mathbf{Sh}(\mathbf{BornCoarse}) \leftrightarrows \mathbf{Sh}^{\{0,1\}}(\mathbf{BornCoarse}) : \mathrm{incl}\,.$$

Note that in order to show that a sheaf in $\mathbf{Sh}(\mathbf{BornCoarse})$ is coarsely invariant it suffices to show that it sends pairs of morphisms in $\mathbf{BornCoarse}$ which are close to each other to pairs of equivalent morphisms in $\mathbf{Spc}^{\mathrm{la}}$.

Remark 3.20 We use the notation $H^{\{0,1\}}$ since this functor resembles the homotopification functor $\mathcal{H}$ in differential cohomology theory [BNV16]. The analogous functor in the construction of the motivic homotopy category in algebraic geometry is discussed, e.g., in Hoyois [Hoy17, Sec. 3.2] and denoted there by L_A.   □

3.3 Flasque Spaces

The following notion is modeled after Higson et al. [HPR96, Sec. 10].

Let X be in $\mathbf{BornCoarse}$.

Definition 3.21 X is flasque if it admits an endomorphism $f : X \to X$ with the following properties:

1. The morphisms f and id_X are close to each other.
2. For every entourage U of X the union $\bigcup_{k \in \mathbb{N}} (f^k \times f^k)(U)$ is again an entourage of X.
3. For every bounded subset B of X there exists k in $\mathbb{N}$ such that $f^k(X) \cap B = \emptyset$.

In this case we will say that flasqueness of X is implemented by f.

Example 3.22 Consider $[0, \infty)$ as a bornological coarse space with the structure induced by the metric; see Example 2.18. Let X be a bornological coarse space. Then the bornological coarse space $[0, \infty) \otimes X$ (see Example 2.32) is flasque. For f we can take the map $f(t, x) := (t + 1, x)$. Let $\mathcal{B}$ denote the bornology of X and note that the bornology $\mathcal{B}_d \times \mathcal{B}$ of $[0, \infty) \otimes X$ is smaller than the bornology of the cartesian product $[0, \infty) \times X$; see Lemma 2.23. Indeed, the bornological coarse space $[0, \infty) \times X$ is not flasque because Condition 3 of Definition 3.21 is violated. But note that $X \ltimes [0, \infty)$ (see Example 2.24) is flasque, too.

Similarly, if $\mathbb{N}_d$ denotes the $\mathbb{N}$ with structures induced form the embedding into $[0, \infty)$, then $\mathbb{N}_d \otimes X$ is flasque for every bornological coarse space X. Flasqueness is again witnessed by the map $f : \mathbb{N}_d \otimes X \to \mathbb{N}_d \otimes X$ given by $f(n, x) := f(n+1, x)$.   □

Let E be in $\mathbf{Sh}^{\{0,1\}}(\mathbf{BornCoarse})$.

Definition 3.23 We say that E vanishes on flasque bornological coarse spaces if $E(X)$ is a final object in $\mathbf{Spc}$ for every flasque bornological coarse space X.

We will say shortly that E vanishes on flasque spaces.

Remark 3.24 If $\emptyset_{\mathbf{Sh}}$ denotes an initial object in $\mathbf{Sh}(\mathbf{BornCoarse})$, then for every E in $\mathbf{Sh}(\mathbf{BornCoarse})$ we have the equivalence of spaces

$$\mathrm{Map}_{\mathbf{Sh}(\mathbf{BornCoarse})}(\emptyset_{\mathbf{Sh}}, E) \simeq * \, .$$

On the other hand, by the sheaf condition (see Definition 3.10) we have the equivalence of spaces

$$\mathrm{Map}_{\mathbf{Sh}(\mathbf{BornCoarse})}(\mathrm{yo}(\emptyset), E) \simeq E(\emptyset) \simeq * \, .$$

We conclude that

$$\mathrm{yo}(\emptyset) \simeq \emptyset_{\mathbf{Sh}} \, ,$$

i.e., that $\mathrm{yo}(\emptyset)$ is an initial object of $\mathbf{Sh}(\mathbf{BornCoarse})$. Hence a sheaf E vanishes on the flasque bornological coarse space X if and only if it is local with respect to the morphism $\mathrm{yo}(\emptyset) \to \mathrm{yo}(X)$. □

Lemma 3.25 *The coarsely invariant sheaves which vanish on flasque spaces form a full localizing subcategory of* $\mathbf{Sh}^{\{0,1\}}(\mathbf{BornCoarse})$.

Proof By Remark 3.24 the category of coarsely invariant sheaves which vanish on flasque spaces is the full subcategory of sheaves which are local with respect to the union of the small sets of morphisms considered in the proof of Lemma 3.19 and the morphisms $\mathrm{yo}(\emptyset) \to \mathrm{yo}(X)$ for all flasque X in $\mathbf{BornCoarse}$. □

We denote the subcategory of coarsely invariant sheaves which vanish on flasque spaces by $\mathbf{Sh}^{\{0,1\},\mathrm{fl}}(\mathbf{BornCoarse})$. We have an adjunction

$$\mathrm{Fl} : \mathbf{Sh}^{\{0,1\}}(\mathbf{BornCoarse}) \leftrightarrows \mathbf{Sh}^{\{0,1\},\mathrm{fl}}(\mathbf{BornCoarse}) : \mathrm{incl} \, .$$

Remark 3.26 The empty bornological coarse space $\emptyset$ is flasque. Therefore we have the equivalence $E(\emptyset) \simeq *$ for any E in $\mathbf{Sh}^{\{0,1\},\mathrm{fl}}(\mathbf{BornCoarse})$. This is compatible with the requirements for E being a sheaf.

If X in $\mathbf{BornCoarse}$ is flasque, then the canonical map $\emptyset \to X$ induces an equivalence $\mathrm{Fl}(\mathrm{yo}(\emptyset)) \to \mathrm{Fl}(\mathrm{yo}(X))$. □

In some applications one considers an apparently more general definition of flasqueness. Let X be in $\mathbf{BornCoarse}$.

Definition 3.27 ([Wri05, Def. 3.10]) X is flasque in the generalized sense if it admits a sequence $(f_k)_{k \in \mathbb{N}}$ of endomorphisms $f_k : X \to X$ with the following properties:

1. $f_0 = \mathrm{id}_X$.
2. The union $\bigcup_{k \in \mathbb{N}} (f_k \times f_{k+1})(\mathrm{diag}_X)$ is an entourage of X.

3. For every entourage U of X the union $\bigcup_{k \in \mathbb{N}} (f_k \times f_k)(U)$ is again an entourage of X.
4. For every bounded subset B of X there exists a k_0 in $\mathbb{N}$ such that for all k in $\mathbb{N}$ with $k \geq k_0$ we have $f_k(X) \cap B = \emptyset$.

If X is flasque in the sense of Definition 3.22 with flasqueness implemented by f, then it is flasque in the generalized sense. We just take $f_k := f^k$ for all k in $\mathbb{N}$.

Let X be a bornological coarse space. In the following lemma $\mathbb{N}_d$ denotes the bornological coarse space with the structures induced by the standard metric. Further, $\iota : X \to \mathbb{N}_d \otimes X$ is the embedding determined by the point 0 in $\mathbb{N}$.

Lemma 3.28 *The following assertions are equivalent:*

1. *X is flasque in the generalized sense.*
2. *There exists a morphism $F : \mathbb{N}_d \otimes X \to X$ such that $F \circ \iota = \mathrm{id}_X$.*

Proof Let X be flasque in the generalized sense. Let $(f_k)_{k \in \mathbb{N}}$ be as in Definition 3.27. Then we define

$$F : \mathbb{N}_d \otimes X \to X \quad F(k, x) := f_k(x).$$

We argue that F is a morphism. We first show that F is proper. Let B be a bounded subset of X. Let k_0 in $\mathbb{N}$ be as in Condition 3.27.4. Then $F^{-1}(B) \subseteq \bigcup_{k=0}^{k_0-1} \{k\} \times f_k^{-1}(B)$. This subset is bounded in $\mathbb{N}_d \otimes X$. We now show that F is controlled. The coarse structure of $\mathbb{N}_d \otimes X$ is generated by the entourages $V := \{(n, n+1) : n \in \mathbb{N}\} \times \mathrm{diag}_X$ and $\mathrm{diag}_\mathbb{N} \times U$ for all entourages U of X. We have $F(V) = \bigcup_{k \in \mathbb{N}} (f_k \times f_{k+1})(\mathrm{diag}_X)$ which is an entourage of X by Condition 3.27.2. Furthermore, $F(U) = \bigcup_{k \in \mathbb{N}} (f_k \times f_k)(U)$ is an entourage of X by Condition 3.27.3. Finally, by Condition 3.27.1 we have $F \circ \iota = \mathrm{id}_X$.

Vice versa, assume that a morphism $F : \mathbb{N}_d \otimes X \to X$ is given such that $F \circ \iota = X$. Then we define family of morphisms $(f_k)_{k \in \mathbb{N}}$ by $f_k(x) := F(k, x)$. Similar considerations as above show that this family satisfies the conditions listed in Definition 3.27. $\square$

Remark 3.29 It is not clear that the condition of being flasque is coarsely invariant. But it is almost obvious that flasqueness in the generalized sense is invariant under coarse equivalences. Let X be flasque in the generalized sense witnessed by the family $(f_k)_{k \in \mathbb{N}}$. Let furthermore $g : X \to Y$ be a coarse equivalence. Then we can choose an inverse $h : Y \to X$ up to closeness. We then define a family of endomorphisms $(f'_k)_{k \in \mathbb{N}}$ of Y by $f'_0 := \mathrm{id}_Y$ and $f'_{k+1} := g \circ f_k \circ h$. It is now easy to check that the family $(f'_k)_{k \in \mathbb{N}}$ satisfies the conditions listed in Definition 3.27. Hence the family $(f'_k)_{k \in \mathbb{N}}$ witnesses that Y is flasque in the generalized sense. $\square$

Let E be in $\mathbf{Sh}^{\{0,1\},\mathrm{fl}}(\mathbf{BornCoarse})$ and X be in $\mathbf{BornCoarse}$.

Lemma 3.30 *If $E \in \mathbf{Sh}^{\{0,1\},\mathrm{fl}}(\mathbf{BornCoarse})$ and X is flasque in the generalized sense, then $E(X) \simeq *$.*

Proof By Lemma 3.28 we have the factorization

$$X \xrightarrow{\iota} \mathbb{N}_d \otimes X \xrightarrow{F} X \tag{3.5}$$

of the identity of X. We apply E and get the factorization

$$E(X) \xrightarrow{E(F)} E(\mathbb{N}_d \otimes X) \xrightarrow{E(\iota)} E(X)$$

of the identity of $E(X)$. Since E vanishes on flasques and $\mathbb{N}_d \otimes X$ is flasque by Example 3.22 we conclude that $\mathrm{id}_{E(X)} \simeq 0$. This implies $E(X) \simeq 0$.   □

3.4   *u*-Continuity and Motivic Coarse Spaces

Finally we will enforce that the Yoneda embedding preserves the colimits considered in Example 2.31.

Let E be in $\mathbf{Sh}^{\{0,1\},\mathrm{fl}}(\mathbf{BornCoarse})$.

Definition 3.31 E is *u*-continuous if for every bornological coarse space $(X, \mathcal{C}, \mathcal{B})$ the natural morphism

$$E(X) \to \lim_{U \in \mathcal{C}} E(X_U)$$

is an equivalence.

Lemma 3.32 *The full subcategory of* $\mathbf{Sh}^{\{0,1\},\mathrm{fl}}(\mathbf{BornCoarse})$ *of u-continuous sheaves is localizing.*

Proof The proof is similar to the proof of Lemma 3.25. We just add the small set of morphisms

$$\operatorname*{colim}_{U \in \mathcal{C}} \mathrm{yo}(X_U) \to \mathrm{yo}(X)$$

for all X in **BornCoarse** to the list of morphisms for which our sheaves must be local.   □

Definition 3.33 The subcategory $\mathbf{Spc}\mathcal{X}$ described in Lemma 3.32 is the subcategory of motivic coarse spaces.

It fits into the localization adjunction

$$U : \mathbf{Sh}^{\{0,1\},\mathrm{fl}}(\mathbf{BornCoarse}) \leftrightarrows \mathbf{Spc}\mathcal{X} : \mathrm{incl} \, .$$

Definition 3.34 We define the functor

$$\text{Yo} := U \circ \text{Fl} \circ H^{\{0,1\}} \circ L \circ \text{yo} : \textbf{BornCoarse} \to \textbf{Spc}\mathcal{X} .$$

Because of Lemma 3.12 we could drop the functor L in this composition.

Every bornological coarse space X represents a motivic coarse space $\text{Yo}(X)$ in **Spc$\mathcal{X}$**, which is called the coarse motivic space associated to X.

By construction we have the following rules:

Corollary 3.35

1. *The ∞-category motivic coarse spaces **Spc$\mathcal{X}$** is presentable and fits into a localization*

$$U \circ \text{Fl} \circ H^{\{0,1\}} \circ L : \textbf{PSh}(\textbf{BornCoarse}) \leftrightarrows \textbf{Spc}\mathcal{X} : \text{incl} .$$

2. *If $(Z, \mathcal{Y})$ is a complementary pair on a bornological coarse space X, then the square*

$$\begin{array}{ccc} \text{Yo}(Z \cap \mathcal{Y}) & \longrightarrow & \text{Yo}(\mathcal{Y}) \\ \downarrow & & \downarrow \\ \text{Yo}(Z) & \longrightarrow & \text{Yo}(X) \end{array}$$

*is cocartesian in **Spc$\mathcal{X}$**.*

3. *If $X \to X'$ is an equivalence of bornological coarse spaces, then $\text{Yo}(X) \to \text{Yo}(X')$ is an equivalence in **Spc$\mathcal{X}$**.*

4. *If X is a flasque bornological coarse space, then $\text{Yo}(X)$ is an initial object of **Spc$\mathcal{X}$**.*

5. *For every bornological coarse space X with coarse structure $\mathcal{C}$ we have*

$$\text{Yo}(X) \simeq \operatorname*{colim}_{U \in \mathcal{C}} \text{Yo}(X_U) .$$

Using Lurie [Lur09, Prop. 5.5.4.20] the ∞-category of motivic coarse spaces has by construction the following universal property:

Corollary 3.36 *For an ∞-category **C** we have an equivalence between* **Fun^{colim}(Spc$\mathcal{X}$, C)** *and the full subcategory of* **Fun^{colim}(PSh(BornCoarse), C)** *of functors which preserve the equivalences or colimits listed in Corollary 3.35.*

The superscript colim stands for small colimit-preserving functors. By the universal property of presheaves [Lur09, Thm. 5.1.5.6] together with the above corollary we get:

Corollary 3.37 *For every large, small cocomplete ∞-category **C** we have an equivalence between* **Fun^{colim}(Spc$\mathcal{X}$, C)** *and the full subcategory of*

Fun(BornCoarse, C) *of functors which satisfy excision, preserve equivalences, annihilate flasque spaces, and which are u-continuous.*

3.5 Coarse Excision and Further Properties

The following generalizes Point 4 of Corollary 3.35. Let X be in **BornCoarse**.

Lemma 3.38 *If X is flasque in the generalized sense, then $\mathrm{Yo}(X)$ is an initial object of* $\mathbf{Spc}\mathcal{X}$.

Proof This follows from the proof of Lemma 3.30. □

Let $(X, \mathcal{C}, \mathcal{B})$ be a bornological coarse space, and let A be a subset of X. We consider the big family $\{A\}$ on X generated by A. By Example 3.15 the inclusion $A \to U[A]$ is an equivalence in **BornCoarse** for every U in $\mathcal{C}$ containing the diagonal of X. By Point 3 of Corollary 3.35 the induced map $\mathrm{Yo}(A) \to \mathrm{Yo}(U[A])$ is an equivalence. Note that

$$\mathrm{Yo}(\{A\}) \simeq \operatorname*{colim}_{U \in \mathcal{C}} \mathrm{Yo}(U[A]).$$

Since the entourages containing the diagonal are cofinal in all entourages we get the following corollary:

Let X be in **BornCoarse**, and let A be a subset of X.

Corollary 3.39 *We have an equivalence* $\mathrm{Yo}(A) \simeq \mathrm{Yo}(\{A\})$.

We now relate the Grothendieck topology τ_χ with the usual notion of coarse excision. Let X be a bornological coarse space, and let (Y, Z) be a pair of subsets of X.

Definition 3.40 (Y, Z) is a coarsely excisive pair if $X = Y \cup Z$ and for every entourage U of X there exists an entourage V of X such that

$$U[Y] \cap U[Z] \subseteq V[Y \cap Z].$$

Let X be a bornological coarse space, and let Y, Z be two subsets of X.

Lemma 3.41 *If (Y, Z) is a coarsely excisive pair, then the square*

$$\begin{array}{ccc} \mathrm{Yo}(Y \cap Z) & \longrightarrow & \mathrm{Yo}(Y) \\ \downarrow & & \downarrow \\ \mathrm{Yo}(Z) & \longrightarrow & \mathrm{Yo}(X) \end{array}$$

is cocartesian in $\mathbf{Spc}\mathcal{X}$.

Proof The pair $(Z, \{Y\})$ is complementary. By Point 2 of Corollary 3.35 the square

$$
\begin{array}{ccc}
\mathrm{Yo}(\{Y\} \cap Z) & \longrightarrow & \mathrm{Yo}(\{Y\}) \\
\downarrow & & \downarrow \\
\mathrm{Yo}(Z) & \longrightarrow & \mathrm{Yo}(X)
\end{array}
$$

is cocartesian. We finish the proof by identifying the respective corners. By Corollary 3.39 we have $\mathrm{Yo}(Y) \simeq \mathrm{Yo}(\{Y\})$. It remains to show that $\mathrm{Yo}(Y \cap Z) \to \mathrm{Yo}(\{Y\} \cap Z)$ is an equivalence. Since

$$U[Y \cap Z] \subseteq U[V[Y] \cap Z]$$

for every two entourages U and V (such that V contains the diagonal) of X we have a map

$$\mathrm{Yo}(\{Y \cap Z\}) \to \mathrm{Yo}(\{V[Y] \cap Z\}) \simeq \mathrm{Yo}(V[Y] \cap Z) \to \mathrm{Yo}(\{Y\} \cap Z) \,.$$

Since (Y, Z) is coarsely excisive, for every entourage U of X we can find an entourage W of X such that $U[Y] \cap Z \subseteq W[Y \cap Z]$. Hence we get an inverse map $\mathrm{Yo}(\{Y\} \cap Z) \to \mathrm{Yo}(\{Y \cap Z\})$. We conclude that both maps in the chain

$$\mathrm{Yo}(Y \cap Z) \to \mathrm{Yo}(\{Y\} \cap Z) \to \mathrm{Yo}(\{Y \cap Z\})$$

are equivalences.   □

Example 3.42 Here we construct an example showing that a pullback in general does not preserve coarsely excisive pairs.

Let X_{max} be the set $\{a, b, w\}$ with the maximal bornological coarse structure. We further consider the coproduct $X := \{a, b\}_{max} \sqcup \{w\}$ in **BornCoarse**.

We consider the subsets $Y := \{a, w\}$ and $Z := \{b, w\}$ of X. Then (Y, Z) is a coarsely excisive pair on X_{max}. The identity of the underlying set X is a morphism $f : X \to X_{max}$. Then $(f^{-1}(Y), f^{-1}(Z)) = (Y, Z)$ is not a coarsely excisive pair on X.   □

Chapter 4
Motivic Coarse Spectra

We will define the category $\mathbf{Sp}\mathcal{X}$ of motivic coarse spectra as the stable version of the category of motivic coarse spaces. We will then obtain a stable version

$$\mathrm{Yo}^s : \mathbf{BornCoarse} \to \mathbf{Sp}\mathcal{X}$$

of the Yoneda functor which turns out to be the universal coarse homology theory. In Sect. 4.1 we introduce this stabilization and discuss the universal property of Yo^s. Section 4.4 contains the definition of the notion of a coarse homology theory and argue that Yo^s is the universal example. In Sect. 4.2 we list some additional properties of Yo^s. Finally, in Sect. 4.3 we consider the coarse homotopy invariance of Yo^s. This is a very useful strengthening of Property 3 of Corollary 3.35. It is bound to the stable context because of the usage of fibre sequences in the proof of Proposition 4.16.

4.1 Stabilization

Before giving the construction of the category of motivic coarse spectra $\mathbf{Sp}\mathcal{X}$ as a stabilization of the category of motivic coarse spaces $\mathbf{Spc}\mathcal{X}$ we demonstrate the method in the case of the construction of the category of spectra $\mathbf{Sp}^{\mathrm{la}}$ from the category of spaces $\mathbf{Spc}^{\mathrm{la}}$.

Remark 4.1 The large ∞-category of small spectra is a presentable stable ∞-category. A reference for it is Lurie [Lur17, Sec. 1]. The ∞-category of spectra can be characterized as the universal presentable stable ∞-category generated by one object. It can be constructed as the stabilization of the category of spaces $\mathbf{Spc}^{\mathrm{la}}$

U. Bunke, A. Engel, *Homotopy Theory with Bornological Coarse Spaces*,
Lecture Notes in Mathematics 2269, https://doi.org/10.1007/978-3-030-51335-1_4

(see (3.2) for a model) . To this end one forms the pointed ∞-category $\mathbf{Spc}^{\mathrm{la}}_{*/}$ of pointed spaces. It has a suspension endofunctor

$$\Sigma : \mathbf{Spc}^{\mathrm{la}}_{*/} \to \mathbf{Spc}^{\mathrm{la}}_{*/}, \quad X \mapsto \mathtt{colim}((* \to *) \leftarrow (* \to X) \to (* \to *)).$$

The category of spectra is then obtained by inverting this functor in the realm of presentable ∞-categories. According to this prescription we set

$$\mathbf{Sp}^{\mathrm{la}} := \mathbf{Spc}^{\mathrm{la}}_{/*}[\Sigma^{-1}] := \mathtt{colim}\{\mathbf{Spc}^{\mathrm{la}}_{*/} \xrightarrow{\Sigma} \mathbf{Spc}^{\mathrm{la}}_{*/} \xrightarrow{\Sigma} \mathbf{Spc}^{\mathrm{la}}_{*/} \xrightarrow{\Sigma} \dots \},$$

where the colimit is taken in the ∞-category $\mathbf{Pr}^L$ of large presentable ∞-categories and left adjoint functors. The category $\mathbf{Sp}^{\mathrm{la}}$ fits into an adjunction

$$\Sigma^{\infty}_{+} : \mathbf{Spc}^{\mathrm{la}} \leftrightarrows \mathbf{Sp}^{\mathrm{la}} : \Omega^{\infty}, \tag{4.1}$$

where Σ^{∞}_{+} is the composition

$$\mathbf{Spc}^{\mathrm{la}} \to \mathbf{Spc}^{\mathrm{la}}_{*/} \to \mathbf{Sp}^{\mathrm{la}}.$$

Here the first functor adds a disjoint base point, and the second functor is the canonical one. The category $\mathbf{Sp}^{\mathrm{la}}$ is characterized by the obvious universal property that precomposition by Σ^{∞}_{+} induces an equivalence

$$\mathbf{Fun}^L(\mathbf{Sp}^{\mathrm{la}}, \mathbf{C}) \simeq \mathbf{Fun}^L(\mathbf{Spc}^{\mathrm{la}}, \mathbf{C}) \simeq \mathbf{C}$$

for every large presentable stable ∞-category $\mathbf{C}$. A reference for the first equivalence is Lurie [Lur17, 1.4.4.5], while the second follows from the universal property of $\mathbf{Spc}^{\mathrm{la}}$.   □

In order to define the ∞-category of motivic coarse spectra we proceed in a similar manner, starting with the presentable category of $\mathbf{Spc}\mathcal{X}$ of motivic coarse spaces. We then let $\mathbf{Spc}\mathcal{X}_{*/}$ be the pointed version of coarse motivic spaces and again consider the suspension endofunctor

$$\Sigma : \mathbf{Spc}\mathcal{X}_{*/} \to \mathbf{Spc}\mathcal{X}_{*/}, \quad X \mapsto \mathtt{colim}((* \to *) \leftarrow (* \to X) \to (* \to *)).$$

Definition 4.2 We define the category of motivic coarse spectra by

$$\mathbf{Sp}\mathcal{X} := \mathbf{Spc}\mathcal{X}_{*/}[\Sigma^{-1}] := \mathtt{colim}\{\mathbf{Spc}\mathcal{X}_{*/} \xrightarrow{\Sigma} \mathbf{Spc}\mathcal{X}_{*/} \xrightarrow{\Sigma} \mathbf{Spc}\mathcal{X}_{*/} \xrightarrow{\Sigma} \dots \},$$

where the colimit is taken in $\mathbf{Pr}^L$.

By construction, $\mathbf{Sp}\mathcal{X}$ is a large presentable stable ∞-category. We have a functor

$$\Sigma_+^{\mathrm{mot}} : \mathbf{Spc}\mathcal{X} \to \mathbf{Spc}\mathcal{X}_{*/} \to \mathbf{Sp}\mathcal{X}$$

which fits into an adjunction

$$\Sigma_+^{\mathrm{mot}} : \mathbf{Spc} \leftrightarrows \mathbf{Sp}\mathcal{X} : \Omega^{\mathrm{mot}}.$$

Definition 4.3 We define the functor

$$\mathrm{Yo}^s := \Sigma_+^{\mathrm{mot}} \circ \mathrm{Yo} : \mathbf{BornCoarse} \to \mathbf{Sp}\mathcal{X}.$$

In particular, every bornological coarse space represents a coarse spectrum

$$\mathrm{Yo}^s(X) := \Sigma_+^{\mathrm{mot}}(\mathrm{Yo}(X)).$$

The functor $\Sigma_+^{\mathrm{mot}} : \mathbf{Spc}\mathcal{X} \to \mathbf{Sp}\mathcal{X}$ has the following universal property [Lur17, Cor. 1.4.4.5]: For every cocomplete stable ∞-category $\mathbf{C}$ precomposition by Σ_+^{mot} provides an equivalence

$$\mathbf{Fun}^{\mathtt{colim}}(\mathbf{Spc}\mathcal{X}, \mathbf{C}) \simeq \mathbf{Fun}^{\mathtt{colim}}(\mathbf{Sp}\mathcal{X}, \mathbf{C}), \tag{4.2}$$

where the super-script $\mathtt{colim}$ stands for small colimit-preserving functors. To be precise, in order to apply [Lur17, Cor. 1.4.4.5] we must assume that $\mathbf{C}$ is presentable and stable. For an extension to all cocomplete stable ∞-categories see the next Lemma 4.4.[1]

If $\mathbf{D}$ is a pointed ∞-category admitting finite colimits, then we can define the colimit

$$\mathbf{Sp}(\mathbf{D}) := \mathtt{colim}(\mathbf{D} \xrightarrow{\Sigma} \mathbf{D} \xrightarrow{\Sigma} \mathbf{D} \xrightarrow{\Sigma} \dots). \tag{4.3}$$

In the lemma below $\mathbf{D}$ is κ-presentable for some regular cardinal κ (in the small universe), and we interpret the colimit in the ∞-category $\mathbf{Pr}_*^{L,\kappa}$ of pointed κ-presentable ∞-categories.

Let $\mathbf{C}$ be a stable ∞-category, and let $\mathbf{D}$ be a pointed ∞-category.

Lemma 4.4 *If $\mathbf{D}$ is presentable and $\mathbf{C}$ is small cocomplete, then the natural functor $\mathbf{D} \to \mathbf{Sp}(\mathbf{D})$ induces an equivalence*

$$\mathbf{Fun}^{\mathtt{colim}}(\mathbf{Sp}(\mathbf{D}), \mathbf{C}) \simeq \mathbf{Fun}^{\mathtt{colim}}(\mathbf{D}, \mathbf{C}).$$

If $\mathbf{C}$ is presentable, then this lemma is shown in Lurie [Lur17, Cor. 1.4.4.5].

---

[1] We thank Denis-Charles Cisinski and Thomas Nikolaus for providing this argument.

Proof We can assume that $\mathbf{D}$ is κ-presentable for some regular cardinal κ. Then we have $\mathbf{D} \simeq \mathrm{Ind}_\kappa(\mathbf{D}^\kappa)$, where $\mathbf{D}^\kappa$ denotes the small category of κ-compact objects in $\mathbf{D}$ and $\mathrm{Ind}_\kappa(\mathbf{D}^\kappa)$ is the free completion of $\mathbf{D}^\kappa$ by κ-filtered colimits. We let $\mathbf{Cat}_{\infty,*}^{L,\kappa}$ be the ∞-category of κ-cocomplete ∞-categories and functors which preserve κ-small colimits (notation $\mathbf{Fun}^{\mathrm{colim},\kappa}(-,-)$). Then $\mathrm{Ind}_\kappa : \mathbf{Cat}_{\infty,*}^{L,\kappa} \to \mathbf{Pr}_*^{L,\kappa}$ satisfies

$$\mathbf{Fun}^{\mathrm{colim}}(\mathrm{Ind}_\kappa(\mathbf{E}), \mathbf{C}) \simeq \mathbf{Fun}^{\mathrm{colim},\kappa}(\mathbf{E}, \mathbf{C})$$

for any cocomplete pointed ∞-category $\mathbf{C}$.

We let $\mathbf{Cat}_\infty^{ex,\kappa}$ be the full subcategory of $\mathbf{Cat}_{\infty,*}^{L,\kappa}$ of stable ∞-categories. We define the functor

$$\mathbf{Sp}_\kappa : \mathbf{Cat}_{\infty,*}^{L,\kappa} \to \mathbf{Cat}_\infty^{ex,\kappa}$$

by Formula (4.3), where the colimit is now interpreted in $\mathbf{Cat}_{\infty,*}^{L,\kappa}$. If $\mathbf{C}$ belongs to $\mathbf{Cat}_\infty^{ex,\kappa}$, then for every $\mathbf{E}$ in $\mathbf{Cat}_\infty^{ex,\kappa}$ the natural functor $\mathbf{E} \to \mathbf{Sp}(\mathbf{E})$ induces an equivalence

$$\mathbf{Fun}^{\mathrm{colim},\kappa}(\mathbf{Sp}_\kappa(\mathbf{E}), \mathbf{C}) \simeq \mathbf{Fun}^{\mathrm{colim},\kappa}(\mathbf{E}, \mathbf{C}) .$$

The functor Ind_κ induces a fully faithful functor (which is even an equivalence if $\kappa > \omega$) $\mathbf{Cat}_{\infty,*}^{L,\kappa} \to \mathbf{Pr}^{L,\kappa}$. Furthermore,

$$\mathrm{Ind}_\kappa \circ \mathbf{Sp}_\kappa \simeq \mathbf{Sp} \circ \mathrm{Ind}_\kappa$$

since Ind_κ is a left-adjoint and therefore commutes with colimits. Consequently, we get a chain of natural equivalences

$$\mathbf{Fun}^{\mathrm{colim}}(\mathbf{Sp}(\mathbf{D}), \mathbf{C}) \simeq \mathbf{Fun}^{\mathrm{colim}}(\mathbf{Sp}(\mathrm{Ind}_\kappa(\mathbf{D}^\kappa)), \mathbf{C})$$

$$\simeq \mathbf{Fun}^{\mathrm{colim}}(\mathrm{Ind}_\kappa(\mathbf{Sp}_\kappa(\mathbf{D}^\kappa)), \mathbf{C})$$

$$\simeq \mathbf{Fun}^{\mathrm{colim},\kappa}(\mathbf{Sp}_\kappa(\mathbf{D}^\kappa), \mathbf{C})$$

$$\simeq \mathbf{Fun}^{\mathrm{colim},\kappa}(\mathbf{D}^\kappa, \mathbf{C})$$

$$\simeq \mathbf{Fun}^{\mathrm{colim}}(\mathrm{Ind}_\kappa(\mathbf{D}^\kappa), \mathbf{C})$$

$$\simeq \mathbf{Fun}^{\mathrm{colim}}(\mathbf{D}, \mathbf{C})$$

$\square$

Our main usage of the stability of an ∞-category $\mathbf{S}$ is the existence of a functor

$$\mathbf{Fun}(\Delta^1, \mathbf{S}) \to \mathbf{Fun}(\mathbb{Z}, \mathbf{S})$$

(where $\mathbb{Z}$ is considered as a poset) which functorially extends a morphism $f : E \to F$ in $\mathbf{S}$ to a cofibre sequence

$$\cdots \to \Sigma^{-1}E \to \Sigma^{-1}F \to \Sigma^{-1}\text{Cofib}(f) \to E \to F \to \text{Cofib}(f) \to \Sigma E \to \Sigma F \to \cdots$$

We will also use the notation $\text{Fib}(f) := \Sigma^{-1}\text{Cofib}(f)$.

If $\mathcal{Y} = (Y_i)_{i \in I}$ is a big family in a bornological coarse space X, then we define the motivic coarse spectrum

$$\text{Yo}^s(\mathcal{Y}) := \underset{i \in I}{\text{colim}}\, \text{Yo}^s(Y_i) . \tag{4.4}$$

Since Σ_+^{mot} preserves colimits we have the equivalence $\text{Yo}^s(\mathcal{Y}) \simeq \Sigma_+^{\text{mot}}(\text{Yo}(\mathcal{Y}))$. The family of inclusions $Y_i \to X$ induces via the universal property of the colimit a canonical morphism

$$\text{Yo}^s(\mathcal{Y}) \to \text{Yo}^s(X) .$$

We will use the notation

$$(X, \mathcal{Y}) := \text{Cofib}(\text{Yo}^s(\mathcal{Y}) \to \text{Yo}^s(X)) . \tag{4.5}$$

Let X be a bornolological coarse space. The stabilized Yoneda functor Yo^s has the following properties:

Corollary 4.5

1. If $\mathcal{Y}$ is a big family in X, then we have a fibre sequence

$$\cdots \to \text{Yo}^s(\mathcal{Y}) \to \text{Yo}^s(X) \to (X, \mathcal{Y}) \to \Sigma\text{Yo}^s(\mathcal{Y}) \to \ldots$$

2. For a complementary pair $(Z, \mathcal{Y})$ on X the natural morphism

$$(Z, Z \cap \mathcal{Y}) \to (X, \mathcal{Y})$$

is an equivalence.
3. If $X \to X'$ is an equivalence of bornological coarse spaces, then $\text{Yo}^s(X) \to \text{Yo}^s(X')$ is an equivalence in $\mathbf{Sp}\mathcal{X}$.
4. If X is flasque, then $\text{Yo}^s(X)$ is a zero object in the stable ∞-category $\mathbf{Sp}\mathcal{X}$. In particular, $\text{Yo}^s(\emptyset)$ is a zero object.
5. For every bornological coarse space X with coarse structure $\mathcal{C}$ we have the equivalence $\text{Yo}^s(X) \simeq \text{colim}_{U \in \mathcal{C}}\, \text{Yo}^s(X_U)$.

Proof The Property 1 is clear from the definition (4.5) of the relative motive and the stability of $\mathbf{Sp}\mathcal{X}$.

The remaining assertions follow from the corresponding assertions of Corollary 3.35 and the fact that Σ^{mot}_+ is a left-adjoint. Hence it preserves colimits and initial objects. In the following we give some details.

Property 2 is an immediate consequence of the fact that the square

$$\begin{array}{ccc} \mathrm{Yo}^s(Z \cap \mathcal{Y}) & \longrightarrow & \mathrm{Yo}^s(\mathcal{Y}) \\ \downarrow & & \downarrow \\ \mathrm{Yo}^s(Z) & \longrightarrow & \mathrm{Yo}^s(X) \end{array}$$ (4.6)

is cocartesian. This follows from Point 2 in Corollary 3.35.

Property 3 immediately follows from Point 3 of Corollary 3.35.

We now show Property 4. If X in **BornCoarse** is flasque, then $\mathrm{Yo}(X)$ is an initial object of $\mathbf{Spc}\mathcal{X}$ by Point 4 of Corollary 3.35. Hence $\mathrm{Yo}^s(X) \simeq \Sigma^{mot}_+(\mathrm{Yo}(X))$ is an initial object of $\mathbf{Sp}\mathcal{X}$. Since $\mathbf{Sp}\mathcal{X}$ is stable an initial object in $\mathbf{Sp}\mathcal{X}$ is a zero object.

Property 5 follows from the u-continuity of Yo stated in Point 5 of Corollary 3.35.

□

4.2   Further Properties of Yo^s

Let X be a bornological coarse space.

Lemma 4.6 *If X is flasque in the generalized sense, then $\mathrm{Yo}^s(X) \simeq 0$.*

Proof This follows from Lemma 3.38 and the fact that Σ^{mot}_+ sends initial objects to zero objects.   □

Let X be in **BornCoarse**, and let A be a subset of X.

Lemma 4.7 *We have an equivalence $\mathrm{Yo}^s(A) \simeq \mathrm{Yo}^s(\{A\})$.*

Proof This follows from Corollary 3.39 and the fact that Σ^{mot}_+ preserves colimits.   □

Let X be a bornological coarse space, and let Y, Z be two subsets of X.

Lemma 4.8 *If (Y, Z) is a coarsely excisive pair on X, then the square*

$$\begin{array}{ccc} \mathrm{Yo}^s(Y \cap Z) & \longrightarrow & \mathrm{Yo}^s(Y) \\ \downarrow & & \downarrow \\ \mathrm{Yo}^s(Z) & \longrightarrow & \mathrm{Yo}^s(X) \end{array}$$

is cocartesian.

Proof This follows from Lemma 3.41 and the fact that Σ^{mot}_+ preserves colimits.   □

Example 4.9 We consider a bornological coarse space X. As we have explained in the Example 3.22 the bornological coarse space $[0, \infty) \otimes X$ is flasque. The pair of subsets

$$((-\infty, 0] \otimes X, [0, \infty) \otimes X)$$

of $\mathbb{R} \otimes X$ is coarsely excisive. By Lemma 4.8 we get the push-out square

$$
\begin{array}{ccc}
\text{Yo}^s(X) & \longrightarrow & \text{Yo}^s([0, \infty) \otimes X) \\
\downarrow & & \downarrow \\
\text{Yo}^s((-\infty, 0] \otimes X) & \longrightarrow & \text{Yo}^s(\mathbb{R} \otimes X)
\end{array}
$$

The lower left and the upper right corners are motives of flasque bornological coarse spaces and hence vanish. Excision therefore provides an equivalence

$$\text{Yo}^s(\mathbb{R} \otimes X) \simeq \Sigma \text{Yo}^s(X). \tag{4.7}$$

Iterating this we get

$$\text{Yo}^s(\mathbb{R}^k \otimes X) \simeq \Sigma^k \text{Yo}^s(X) \tag{4.8}$$

for all $k \geq 0$. □

Example 4.10 The following is an application of the stability of the category of coarse motivic spectra. We consider a family $(X_i)_{i \in I}$ bornological coarse spaces and form the free union (Definition 2.27) $X := \bigsqcup_{i \in I}^{\text{free}} X_i$.

We fix an element $i \in I$ and set $I' := I \setminus \{i\}$. Then $(X_i, \bigsqcup_{j \in I'}^{\text{free}} X_j)$ is a coarsely excisive pair on X. Since the intersection of the two entries is disjoint and $\text{Yo}^s(\emptyset) \simeq 0$, excision (Lemma 4.8) gives an equivalence

$$\text{Yo}^s(X) \simeq \text{Yo}^s(X_i) \oplus \text{Yo}^s\left(\bigsqcup_{j \in I'}^{\text{free}} X_j\right).$$

In particular we obtain a projection

$$p_i : \text{Yo}^s(X) \to \text{Yo}^s(X_i). \tag{4.9}$$

Note that this projection does not come from a morphism in **BornCoarse**. We can combine the projections p_i for all i in I to a morphism

$$p : \mathrm{Yo}^s(X) \to \prod_{j \in I} \mathrm{Yo}^s(X_j) . \tag{4.10}$$

This map will play a role in the definition of additivity of a coarse homology theory later. If I is finite, then it is an equivalence by excision.   □

Excision can deal with finite decompositions. The following lemma investigates what happens in the case of inifinite coproducts.

Let $\mathcal{A}$ be a set of objects in **BornCoarse**.

Definition 4.11 We let $\mathbf{Sp}\mathcal{X}\langle\mathcal{A}\rangle$ denote the minimal cocomplete stable full subcategory of $\mathbf{Sp}\mathcal{X}$ containing $\mathrm{Yo}^s(\mathcal{A})$.

We let $\mathcal{A}_{\mathrm{disc}}$ be the set of discrete bornological coarse spaces (see Definition 2.14).

Let $(X_i)_{i \in I}$ be a family in **BornCoarse**. Then we have a canonical maps

$$\bigoplus_{i \in I} \mathrm{Yo}^s(X_i) \to \mathrm{Yo}^s(\bigsqcup_{i \in I} X_i) , \quad \bigoplus_{i \in I} \mathrm{Yo}^s(X_i) \to \mathrm{Yo}^s(\overset{\mathrm{mixed}}{\underset{i \in I}{\bigsqcup}} X_i) \tag{4.11}$$

induced by the inclusions of the components X_i into the coproduct or the mixed union.

Lemma 4.12 *The fibres of the maps in* (4.11) *belong to* $\mathbf{Sp}\mathcal{X}\langle\mathcal{A}_{\mathrm{disc}}\rangle$.

Proof We consider the case of a mixed union. Let $X := \bigsqcup_{i \in I}^{\mathrm{mixed}} X_i$. Let $\mathcal{C}$ denote the coarse structure of X and recall that we have the equivalence

$$\mathrm{Yo}^s(X) \simeq \underset{U \in \mathcal{C}}{\mathrm{colim}}\, \mathrm{Yo}^s(X_U) .$$

Let now U in $\mathcal{C}$ be given such that it contains the diagonal. Then there exists a finite subset J of I and entourages U_j in $\mathcal{C}_j$ for all j in J such that

$$U = \bigcup_{j \in J} U_j \cup \bigcup_{i \in I \setminus J} \mathrm{diag}(X_i) .$$

We conclude that

$$X_U \cong \bigsqcup_{j \in J} X_{j,U_j} \sqcup \overset{\mathrm{mixed}}{\underset{i \in I \setminus J}{\bigsqcup}} X_{i,\mathrm{disc}} ,$$

where $X_{i,\mathrm{disc}} := (X_i, \mathcal{C}\langle\emptyset\rangle, \mathcal{B}_i)$. We conclude that

$$\mathrm{Yo}^s(X_U) \simeq \bigoplus_{j \in J} \mathrm{Yo}^s(X_{j,U_j}) \oplus \overset{\text{mixed}}{\mathrm{Yo}^s\Big(\bigsqcup_{i \in I \setminus J} X_{i,\mathrm{disc}}\Big)}.$$

Let not J' be a second finite subset of I such that $J \subseteq J'$. Then by excision we have a decomposition

$$\overset{\text{mixed}}{\mathrm{Yo}^s\Big(\bigsqcup_{i \in I \setminus J} X_{i,\mathrm{disc}}\Big)} \simeq \overset{\text{mixed}}{\mathrm{Yo}^s\Big(\bigsqcup_{i \in I \setminus J'} X_{i,\mathrm{disc}}\Big)} \oplus \bigoplus_{i \in J' \setminus J} \mathrm{Yo}^s(X_{i,\mathrm{disc}}).$$

In particular we have a projection

$$\overset{\text{mixed}}{\mathrm{Yo}^s\Big(\bigsqcup_{i \in I \setminus J} X_{i,\mathrm{disc}}\Big)} \to \overset{\text{mixed}}{\mathrm{Yo}^s\Big(\bigsqcup_{i \in I \setminus J'} X_{i,\mathrm{disc}}\Big)}.$$

We now take the colimit over $\mathcal{C}$ in two stages such that the outer colimit increases J and the inner colimits runs over the entourages with a fixed J. Then we get a fibre sequence

$$\Sigma^{-1} R \to \bigoplus_{i \in I} \mathrm{Yo}^s(X_i) \to \mathrm{Yo}^s(X) \to R,$$

where

$$R := \underset{J \subseteq I \, finite}{\mathrm{colim}} \, \overset{\text{mixed}}{\mathrm{Yo}^s\Big(\bigsqcup_{i \in I \setminus J} X_{i,\mathrm{disc}}\Big)} \tag{4.12}$$

is a remainder term as claimed by the lemma.

The above argument works word for word also in the case of the coproduct $X := \bigsqcup_{i \in I} X_i$. In this case the remainder term will be

$$\underset{J}{\mathrm{colim}} \, \mathrm{Yo}^s\Big(\bigsqcup_{i \in I \setminus J} X_{i,\mathrm{disc}}\Big). \tag{4.13}$$

$\square$

4.3   Homotopy Invariance

In order to define the notion of homotopy in the context of bornological coarse spaces we introduce appropriate cylinder objects (see Mitchener [Mit10, Sec. 3] for a similar construction). A coarse cylinder on X depends on the choice $p = (p_-, p_+)$ of two bornological (Definition 2.8) maps $p_+ : X \to [0, \infty)$ and $p_- : X \to (-\infty, 0]$, where the rays have the usual bornology of bounded subsets.

Let $(X, \mathcal{C}, \mathcal{B})$ be a bornological coarse space. Furthermore, we equip $\mathbb{R}$ with the metric bornological coarse structure $\mathcal{C}_d$ and $\mathcal{B}_d$. We consider the bornological coarse space $\mathbb{R} \otimes X$ introduced in Example 2.32.

Remark 4.13 Using $\mathbb{R} \otimes X$ instead of the cartesian product $\mathbb{R} \times X$ has the effect that the projections $\mathbb{R} \otimes X \to \mathbb{R}$ (for unbounded X) and $\mathbb{R} \otimes X \to X$ are not morphisms of bornological coarse spaces since they are not proper. On the other hand, our definition guarantees that the inclusion $X \to \mathbb{R} \otimes X$ given by $\{0\}$ in $\mathbb{R}$ is an inclusion of a bornological coarse subspace. A further reason for using $\otimes$ instead of $\times$ is that we want the half cylinders in $\mathbb{R} \otimes X$ to be flasque, see Example 3.22. This property will be used in an essential manner in the proof of Proposition 4.16 below.   □

Let X be a bornological coarse space, and let $p_+ : X \to [0, \infty)$ and $p_- : X \to (-\infty, 0]$ be bornological maps.

Definition 4.14   The coarse cylinder $I_p X$ is defined as the subset

$$I_p X := \{(t, x) \in \mathbb{R} \times X : p_-(x) \le t \le p_+(x)\} \subseteq \mathbb{R} \otimes X$$

equipped with the induced bornological coarse structure.

Lemma 4.15   *The projection* $\pi : I_p X \to X$ *is a morphism of bornological coarse spaces.*

Proof   The projection is clearly controlled. We must show that it is proper. Let B be a bounded subset of X. Since the maps $p_\pm$ are bornological there exists a positive real number C such that $|p_\pm(x)| \le C$ for all x in B. We then have $\pi^{-1}(B) \subseteq [-C, C] \times B \cap I_p X$, and this subset is bounded.   □

Proposition 4.16   *For a coarse cylinder* $I_p X$ *on* X *the projection* $I_p X \to X$ *induces an equivalence*

$$\mathrm{Yo}^s(I_p X) \to \mathrm{Yo}^s(X).$$

Proof   We consider the bornological coarse subspaces of $\mathbb{R} \otimes X$

$$W := (-\infty, 0] \times X \cup I_p(X), \qquad Z := [0, \infty) \times X \cap I_p(X)$$

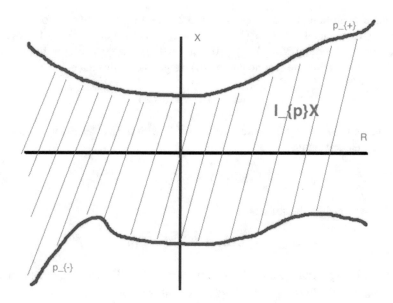

Coarse cylinder

Coarse cylinder

and the big family $\{Y\}$ generated by the subset $Y := (-\infty, 0] \times X$, see Example 3.3.
Then $(Z, \{Y\})$ is a complementary pair on W. By Point 1 of Corollary 4.5 we have
fibre sequences

$$\cdots \to \mathrm{Yo}^s(\{Y\}) \to \mathrm{Yo}^s(W) \to (W, \{Y\}) \to \cdots$$

and

$$\cdots \to \mathrm{Yo}^s(Z \cap \{Y\}) \to \mathrm{Yo}^s(Z) \to (Z, Z \cap \{Y\}) \to \cdots,$$

where we use the construction in Example 3.7 in the second case. By Point 2 of
Corollary 4.5 we have the excision equivalence

$$(Z, Z \cap \{Y\}) \simeq (W, \{Y\}). \tag{4.14}$$

We now observe that Y and W are flasque. In both cases flasqueness is
implemented by the self maps given by the restriction of the map

$$f : \mathbb{R} \otimes X \to \mathbb{R} \otimes X, \quad (t, x) \mapsto (t - 1, x).$$

The three conditions listed in Definition 3.21 are easy to verify. It follows from Point 4 of Corollary 4.5 that $\mathrm{Yo}^s(W) \simeq 0$, and from Point 4 of Corollary 4.5 together with Lemma 3.39 that $\mathrm{Yo}^s(\{Y\}) \simeq 0$. The first fibre sequence now implies that $(W, \{Y\}) \simeq 0$. Using the second and (4.14) we get the equivalence

$$\mathrm{Yo}^s(Z \cap \{Y\}) \simeq \mathrm{Yo}^s(Z) \,.$$

Note that $Z \cap \{Y\} \simeq \{X\}$, where we identify X with the subset $\{0\} \times X$ of Z. We thus get the equivalence

$$\mathrm{Yo}^s(X) \simeq \mathrm{Yo}^s(Z) \,. \tag{4.15}$$

We now consider the subsets

$$V := I_p X \cup [0, \infty) \times X \,, \quad U := [0, \infty) \times X$$

of $\mathbb{R} \otimes X$. We again have a complementary pair $(I_p X, \{U\})$ in V. Argueing as above and using that V and U are flasque we get an equivalence

$$\mathrm{Yo}^s(I_p X \cap \{U\}) \simeq \mathrm{Yo}^s(I_p X) \,.$$

We now observe that $I_p X \cap U = Z$. This implies by Point 4 of Corollary 4.5 that

$$\mathrm{Yo}^s(I_p X \cap \{U\}) \simeq \mathrm{Yo}^s(Z) \,.$$

Combining this with the equivalence (4.15) we get the equivalence

$$\mathrm{Yo}^s(X) \simeq \mathrm{Yo}^s(I_p X)$$

induced by the inclusion as the zero section. This is clearly an inverse of the projection.   □

We consider a coarse cylinder $I_p X$ over X. We can consider the maps of sets

$$i_\pm : X \to I_p X \,, \quad i_\pm(x) = (p_\pm(x), x) \,.$$

These maps are morphisms if the bornological maps $p_+ : X \to [0, \infty)$ and $p_- : X \to (-\infty, 0]$ are in addition controlled.

Let X and X' be bornological coarse spaces and $f_+, f_- : X \to X'$ be two morphisms.

Definition 4.17 We say that f_+ and f_- are homotopic to each other if there exist a pair of controlled and bornological maps $p = (p_+, p_-)$ and a morphism $h : I_p X \to X'$ such that $f_\pm = h \circ i_\pm$.

Observe that $\pi \circ i_{\pm} = \mathrm{id}_X$. Since $\mathrm{Yo}^s(\pi)$ is an equivalence, the functor Yo^s sends i_+ and i_- to equivalent maps. We have the following consequence.

Let X and X' be bornological coarse spaces and $f_+, f_- : X \to X'$ be two morphisms.

Corollary 4.18 *If f_+ and f_- are homotopic, then $\mathrm{Yo}^s(f_+)$ and $\mathrm{Yo}^s(f_-)$ are equivalent.*

Let $f : X \to X'$ be a morphism between bornological coarse spaces.

Definition 4.19 We say that f is a homotopy equivalence if there exist a morphism $g : X' \to X$ such that the compositions $f \circ g$ and $g \circ f$ are homotopic to the respective identities.

Corollary 4.20 *The functor Yo^s sends homotopy equivalences to equivalences.*

It is easy to see that if f_+ and f_- are close to each other, then they are homotopic. By Point 3 of Corollary 4.5 the functor Yo^s sends close morphisms to equivalent morphisms. But homotopy is a much weaker relation. The Corollary 4.20 can be considered as a vast improvement of Point 3 of Corollary 4.5.

Example 4.21 Let n be in $\mathbb{N}$ such that $n \geq 1$. For r in $(0, \infty)$ we let $B^n(0, r)$ denote the closed Euclidean ball in $\mathbb{R}^n$ at the origin of radius r. Let I be some set and $r : I \to (0, \infty)$ be some function. We consider the set

$$X := \bigsqcup_{i \in I} B^n(0, r(i)) .$$

We define a coarse structure on X to be generated by the family of entourages $(\tilde{U}_r)_{r \geq 0}$ on X, where $\tilde{U}_r := \bigcup_{i \in I} U_{r,i}$, and where $U_{r,i}$ is the entourage U_r (see Example 2.1) considered as an entourage of the component of X with index i in I. In other words, we consider X as a quasi-metric space where all components have the induced Euclidean metric and their mutual distance is infinite. We let the bornology on X be generated by the subsets $B^n(0, r(i))$ (the component with index i) for all i in I. The bornology and the coarse structure are compatible. In this way we have described a bornological coarse space X.

We consider I with the discrete coarse structure and the minimal bornology. Then we have a natural inclusion

$$\iota : I \hookrightarrow X$$

as the centers of the balls.

We claim that ι is a homotopy equivalence with inverse the projection $\pi : X \to I$. Note that $\pi \circ \iota \cong \mathrm{id}_I$. We must exhibit a homotopy between $\iota \circ \pi$ and id_X. We let (i, x) denote the point $x \in B^n(0, r(i))$ of the i'th component. We consider the functions $p_- := 0$ and $p_+(i, x) := \|x\|$ and form the coarse cylinder $I_p X$ associated to the pair $p := (p_-, p_+)$. One can check that

$$h : I_p X \rightarrow X , \quad h(t, (i, x)) := \begin{cases} (i, x - t \frac{x}{\|x\|}) & x \neq 0 \\ (i, 0) & x = 0 \end{cases}$$

is a morphism. Since the functions $p_\pm$ are controlled this morphism is a homotopy from the composition $\iota \circ \pi$ and id_X as required.

We conclude that

$$\mathrm{Yo}^s(I) \simeq \mathrm{Yo}^s(X) .$$

If the function r is unbounded, then ι and π are not coarse equivalences.   □

4.4   Axioms for a Coarse Homology Theory

Let $\mathbf{C}$ be a cocomplete stable ∞-category, and let

$$E : \mathbf{BornCoarse} \rightarrow \mathbf{C}$$

be a functor. For a big family $\mathcal{Y} = (Y_i)_{i \in I}$ on a bornological coarse space X we set

$$E(\mathcal{Y}) := \operatorname*{colim}_{i \in I} E(Y_i)$$

and

$$E(X, \mathcal{Y}) := \mathrm{Cofib}(E(\mathcal{Y}) \rightarrow E(X)) .$$

Definition 4.22 E is a coarse homology theory if it has the following properties:

1. (excision) For every complementary pair $(Z, \mathcal{Y})$ on a bornological coarse space X the natural morphism

$$E(Z, Z \cap \mathcal{Y}) \rightarrow E(X, \mathcal{Y})$$

   is an equivalence.
2. (coarse invariance) If $X \rightarrow X'$ is an equivalence of bornological coarse spaces, then $E(X) \rightarrow E(X')$ is an equivalence in $\mathbf{C}$.
3. (vanishing on flasques) If X is a flasque bornological coarse space, then $0 \simeq E(X)$.

4. (*u*-continuity) For every bornological coarse space X with coarse structure $\mathcal{C}$ we have the equivalence

$$E(X) \simeq \operatorname*{colim}_{U \in \mathcal{C}} E(X_U)$$

induced by the collection of canonical morphisms $X_U \to X$.

Remark 4.23 The excision axiom is equivalent to the requirement, that for every complementary pair $(Z, \mathcal{Y})$ on a bornological coarse space X the square

$$\begin{array}{ccc} E(Z \cap \mathcal{Y}) & \longrightarrow & E(Z) \\ \downarrow & & \downarrow \\ E(\mathcal{Y}) & \longrightarrow & E(X) \end{array}$$

is a push-out square. □

As a consequence of Corollary 3.37 and (4.2) the ∞-category $\mathbf{Sp}\mathcal{X}$ of motivic coarse spectra has the following universal property:

Corollary 4.24 *For every large, small cocomplete stable ∞-category $\mathbf{C}$ precomposition by Yo^s induces an equivalence between $\mathbf{Fun}^{\mathrm{colim}}(\mathbf{Sp}\mathcal{X}, \mathbf{C})$ and the full subcategory of $\mathbf{Fun}(\mathbf{BornCoarse}, \mathbf{C})$ of $\mathbf{C}$-valued coarse homology theories.*

Note that $\mathtt{colim}$ stands for the small colimit-preserving functors. In particular, Yo^s is a $\mathbf{Sp}\mathcal{X}$-valued coarse homology theory.

Usually the targets $\mathbf{C}$ of our coarse homology theories are small and admit all very small colimits. To transfer motivic results to such coarse homology theories using Corollary 4.24 we employ the following trick.

There exists a fully faithful, very small colimit preserving inclusion $i : \mathbf{C} \to \mathbf{C}^{\mathrm{la}}$ of $\mathbf{C}$ into a large stable ∞-category which admits all small colimits, see Remark 4.27.

Here are two typical arguments using this trick.

If $E : \mathbf{BornCoarse} \to \mathbf{C}$ is a coarse homology theory, then we define

$$E^{\mathrm{la}} := i \circ E : \mathbf{BornCoarse} \to \mathbf{C}^{\mathrm{la}} .$$

Because i preserves colimits, it is again a coarse homology theory. By Corollary 4.24 it corresponds essentially uniquely to a small colimit preserving functor $\mathbf{Sp}\mathcal{X} \to \mathbf{C}^{\mathrm{la}}$ for which we use the same symbol. For X in $\mathbf{BornCoarse}$ we then have $i(E(X)) \simeq E^{\mathrm{la}}(\mathrm{Yo}^s(X))$.

Assume that $f : X \to Y$ is a morphism in $\mathbf{BornCoarse}$.

Corollary 4.25 *If $\mathrm{Yo}^s(f) \colon \mathrm{Yo}^s(X) \to \mathrm{Yo}^s(Y)$ is an equivalence, then $E(f) \colon E(X) \to E(Y)$ is an equivalence.*

Proof We use that i is fully faithful. □

Assume that $E \to E'$ is a natural transformation of coarse homology theories, and let X be in **BornCoarse**.

Corollary 4.26 *If $E^{\mathrm{la}}(\mathrm{Yo}^s(X)) \to E'^{,\mathrm{la}}(\mathrm{Yo}^s(X))$ is an equivalence, then $E(X) \to E'(X)$ is an equivalence.*

Proof We use that $E(X) \simeq E^{\mathrm{la}}(\mathrm{Yo}^s(X))$ and again that i is fully faithful.   □

Remark 4.27 The construction of such an embedding i is standard. We let **Sp**^{la} be the category of large spectra and consider the stable Yoneda embedding

$$\mathrm{yo}^s : \mathbf{C} \to \mathbf{Fun}(\mathbf{C}^{\mathrm{op}}, \mathbf{Sp}^{\mathrm{la}}) \ .$$

We let S be the small set of the canonical morphisms

$$\operatorname*{colim}_{I} \mathrm{yo}^s(D) \to \mathrm{yo}^s(\operatorname*{colim}_{I} D)$$

in $\mathbf{Fun}(\mathbf{C}^{\mathrm{op}}, \mathbf{Sp}^{\mathrm{la}})$ for all diagrams $D : I \to \mathbf{C}$ indexed by very small categories I. Since $\mathbf{Fun}(\mathbf{C}^{\mathrm{op}}, \mathbf{Sp}^{\mathrm{la}})$ is presentable we can define $\mathbf{C}^{\mathrm{la}}$ as the category of S-local objects in $\mathbf{Fun}(\mathbf{C}^{\mathrm{op}}, \mathbf{Sp}^{\mathrm{la}})$. It fits into a adjunction

$$L : \mathbf{Fun}(\mathbf{C}^{\mathrm{op}}, \mathbf{Sp}^{\mathrm{la}}) \leftrightarrows \mathbf{C}^{\mathrm{la}} : \mathrm{incl} \ .$$

Note that $\mathrm{yo}^s(C)$ is S-local for every object C of $\mathbf{C}$. This follows from

$$\mathrm{map}_{\mathbf{Fun}(\mathbf{C}^{\mathrm{op}}, \mathbf{Sp}^{\mathrm{la}})}(\mathrm{yo}^s(\operatorname*{colim}_{I} D), \mathrm{yo}^s(C)) \simeq \mathrm{map}_{\mathbf{C}}(\operatorname*{colim}_{I} D, C)$$

$$\simeq \operatorname*{lim}_{I} \mathrm{map}_{\mathbf{C}}(D, C)$$

$$\simeq \operatorname*{lim}_{I} \mathrm{map}_{\mathbf{Fun}(\mathbf{C}^{\mathrm{op}}, \mathbf{Sp}^{\mathrm{la}})}(\mathrm{yo}^s(D), \mathrm{yo}^s(C))$$

$$\simeq \mathrm{map}_{\mathbf{Fun}(\mathbf{C}^{\mathrm{op}}, \mathbf{Sp}^{\mathrm{la}})}(\operatorname*{colim}_{I} \mathrm{yo}^s(D), \mathrm{yo}^s(C)) \ ,$$

where $\mathrm{map}(-, -)$ stands for the mapping spectrum in the respective stable ∞-category, and $D : I \to \mathbf{C}$ is any diagram with a very small index category.

We now set $i := \mathrm{yo}^s : \mathbf{C} \to \mathbf{C}^{\mathrm{la}}$. This immediately implies that i is fully faithful. We finally observe that i preserves very small colimits. Let C be in $\mathbf{C}^{\mathrm{la}}$. Then we have the following chain of equivalences

$$\mathrm{map}_{\mathbf{C}^{\mathrm{la}}}(i(\operatorname*{colim}_{I} D), C) \simeq \mathrm{map}_{\mathbf{Fun}(\mathbf{C}^{\mathrm{op}}, \mathbf{Sp}^{\mathrm{la}})}(\mathrm{yo}^s(\operatorname*{colim}_{I} D), C)$$

$$\overset{!}{\simeq} \mathrm{map}_{\mathbf{Fun}(\mathbf{C}^{\mathrm{op}}, \mathbf{Sp}^{\mathrm{la}})}(\operatorname*{colim}_{I} \mathrm{yo}^s(D), C)$$

$$\simeq \mathrm{map}_{\mathbf{C}^{\mathrm{la}}}(L(\operatorname*{colim}_{I} \mathrm{yo}^s(D)), C)$$

$$\simeq \mathrm{map}_{\mathbf{C}^{\mathrm{la}}}(\operatorname*{colim}_{I}^{\mathbf{C}^a} \mathrm{yo}^s(D), C)$$

where $\operatorname{colim}_I \mathrm{yo}^s(D)$ is interpreted in $\mathbf{Fun}(\mathbf{C}^{\mathrm{op}}, \mathbf{Sp}^{\mathrm{la}})$, and $\operatorname{colim}_I^{\mathbf{C}^a} \mathrm{yo}^s(D)$ is interpreted in $\mathbf{C}^{\mathrm{la}}$. For the equivalence marked by ! we use the assumption that C is S-local.   □

Let $E : \mathbf{BornCoarse} \to \mathbf{C}$ be a coarse homology theory.

Corollary 4.28 *The following statements are true:*

1. *If A is a subset of a bornological coarse space with the induced bornological coarse structure, then the natural morphism induces an equivalence $E(A) \simeq E(\{A\})$.*
2. *If X is a bornological coarse space which is flasque in the generalized sense (Definition 3.27), then $E(X) \simeq 0$.*
3. *E sends homotopic morphisms (Definition 4.17) to equivalent morphisms.*
4. *E is coarsely excisive, i.e., for a coarsely excisive pair (Y, Z) (Definition 3.40) on X the square*

$$
\begin{array}{ccc}
E(Y \cap Z) & \longrightarrow & E(Y) \\
\downarrow & & \downarrow \\
E(Z) & \longrightarrow & E(X)
\end{array}
$$

is cocartesian.

Proof We use Corollary 4.25 in order to transfer the motivic relations relations between values of E.

The Assertion 2 now follows from Lemma 4.6. The Assertion 3 follows from Corollary 4.20. The Assertion 4 is a consequence of Lemma 4.8. And finally, the Assertion 1 follows from Lemma 4.7.   □

Chapter 5
Merging Coarse and Uniform Structures

If a bornological coarse space has an additional compatible uniform structure and a big family of subsets, then we can define a new bornological coarse structure called the hybrid structure (introduced by Wright [Wri05, Sec. 5]). This new structure will allow us to coarsely decompose simplicial complexes into cells and therefore to perform inductive arguments by dimension. Such arguments are at the core of proofs of the coarse Baum–Connes conjecture [Wri05], but also an important ingredient in proofs of the Farell–Jones conjecture in various cases [BLR08, BL11].

In Sect. 5.1 we introduce the hybrid coarse structure. Our main technical results are the decomposition theorem and the homotopy theorem shown in Sects. 5.2 and 5.3. These results will be applied in Sect. 5.5 in order to decompose the motives of simplicial complexes. In Sect. 5.4 we provide sufficient conditions for the flasqueness of hybrid spaces which will be used to study coarsening spaces in Sect. 5.6.

In the subsequent papers [BEb] and [BEKW20] we formalize bornological coarse spaces with additional uniform structures by introducing the category of uniform bornological coarse spaces **UBC**.

5.1 The Hybrid Structure

We start with recalling the notion of a uniform structure.

Let X be a set.

Definition 5.1 A uniform structure on X is a subset $\mathcal{T}$ of $\mathcal{P}(X \times X)$ satisfying following conditions

1. For every W in $\mathcal{T}$ we have $\mathrm{diag}_X \subseteq W$.
2. $\mathcal{T}$ is closed under taking supersets.
3. $\mathcal{T}$ is closed under taking inverses.

© The Editor(s) (if applicable) and The Author(s), under exclusive licence
to Springer Nature Switzerland AG 2020
U. Bunke, A. Engel, *Homotopy Theory with Bornological Coarse Spaces*,
Lecture Notes in Mathematics 2269, https://doi.org/10.1007/978-3-030-51335-1_5

4. $\mathcal{T}$ is closed under finite intersections.
5. For every W in $\mathcal{T}$ there exists W' in $\mathcal{T}$ such that $W' \circ W' \subseteq W$.

The elements of $\mathcal{T}$ will be called uniform entourages. We will occasionally use the term coarse entourage for elements of a coarse structure on X in order to clearly distinguish them from uniform entourages.

A uniform space is a pair $(X, \mathcal{T})$ of a set X with a uniform structure $\mathcal{T}$.

Let $\mathcal{T}$ be a uniform structure on a set X.

Definition 5.2 A uniform structure $\mathcal{T}$ is Hausdorff if $\bigcap_{W \in \mathcal{T}} W = \mathrm{diag}(X)$.

Example 5.3 If (X, d) is a metric space, then we can define a uniform structure $\mathcal{T}_d$. It is the smallest uniform structure containing the subsets U_r for r in $(0, \infty)$ given by (2.1). The uniform structure $\mathcal{T}_d$ is Hausdorff.   □

Let X be a set with a uniform structure $\mathcal{T}$ and a coarse structure $\mathcal{C}$.

Definition 5.4 We say that $\mathcal{T}$ and $\mathcal{C}$ are compatible if $\mathcal{T} \cap \mathcal{C} \neq \emptyset$.

In other words, a uniform and a coarse structure are compatible if there exist a controlled uniform entourage.

Example 5.5 If (X, d) is a metric space, then the coarse structure $\mathcal{C}_d$ introduced in Example 2.18 and the uniform structure $\mathcal{T}_d$ from Example 5.3 are compatible.   □

Example 5.6 Let X be a set. The minimal coarse structure $\mathcal{C}_{min}$ on X is only compatible with the discrete uniform structure $\mathcal{T}_{\mathrm{disc}} = \mathcal{P}(X \times X)$. On the other hand, the maximal coarse structure $\mathcal{C}_{max}$ is compatible with every uniform structure on X.   □

Let $(X, \mathcal{C}, \mathcal{B})$ be a bornological coarse space equipped with an additional compatible uniform structure $\mathcal{T}$ and with a big family $\mathcal{Y} = (Y_i)_{i \in I}$. We consider $\mathcal{T}$ as a filtered poset using the opposite of the inclusion relation

$$U \leq U' := U' \subseteq U .$$

Let I, J be partially ordered sets, and let $\phi : I \to J$ be a map between the underlying sets.

Definition 5.7 The map ϕ is called a function if it is order preserving. It is cofinal, if in addition for every j in J there exists i in I such that $j \leq \phi(i)$.

Example 5.8 We assume that the uniform structure $\mathcal{T}_d$ is induced by a metric d (see Example 5.3). For a positive real number r we can consider the uniform entourages U_r as defined in (2.1). We equip $(0, \infty)$ with the usual order and let $(0, \infty)^{\mathrm{op}}$ denote the same set with the opposite order. Consider a function $\phi : I \to (0, \infty)^{\mathrm{op}}$. Then $I \ni i \mapsto U_{\phi(i)} \in \mathcal{T}_d$ is cofinal if and only if $\lim_{i \in I} \phi(i) = 0$.   □

We consider $\mathcal{P}(X \times X)$ as a poset with the inclusion relation and we consider a function $\phi : I \to \mathcal{P}(X \times X)^{\mathrm{op}}$, where I is some poset.

Definition 5.9 The function ϕ is $\mathcal{T}$-admissible if for every U in $\mathcal{T}$ there exists i in I such that $\phi(i) \subseteq U$.

Note that we do not require that ϕ takes values in $\mathcal{T}$.

Let $\mathcal{Y} = (Y_i)_{i \in I}$ be a filtered family of subsets of a set X. For a function $\phi : I \to \mathcal{P}(X \times X)^{\mathrm{op}}$ we consider the following subset of $X \times X$:

$$U_\phi := \{(x, x') \in X \times X : (\forall i \in I : x, x' \in Y_i \text{ or } (x, x') \in \phi(i) \setminus (Y_i \times Y_i))\}. \tag{5.1}$$

Let us spell this out in words: a pair (x, x') belongs to U_ϕ if and only if for every i in I both entries x, x' belong to Y_i or, if not, the pair (x, x') belongs to the subset $\phi(i)$ of $X \times X$.

Let $(X, \mathcal{C}, \mathcal{B})$ be a bornological coarse space equipped with a compatible uniform structure $\mathcal{T}$ and a big family $\mathcal{Y} = (Y_i)_{i \in I}$.

Definition 5.10 We define the hybrid coarse structure on X by

$$\mathcal{C}_h := \mathcal{C}\langle\{U_\phi \cap V : V \in \mathcal{C} \text{ and } \phi : I \to \mathcal{P}(X \times X)^{\mathrm{op}} \text{ is } \mathcal{T}\text{-admissible}\}\rangle.$$

and set

$$X_h := (X, \mathcal{C}_h, \mathcal{B}).$$

By definition we have the inclusion $\mathcal{C}_h \subseteq \mathcal{C}$. This implies in particular that $\mathcal{C}_h$ is compatible with the bornological structure $\mathcal{B}$. The triple $(X, \mathcal{C}_h, \mathcal{B})$ is thus a bornological coarse space. Furthermore, the identity of the underlying sets is a morphism of bornological coarse spaces

$$(X, \mathcal{C}_h, \mathcal{B}) \to (X, \mathcal{C}, \mathcal{B}).$$

Note that the hybrid coarse structure depends on the original coarse structure $\mathcal{C}$, the uniform structure $\mathcal{T}$, and the big family $\mathcal{Y}$.

Remark 5.11 If the uniform structure $\mathcal{T}$ is determined by a metric and we have $I \cong \mathbb{N}$ as partially ordered sets, then we can describe the hybrid structure $\mathcal{C}_h$ as follows:

A subset U of $X \times X$ belongs to $\mathcal{C}_h$ if U in $\mathcal{C}$ and if for every ε in $(0, \infty)$ there exists an i in I such that $U \subseteq (Y_i \times Y_i) \cup U_\varepsilon$.

It is clear that any hybrid entourage satisfies this condition. In the other direction, it is easy to check that a subset U satisfying the condition above is contained in $U \cap U_\phi$ for an appropriate ϕ. In order to construct such a ϕ one uses the cofinal family $(U_{1/n})_{n \in \mathbb{N} \setminus \{0\}}$ in $\mathcal{T}$ and that $I = \mathbb{N}$ is well-ordered. $\qquad\square$

Example 5.12 Let $(X, \mathcal{C}, \mathcal{B})$ be a bornological coarse space equipped with an additional uniform structure $\mathcal{T}$. Then we can define the hybrid structure associated to $\mathcal{C}$, $\mathcal{T}$ and the big family $\mathcal{B}$ (see Example 3.4). This hybrid structure is called the $\mathcal{C}_0$-structure by Wright [Wri05, Def. 2.2].   □

We now consider functoriality of the hybrid structure.

Let $(X, \mathcal{T})$ and $(X', \mathcal{T}')$ be uniform spaces, and let $f : X \to X'$ be a map of sets.

Definition 5.13 The map f is uniformly continuous if for every V' in $\mathcal{T}'$ there exists V in $\mathcal{T}$ such that $(f \times f)(V) \subseteq V'$.

Let X, X' be bornological coarse spaces which come equipped with uniform structures $\mathcal{T}$ and $\mathcal{T}'$ and big families $\mathcal{Y} = (Y_i)_{i \in I}$ and $\mathcal{Y}' = (Y'_{i'})_{i' \in I'}$. Let $f : X \to X'$ be a morphism of bornological coarse spaces. In the following we introduce the condition of being compatible in order to express that f behaves well with respect to the additional structures. The main motivation for imposing the conditions is to ensure Lemma 5.15 below.

Definition 5.14 The morphism f is compatible if the following conditions are satisfied:

1. f is uniformly continuous.
2. For every i in I there exists i' in I' such that $f(Y_i) \subseteq Y'_{i'}$.

Lemma 5.15 *If f is compatible, then it is a morphism $f : X_h \to X'_h$.*

Proof It is clear that f is proper since we have not changed the bornological structures.

We must check that f is controlled with respect to the hybrid structures.

Let $\phi : I \to \mathcal{P}(X \times X)^{\mathrm{op}}$ be a $\mathcal{T}$-admissible function, and let V be a coarse entourage of X. We then consider the entourage $U_\phi \cap V$ of X_h. We have

$$(f \times f)(U_\phi \cap V) \subseteq (f \times f)(U_\phi) \cap (f \times f)(V).$$

Since $(f \times f)(V)$ is an entourage of X' it suffices to show that there exists a suitable $\mathcal{T}'$-admissible function $\phi' : I' \to \mathcal{P}(X' \times X')^{\mathrm{op}}$ such that we have $(f \times f)(U_\phi) \subseteq U_{\phi'}$. Since f is compatible we can choose a map of sets $\lambda : I \to I'$ such that $Y_i \subseteq Y'_{\lambda(i)}$ for all i in I. We now define a map of sets $\tilde{\phi}' : I' \to \mathcal{P}(X' \times X')$ by

$$\tilde{\phi}'(i') := \bigcap_{i \in \lambda^{-1}(i')} (f \times f)(\phi(i)).$$

Note that the empty intersection of subsets of X' is by convention equal to $X' \times X'$. Then we construct the order preserving function $\phi' : I' \to \mathcal{P}(X' \times X')^{\mathrm{op}}$ by

$$\phi'(i') := \bigcap_{j' \leq i'} \tilde{\phi}'(j').$$

We claim that ϕ' is $\mathcal{T}'$-admissible. Let W' in $\mathcal{T}'$ be given. Since f is uniformly continuous we can choose W in $\mathcal{T}$ such that $(f \times f)(W) \subseteq W'$. Since ϕ is $\mathcal{T}$-admissible we can choose i in I such that $\phi(i) \subseteq W$. Then $\phi'(\lambda(i)) \subseteq W'$ by construction.

Assume that (x, y) is in U_ϕ. We must show that $(f(x), f(y)) \in U_{\phi'}$. We consider i' in I' and have the following two cases:

1. $(f(x), f(y)) \in Y'_{i'} \times Y'_{i'}$.
2. $(f(x), f(y)) \notin Y'_{i'} \times Y'_{i'}$. In this case we must show that $(f(x), f(y)) \in \phi'(i')$. Let j' in I' be any element such that $j' \leq i'$. Then we must show that $(f(x), f(y)) \in \tilde{\phi}'(j')$. We have two subcases:

   (a) j' is in the image of λ. For any choice of a preimage j in $\lambda^{-1}(j')$ we have $(x, y) \notin Y_j \times Y_j$ since otherwise $(f(x), f(y)) \in Y'_{j'} \times Y'_{j'} \subseteq Y'_{i'} \times Y'_{i'}$. Hence for all j in $\lambda^{-1}(j')$ we have $(x, y) \in \phi(j)$. Then $(f(x), f(y)) \in \tilde{\phi}'(j')$ as required.
   (b) j' is not in the image of λ. Then $(f(x), f(y)) \in \tilde{\phi}'(j') = X' \times X'$.

All the above shows that f is controlled, which finishes this proof. □

Example 5.16 A typical example of a bornological coarse space with a C_0-structure is the cone $\mathcal{O}(Y)$ over a uniform space $(Y, \mathcal{T}_Y)$. We consider Y as a bornological coarse space with the maximal structures. We then form the bornological coarse space

$$([0, \infty) \times Y, \mathcal{C}, \mathcal{B}) := [0, \infty) \otimes Y ,$$

see Example 2.32. The metric induces a uniform structure $\mathcal{T}_d$ on $[0, \infty)$, and we form the product uniform structure $\mathcal{T} := \mathcal{T}_d \times \mathcal{T}_Y$ on $[0, \infty) \times Y$. The cone over $(Y, \mathcal{T}_Y)$ is now defined as

$$\mathcal{O}(Y) := ([0, \infty) \otimes Y, C_0, \mathcal{B}) ,$$

where one should not overlook that we use the C_0-structure.

Note that instead of all bounded subsets we could also take the big family

$$\mathcal{Y} := ([0, n] \times Y)_{n \in \mathbb{N}} .$$

The resulting hybrid structure is equal to the C_0-structure since $\mathcal{Y}$ is cofinal in $\mathcal{B}$.

Note that $[0, \infty) \otimes Y$ is flasque by Example 3.22, but the cone $\mathcal{O}(Y)$ it is not flasque in general since it has the smaller C_0-structure.

Let $\mathbf{U}$ denote the category of uniform spaces and uniformly continuous maps. It follows from Lemma 5.15 that the cone construction determines a functor

$$\mathcal{O} : \mathbf{U} \to \mathbf{BornCoarse}$$

which is of great importance in coarse algebraic topology, see [BEb] and
[BEKW20].   □

Let $(X, \mathcal{C}, \mathcal{B})$ be a bornological coarse space equipped with a compatible uniform
structure $\mathcal{T}$ and a big family $\mathcal{Y} = (Y_i)_{i \in I}$. We consider furthermore a subset Z of
X. It has an induced bornological coarse structure $(\mathcal{C}_{|Z}, \mathcal{B}_{|Z})$, an induced uniform
structure $\mathcal{T}_{|Z}$, and a big family $Z \cap \mathcal{Y} := (Z \cap Y_i)_{i \in I}$. We let $(\mathcal{C}_{|Z})_h$ denote the
hybrid coarse structure on Z defined by these induced structures. Let furthermore
$(\mathcal{C}_h)_{|Z}$ be the coarse structure on Z induced from the hybrid coarse structure $\mathcal{C}_h$ on
X.

Lemma 5.17 *We have an equality* $(\mathcal{C}_h)_{|Z} = (\mathcal{C}_{|Z})_h$.

Proof It is clear that $(\mathcal{C}_h)_{|Z} \subseteq (\mathcal{C}_{|Z})_h$.

Let now $\phi : I \to \mathcal{P}(Z \times Z)^{\mathrm{op}}$ be $\mathcal{T}_{|Z}$-admissible, and let V in $\mathcal{C}_{|Z}$ and U_ϕ be as
in (5.1). We must show that $U_\phi \cap V \in (\mathcal{C}_h)_{|Z}$. Let ψ be the composition of ϕ with
the inclusion $\mathcal{P}(Z \times Z)^{\mathrm{op}} \to \mathcal{P}(X \times X)^{\mathrm{op}}$. Then ψ is $\mathcal{T}$-admissible and

$$U_\phi \cap V = (U_\psi \cap V) \cap (Z \times Z),$$

hence $U_\phi \cap V \in (\mathcal{C}_h)_{|Z}$.   □

Remark 5.18 In this proof it is important to allow that $\mathcal{T}$-admissible functions can
have values which do not belong to $\mathcal{T}$. In fact, the inclusion $\mathcal{P}(Z \times Z) \to \mathcal{P}(X \times X)$
in general does not map $\mathcal{T}_{|Z}$ to $\mathcal{T}$.   □

5.2 Decomposition Theorem

In this subsection we show the decomposition theorem. It roughly states that a
uniform decomposition of a space with a hybrid structure associated to a family $\mathcal{Y}$
induces a decomposition of its motivic coarse spectrum relative to this family. The
precise formulation is Theorem 5.22. As a consequence we deduce an excisiveness
result for the cone functor.

5.2.1 Uniform Decompositions and Statement of the Theorem

We consider a uniform space $(X, \mathcal{T})$. For a uniform entourage U in $\mathcal{T}$ we write
$\mathcal{T}_{\subseteq U}$ (or $\mathcal{P}(X \times X)^{\mathrm{op}}_{\subseteq U}$, respectively) for the partially ordered subset of uniform
entourages (or subsets, respectively) which are contained in U.

Let (Y, Z) be a pair of two subsets of X.

Definition 5.19 The pair (Y, Z) is called a uniform decomposition if:

1. $X = Y \cup Z$.
2. There exists V in $\mathcal{T}$ and a function $s : \mathcal{P}(X \times X)^{\mathrm{op}}_{\subseteq V} \to \mathcal{P}(X \times X)^{\mathrm{op}}$ such that:

   (a) The restriction $s_{|\mathcal{T}_{\subseteq V}} : \mathcal{T}_{\subseteq V} \to \mathcal{P}(X \times X)^{\mathrm{op}}$ is $\mathcal{T}$-admissible.
   (b) For every W in $\mathcal{P}(X \times X)^{\mathrm{op}}_{\subseteq V}$ we have the relation

$$W[Y] \cap W[Z] \subseteq s(W)[Y \cap Z]. \tag{5.2}$$

Remark 5.20 Note that the condition in Definition 5.19 is apparently stronger than the condition (compare with Definition 3.40 of a coarsely excisive decomposition) that for every W in $\mathcal{P}(X \times X)^{\mathrm{op}}_{\subseteq V}$ there exists W' in $\mathcal{P}(X \times X)$ such that $W[Y] \cap W[Z] \subseteq W'[Y \cap Z]$. First of all W' must become small if W becomes small. And even stronger, the choice of W' for given W must be expressible by a function as stated. □

Example 5.21 Assume that (X, d) metric space such that the restriction of d to every component of X is a path-metric. We further assume that there exists a constant $c > 0$ such that the distance between every two components of X is greater than c. We consider a pair (Y, Z) of closed subsets such that $Y \cup Z = X$. If we equip X with the uniform structure $\mathcal{T}_d$ induced by the metric, then (Y, Z) is a uniform decomposition.

Here is the argument:

For every r in $[0, \infty)$ we define the subset

$$\bar{U}_r := \{(x, y) \in X \times X : d(x, y) \le r\}$$

of $X \times X$. Note that $\bar{U}_r$ is a uniform entourage provided r is in $r > 0$. We define the uniform entourage

$$V := \bar{U}_c \cap \bar{U}_1,$$

where c is as above. We consider the subset

$$Q := \{0\} \cup \{n^{-1} : n \in \mathbb{N} \ \& \ n \ne 0\}$$

of $[0, 1]$. For $W \in \mathcal{P}(X \times X)^{\mathrm{op}}_{\subseteq V}$ we let $d(W)$ in Q be the minimal element such that $W \subseteq \bar{U}_{d(W)}$. Note that $d(W)$ exists, and if W in $\mathcal{T}_V$ is such that $W' \le W$ (i.e., $W \subseteq W'$) then $d(W) \le d(W')$. We then define the function

$$s : \mathcal{P}(X \times X)^{\mathrm{op}}_{\subseteq V} \to \mathcal{P}(X \times X)^{\mathrm{op}}, \quad s(W) := \bar{U}_{d(W)}.$$

Since $s(\bar{U}_r) = \bar{U}_r$ for r in Q this function is $\mathcal{T}_d$-admissible.

Let us check that this function s satisfies (5.2). We consider an element x of $W[Y] \cap W[Z]$. Since $Y \cup Z = X$ we have $x \in Y$ or $x \in Z$. Without loss of generality we can assume that $x \in Y$. Then there exists z in Z such that $(x, z) \in W$. We choose a path γ from x to z realizing the distance from x to z (note that $W \subseteq V \subseteq \bar{U}_c$ and therefore x and z are in the same path component of X). Let now x' be the first point on the path γ with $x' \in Y \cap Z$. Such a point exists since Y and Z are both closed, $x \in Y$ and $z \in Z$. Then $(x, x') \in s(W)$, which finishes the proof of (5.2).   □

We consider a bornological coarse space $(X, \mathcal{C}, \mathcal{B})$ with an additional compatible uniform structure $\mathcal{T}$, big family $\mathcal{Y} = (Y_i)_{i \in I}$, and uniform decomposition Y, Z. We let X_h denote the associated bornological coarse space with the hybrid structure. Moreover, we let $Y_h{}^1$ and Z_h denote the subsets Y and Z equipped with the bornological coarse structure induced from X_h. By Lemma 5.17 this induced structure coincides with the hybrid structure defined by the induced uniform structures and big families on these subsets.

Recall the pair notation (4.5).

Theorem 5.22 (Decomposition Theorem) *We assume that $I = \mathbb{N}$ and that the uniform structure on X is Hausdorff. Then the following square in $\mathbf{Sp}\mathcal{X}$ is cocartesian*

$$
\begin{array}{ccc}
((Y \cap Z)_h, Y \cap Z \cap \mathcal{Y}) & \longrightarrow & (Y_h, Y \cap \mathcal{Y}) \\
\downarrow & & \downarrow \\
(Z_h, Z \cap \mathcal{Y}) & \longrightarrow & (X_h, \mathcal{Y})
\end{array}
\tag{5.3}
$$

The proof of the theorem will be given in the next section.

Remark 5.23 The assumption in the decomposition theorem that $I = \mathbb{N}$ seems to be technical. It covers our main applications to cones or the coarsening spaces. Therefore we have not put much effort into a generalization.   □

5.2.2   Proof of the Decomposition Theorem

We retain the notation from the statement of the theorem.

In order to simplify the notation we omit the subscript h at Y and Z. All subsets of X occurring in the following argument are considered with the induced bornological coarse structure from X_h.

In general we use the symbol I for the index set, but at the place where the assumption $I = \mathbb{N}$ is relevant, we use $\mathbb{N}$ instead.

---

[1]This should not be confused with the notation Y_i for the members of the big family $\mathcal{Y}$.

For every i in I we have a push-out square (compare with (4.6) and see also (3.1) for notation)

$$
\begin{array}{ccc}
\mathrm{Yo}^s(\{Y \cup Y_i\} \cap (Z \cup Y_i)) & \longrightarrow & \mathrm{Yo}^s(\{Y \cup Y_i\}) \\
\downarrow & & \downarrow \\
\mathrm{Yo}^s(Z \cup Y_i) & \longrightarrow & \mathrm{Yo}^s(X_h)
\end{array}
$$

associated to the complementary pair $(Z \cup Y_i, \{Y \cup Y_i\})$. Taking the cofibre of the obvious map from the push-out square

$$
\begin{array}{ccc}
\mathrm{Yo}^s(Y_i) & \longrightarrow & \mathrm{Yo}^s(Y_i) \\
\downarrow & & \downarrow \\
\mathrm{Yo}^s(Y_i) & \longrightarrow & \mathrm{Yo}^s(Y_i)
\end{array}
$$

we get the push-out square (with self-explaining pair notation)

$$
\begin{array}{ccc}
(\{Y \cup Y_i\} \cap (Z \cup Y_i), Y_i) & \longrightarrow & (\{Y \cup Y_i\}, Y_i) \\
\downarrow & & \downarrow \\
(Z \cup Y_i, Y_i) & \longrightarrow & (X_h, Y_i)
\end{array}
$$

We now form the colimit over i in I and get a push-out square

$$
\begin{array}{ccc}
(\{Y \cup \mathcal{Y}\} \cap (Z \cup \mathcal{Y}), \mathcal{Y}) & \longrightarrow & (\{Y \cup \mathcal{Y}\}, \mathcal{Y}) \\
\downarrow & & \downarrow \\
(Z \cup \mathcal{Y}, \mathcal{Y}) & \longrightarrow & (X_h, \mathcal{Y})
\end{array}
$$

Note that the combination of $\{-\}$ and $\mathcal{Y}$ indicates a double colimit. In the following we show that this square is equivalent to the square (5.3) by identifying the corresponding corners. In Remark 5.24 below we will verify the equivalence

$$(Z \cup \mathcal{Y}, \mathcal{Y}) \simeq (Z, Z \cap \mathcal{Y}). \tag{5.4}$$

Using Corollary 3.39 and this remark again (for Y in place of Z) we get the equivalence

$$(\{Y \cup \mathcal{Y}\}, \mathcal{Y}) \simeq (Y \cup \mathcal{Y}, \mathcal{Y}) \simeq (Y, Y \cap \mathcal{Y}).$$

The core of the argument is to observe the first equivalence in the chain

$$(\{Y \cup \mathcal{Y}\} \cap (Z \cup \mathcal{Y}), \mathcal{Y}) \simeq (\{Y \cap Z\} \cap Z \cup \mathcal{Y}, \mathcal{Y}) \simeq ((Y \cap Z) \cup \mathcal{Y}, \mathcal{Y}) \simeq (Y \cap Z, Y \cap Z \cap \mathcal{Y}),$$

where the remaining equivalences are other instances of the cases considered above.
We must show that

$$\mathrm{Yo}^s\left(\{Y \cup \mathcal{Y}\} \cap (Z \cup \mathcal{Y})\right) \simeq \mathrm{Yo}^s\left(\{Y \cap Z\} \cap Z \cup \mathcal{Y}\right). \tag{5.5}$$

Note that both sides are filtered colimits over families of motives of subsets of X_h.
We will show that these families of subsets dominate each other in a cofinal way.
This then implies the equivalence of the colimits.

One direction is easy. Let V be a coarse entourage of X_h containing the diagonal
and let i be in I. Then clearly

$$V[Y \cap Z] \cap Z \cup Y_i \subseteq V[Y \cup Y_i] \cap (Z \cup Y_i).$$

The other direction is non-trivial.

From now on we write $\mathbb{N}$ instead of I and choose i in $\mathbb{N}$. We furthermore consider
a coarse entourage U of X and a $\mathcal{T}$-admissible function

$$\phi : \mathbb{N} \to \mathcal{P}(X \times X)^{\mathrm{op}}.$$

This data defines a coarse entourage

$$V := U \cap U_\phi$$

of X_h. We must show that there exists j in I and an entourage W of X_h such that

$$V[Y \cup Y_i] \cap (Z \cup Y_i) \subseteq W[Y \cap Z] \cap Z \cup Y_j.$$

The uniformity of the decomposition (Y, Z) provides a uniform entourage C in
$\mathcal{T}$ (called V in Definition 5.19) and a $\mathcal{T}$-admissible function

$$s : \mathcal{P}(X \times X)^{\mathrm{op}}_{\subseteq C} \to \mathcal{P}(X \times X)^{\mathrm{op}}.$$

Since the uniform and coarse structures of X are compatible we can choose an
entourage E which is uniform and coarse. Using that s is $\mathcal{T}$-admissible we can
assume (after decreasing C if necessary) that s takes values in $\mathcal{P}(X \times X)^{\mathrm{op}}_{\subseteq E}$, so
in particular in coarse entourages. Without loss of generality we can furthermore
assume that $\mathrm{diag}(X) \subseteq s(W)$ for every W in $\mathcal{P}(X \times X)^{\mathrm{op}}_{\subseteq C}$ (otherwise replace s
by $s \cup \mathrm{diag}(X)$).

We define the function

$$\phi_C : \mathbb{N} \to \mathcal{P}(X \times X)^{\mathrm{op}}_{\subseteq C}, \quad \phi_C(i) := (C \cap \phi(i)) \cup \mathrm{diag}(X).$$

We furthermore define the maps

$$k' : \mathbb{N} \to \mathbb{N} \cup \{\infty\}, \quad \psi : \mathbb{N} \to \mathcal{P}(X \times X)^{\mathrm{op}}$$

as follows. For $\ell \in \mathbb{N}$ we consider the set

$$N(\ell) := \left\{ k' \in \mathbb{N} \ : \ \left(\forall k \in \mathbb{N} \mid k \leq k' \Rightarrow Y_k \times s(\phi_C(k-1))^{-1}[Y_k] \subseteq Y_\ell \times Y_\ell\right) \right\}.$$

1. If $N(\ell)$ is not empty and bounded, then we set

$$k'(\ell) := \max N(\ell), \quad \psi(\ell) := s(\phi_C(k'(\ell))).$$

2. If $N(\ell)$ is unbounded, then we set

$$k'(\ell) := \infty, \quad \psi(\ell) := \mathrm{diag}(X).$$

3. If $N(\ell)$ is empty (this can only happen for small ℓ), then we set

$$k'(\ell) := 0, \quad \psi(\ell) := X \times X.$$

We show that ψ is a $\mathcal{T}$-admissible function.

We first observe that k' is monotonously increasing since the family $(Y_\ell)_{\ell \in I}$ is so. Since s and ϕ_C are functions, ψ is also a function.

Let Q in $\mathcal{T}$ be given. We must show that there exists ℓ in $\mathbb{N}$ such that $\psi(\ell) \subseteq Q$. Since $s_{|\mathcal{T}_{\subseteq C}}$ is $\mathcal{T}$-admissible we can find W in $\mathcal{T}_{\subseteq C}$ such that $s(W) \subseteq Q$. Since ϕ_C is $\mathcal{T}$-admissible we can chose i in $\mathbb{N}$ such that $\phi_C(i) \subseteq W$. Using that $\mathcal{Y}$ is big and that s is bounded by E we see that $\lim_{\ell \to \infty} k'(\ell) = \infty$. Hence we can choose ℓ in $\mathbb{N}$ so large that $k'(\ell) \geq i$. Then $\psi(\ell) \subseteq Q$. This finishes the proof that ψ is $\mathcal{T}$-admissible.

We now fix i' in $\mathbb{N}$ so large that $i \leq i'$ and $\phi(i') \subseteq C$. This is possible by the $\mathcal{T}$-admissibility of ϕ. We then set

$$W := (\psi(i') \cap U_\psi) \cup \mathrm{diag}(X) \in \mathcal{C}_h.$$

Using that $\mathcal{Y}$ is big we further choose j in $\mathbb{N}$ such that $i' \leq j$ and $U[Y_{i'}] \subseteq Y_j$.

The following chain of inclusions finishes the argument:

$$V[Y \cup Y_i] \cap (Z \cup Y_i) \subseteq V[Y \cup Y_{i'}] \cap (Z \cup Y_{i'}) \subseteq W[Y \cap Z] \cap Z \cup Y_j.$$

The first inclusion is clear since $i \leq i'$ and therefore $Y_i \subseteq Y_{i'}$. It remains to show the second inclusion.

We show the claim by the following case-by-case discussion. Let x be in $V[Y \cup Y_{i'}] \cap (Z \cup Y_{i'})$.

1. If $x \in Y_{i'}$, then $x \in Y_j$ since $Y_{i'} \subseteq U[Y_{i'}] \subseteq Y_j$ by the choice of j.
2. Assume now that $x \in Z \backslash Y_{i'}$. Then there exists b in $Y \cup Y_{i'}$ such that $(x, b) \in V$.

    (a) If $b \in Y_i$, then $x \in U[Y_{i'}] \subseteq Y_j$.

(b) Otherwise $b \in Y \setminus Y_i$. In view of the definition (5.1) of U_ϕ we then have $(x, b) \in \phi(i')$. By our choice of i' this also implies $(x, b) \in \phi_C(i')$. We distinguish two cases:

- There exists k in $\mathbb{N}$ such that $(x, b) \in Y_k \times Y_k$. We can assume that k is minimal with this property. Then $(x, b) \in \phi_C(k - 1)$, where we set $\phi_C(-1) := C$. Consequently, $x \in \phi_C(k - 1)[Y] \cap Z$, and by the defining property of s there exists c in $Y \cap Z$ such that $(x, c) \in s(\phi_C(k - 1))$. For every k' in $\mathbb{N}$ we then have exactly one the following two cases:

  – $k' \leq k - 1$: Then $(x, c) \in s(\phi_C(k'))$.
  – $k' \geq k$: Then $(x, c) \in Y_k \times s(\phi_C(k - 1))^{-1}[Y_k]$.

  Let now ℓ in $\mathbb{N}$ be given. We let $k' := k'(\ell)$ be as in the definition of ψ. If $(x, c) \notin \psi(\ell) = s(\phi_C(k'))$, then $k' \geq k$ and we have case B:

  $$(x, c) \in Y_k \times s(\phi_C(k - 1))^{-1}[Y_k] \subseteq Y_\ell \times Y_\ell,$$

  where the last inclusion holds true by the definition of $k'(\ell)$. This implies that $x \in W[Y \cap Z] \cap Z$.

- For every k in $\mathbb{N}$ we have $(x, b) \notin Y_k \times Y_k$. Then we have

  $$(x, b) \in \bigcap_{k \in \mathbb{N}} \phi(k) = \mathtt{diag}(X)$$

  (the last equality holds true since ϕ is $\mathcal{T}$-admissible and $\mathcal{T}$ is Hausdorff by assumption), i.e., $x = b$. It follows that $x \in Z \cap Y$, hence $x \in W[Y \cap Z] \cap Z$.

The proof is therefore finished.

Remark 5.24 We have to verify the equivalence (5.4).

We retain the notation introduced for Theorem 5.22. We have the following chain of equivalences:

$(Z, Z \cap \mathcal{Y})$

$\simeq \underset{i \in I}{\mathrm{colim}}(Z, Z \cap Y_i)$ definition

$\simeq \underset{i \in I}{\mathrm{colim}}(Z, Z \cap \{Y_i\})$ coarse invariance

$\simeq \underset{i \in I}{\mathrm{colim}} \underset{j \in I, j \geq i}{\mathrm{colim}}(Z \cup Y_j, (Z \cup Y_j) \cap \{Y_i\})$ excision

$\simeq \underset{i \in I}{\mathrm{colim}} \underset{j \in I, j \geq i}{\mathrm{colim}} \underset{U \in \mathcal{C}_h}{\mathrm{colim}}(Z \cup Y_j, (Z \cup Y_j) \cap U[Y_i])$ definition

$\simeq \underset{i \in I}{\mathrm{colim}} \underset{U \in \mathcal{C}_h}{\mathrm{colim}} \underset{j \in I, j \geq i, Y_j \supseteq U[Y_i]}{\mathrm{colim}}(Z \cup Y_j, U[Y_i])$ $\mathcal{Y}$ is big and Lemma 4.7

$$\simeq \operatorname*{colim}_{i \in I} \operatorname*{colim}_{j \in I, j \geq i}(Z \cup Y_j, Y_i) \qquad \text{coarse invariance, Corollary 4.5.3}$$

$$\simeq \operatorname*{colim}_{i \in I}(Z \cup Y_i, Y_i) \qquad \text{cofinality}$$

$$\simeq (Z \cup \mathcal{Y}, \mathcal{Y}) \qquad \text{definition}$$

This verifies the equivalence (5.4). □

5.2.3 Excisiveness of the Cone-at-Infinity

Recall the cone functor $\mathcal{O} \colon \mathbf{U} \to \mathbf{BornCoarse}$ from Example 5.16.

If $f \colon (X, \mathcal{T}_X) \to (X', \mathcal{T}_{X'})$ is a morphism of uniform spaces, then by Lemma 5.15 we get an induced morphism $\mathcal{O}(f) \colon \mathcal{O}(X) \to \mathcal{O}(X')$ in $\mathbf{BornCoarse}$. We can thus define a functor

$$\mathcal{O}^{\infty} \colon \mathbf{U} \to \mathbf{Sp}\mathcal{X}, \quad (X, \mathcal{T}_X) \mapsto (\mathcal{O}(X), \mathcal{Y}), \tag{5.6}$$

where we use

$$\mathcal{Y} := ([0, n] \times X)_{n \in \mathbb{N}} \tag{5.7}$$

as the big family.

Corollary 5.25 *If the uniform structure on X is Hausdorff and (Y, Z) is a uniform decomposition of X, then the square*

$$
\begin{array}{ccc}
\mathcal{O}^{\infty}(Y \cap Z) & \longrightarrow & \mathcal{O}^{\infty}(Y) \\
\downarrow & & \downarrow \\
\mathcal{O}^{\infty}(Z) & \longrightarrow & \mathcal{O}^{\infty}(X)
\end{array}
$$

is cocartesian.

Proof Note that $([0, \infty) \times Y, [0, \infty) \times Z)$ is a uniform decomposition of $[0, \infty) \times X$ with the product uniform structure (which is again Hausdorff), and the big family $\mathcal{Y}$ in (5.7) is indexed by $\mathbb{N}$. □

5.3 Homotopy Theorem

In this subsection we state and prove the homotopy theorem. It asserts that the motivic coarse spectrum of a bornological coarse space with the hybrid structure associated to a uniform structure and big family is homotopy invariant. Here the notion of homotopy is compatible with the uniform structure. The idea is to relate this notion of homotopy with the notion of coarse homotopy introduced in Sect. 4.3. The details are surprisingly complicated.

5.3.1   Statement of the Theorem

Let X be a bornological coarse space with an additional compatible uniform structure $\mathcal{T}$ and a big family $\mathcal{Y} = (Y_i)_{i \in I}$. We consider the unit interval $[0, 1]$ as a bornological coarse space with the structures induced by the metric (hence with the maximal structures) and the compatible metric uniform structure. We abbreviate $IX := [0, 1] \ltimes X$, see Example 2.24. The projection $IX \to X$ is a morphism. The bornological coarse space IX has a big family $I\mathcal{Y} := ([0, 1] \times Y_i)_{i \in I}$ and a compatible product uniform structure. We let $(IX)_h$ denote the corresponding hybrid bornological coarse space (Definition 5.10). The projection is compatible (Definition 5.14) so that by Lemma 5.15 it is also a morphism

$$(IX)_h \to X_h . \tag{5.8}$$

Let X be a bornological coarse space with a compatible uniform structure $\mathcal{T}$ and a big family $\mathcal{Y} = (Y_i)_{i \in I}$.

Theorem 5.26 (Homotopy Theorem)   *Assume:*

1. $I = \mathbb{N}$.
2. For every bounded subset B of X there exists i in I such that $B \subseteq Y_i$.

Then the projection (5.8) *induces an equivalence*

$$\mathrm{Yo}^s((IX)_h) \to \mathrm{Yo}^s(X_h) .$$

5.3.2   Proof of the Homotopy Theorem

The rough idea is to identify $(IX)_h$ with a coarse cylinder over X_h and then to apply Proposition 4.16. In the following we make this idea precise. Point 5 of Corollary 4.5 gives an equivalence

$$\mathrm{Yo}^s((IX)_h) \simeq \operatorname*{colim}_{U \in I\mathcal{C}_h} \mathrm{Yo}^s((IX)_U) ,$$

where the colimit runs over the poset $I\mathcal{C}_h$ of the entourages of the hybrid coarse structure of $(IX)_h$. See also Example 2.31 for notation.

We consider $[1, \infty)$ as a bornological coarse space with the structures induced from the metric. Let $q : X \to [1, \infty)$ be a function. Then we define the pair of functions $p := (q, 0)$ and consider the cylinder $I_p X$. We define the maps of sets

$$f : I_p X \to [0, 1] \times X, \quad f(t, x) := (t/q(x), x)$$

and

$$g : [0, 1] \times X \to I_p X, \quad g(t, x) := (tq(x), x)$$

by rescaling the time variable. They are inverse to each other.

By Lemma 5.27, for every coarse entourage V of X we can find a function q as above such that $q : X_V \to [1, \infty)$ is bornological and controlled and

$$\lim_{i \in I} \sup_{x \in X \setminus Y_i} q(x)^{-1} = 0. \tag{5.9}$$

Note that we take the supremum for subsets in the ordered set $[0, \infty]$, and that the supremum of the empty set is equal to 0. This is relevant if there exists an i in I with $Y_i = X$.

It suffices to show:

1. For any entourage V_h in $\mathcal{C}_h$ and controlled and bornological function $q : X_{V_h} \to [1, \infty)$ with (5.9) there exists an entourage U_h in $I\mathcal{C}_h$ such that $f : I_p X_{V_h} \to (IX)_{U_h}$ is a morphism.
2. For every U_h in $I\mathcal{C}_h$ there exists V_h' in $\mathcal{C}_h$ and a controlled and bornological function $q' : X_{V_h'} \to [1, \infty)$ with (5.9) such that $g' : (IX)_{U_h} \to I_{p'} X_{V_h'}$ is a morphism.

We then choose U_h' for V_h' similarly. We can assume that $V_h \subseteq V_h'$ and $U_h \subseteq U_h'$. We obtain a commuting diagram

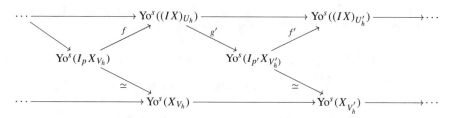

where the upper and lower horizontal morphisms are induced by the identities of the underlying sets, and the morphisms indicated as equivalences are the projections, see Proposition 4.16. The diagram induces by a cofinality consideration the middle equivalence between the colimits in the following chain

$$\mathrm{Yo}^s(X_h) \simeq \operatorname*{colim}_{V_h \in \mathcal{C}_h} \mathrm{Yo}^s(X_{V_h}) \simeq \operatorname*{colim}_{U_h \in I\mathcal{C}_h} \mathrm{Yo}^s((IX)_{U_h}) \simeq \mathrm{Yo}^s((IX)_h).$$

The maps f and g are clearly proper. So we must discuss under which conditions they are controlled.

We now take advantage that we can explicitly parametrize a cofinal set of coarse entourages of $I\mathcal{C}_h$ in the form $U_h := IV \cap V_\phi$. In detail, we take a coarse entourage V of X and define the entourage $IV := [0, 1] \times [0, 1] \times V$ of the product $[0, 1] \times X$. We also choose a pair $\phi = (\kappa, \psi)$ of a function $\kappa : I \to (0, \infty)$ with $\lim_{i \in I} \kappa(i) = 0$ and a $\mathcal{T}$-admissible function $\psi : I \to \mathcal{P}(X \times X)^{\mathrm{op}}$. Using the metric on $[0, 1]$ the function ϕ determines a function $I \to \mathcal{P}(IX \times IX)^{\mathrm{op}}$ which is admissible with respect to the product uniform structure on IX. In particular, ϕ defines a subset $V_\phi \subseteq \mathcal{P}(IX \times IX)$ as in (5.1). In detail, $((s, x), (t, y)) \in V_\phi$ if for every i in I we have $(x, y) \in Y_i \times Y_i$ or $|s - t| \leq \kappa(i)$ and $(x, y) \in V \cap \psi(i)$.

We will now analyse f. We fix the data for the entourage V_h. To this end we fix V in $\mathcal{C}$ and a $\mathcal{T}$-admissible function $\psi : I \to \mathcal{P}(X \times X)^{\mathrm{op}}$. Then we set $V_h := V \cap V_\psi$ in $\mathcal{C}_h$. We furthermore fix e in $(0, \infty)$ and let U_e be the corresponding metric entourage on $\mathbb{R}$. Then $\tilde{V}_h := U_e \times V_h$ (we omit to write the restriction from $\mathbb{R} \times X$ to $I_p X$) generates the coarse structure of $I_p X_{V_h}$. Our task is now to find an entourage U_h in $I\mathcal{C}_h$ such that $(f \times f)(\tilde{V}_h) \subseteq U_h$.

Since $q : X_{V_h} \to [1, \infty)$ is controlled and $V_h \subseteq V$ there exists a constant C in $(0, \infty)$ such that $(x, y) \in V$ implies that $|q(x) - q(y)| \leq C$. We define the function $\kappa : \mathbb{N} \to [0, \infty)$ by

$$\kappa(i) := (C + e) \cdot \sup_{x \in X \setminus Y_i} q(x)^{-1}.$$

By (5.9) we have $\lim_{i \in I} \kappa(i) = 0$. We now set $\phi := (\kappa, \psi)$ and define $U_h := IV \cap V_\phi$.

We now show that $(f \times f)(\tilde{V}_h) \subseteq U_h$. We assume that $((s, x), (t, y))$ is in $\tilde{V}_h$. Then $|t - s| \leq e$ and $(x, y) \in V \cap V_\psi$. For every i in I we have two cases:

1. If $(x, y) \in V \cap (Y_i \times Y_i)$. Then $((s, x), (t, y)) \in ([0, 1] \times Y_i) \times ([0, 1] \times Y_i)$.
2. Otherwise, $(x, y) \in V \cap \phi(i)$ and w.l.o.g $x \notin Y_i$. Then

$$|s/q(x) - t/q(y)| \leq e/q(x) + t(|q(x) - q(y)|)q(x)^{-1}q(y)^{-1}$$

$$\leq (e + C)q(x)^{-1} \leq \kappa(i).$$

This shows that $(f(s, x), f(t, y)) \in U_h$ as required.

We will now investigate g. We choose a coarse entourage V of X and $\phi = (\kappa, \psi)$ as above. These choices define the entourage $U_h := IV \cap V_\phi$ in $I\mathcal{C}_h$. We must find V_h' in $\mathcal{C}_h$ and a function q' appropriately. We take $V_h' := V \cap V_\psi$. In order to find q' we use the following lemma which we will prove afterwards.

Lemma 5.27 *For every coarse entourage V of X there exists a controlled and bornological function $q : X_V \to (0, \infty)$ such that the sets of real numbers*

$$\left\{ q(x) \cdot \left(\inf_{i \in I : x \notin Y_i} \kappa(i) \right) : x \in X \right\} \text{ and } \left\{ q(x) : x \in \bigcap_{i \in I} Y_i \right\}$$

are bounded by some positive real number D and (5.9) holds.

Let q' (called q in the lemma) and D be as in the lemma. Since q' is controlled we can choose C in $(0, \infty)$ such that $(x, y) \in V$ implies that $|q'(x) - q'(y)| \leq C$. We define the entourage $\tilde{V}_h := U_{C+D} \times V'_h$ of $I_{p'} X_{V'_h}$. Assume that $((s, x), (t, y))$ belongs to U_h. For every i in I we distinguish two cases:

1. $(x, y) \in V \cap (Y_i \times Y_i)$.
2. Otherwise we can w.l.o.g. assume that $x \notin Y_i$. Then

$$\left| sq'(x) - tq'(y) \right| \leq q'(x) \cdot \left(\inf_{i \in I : x \notin Y_i} \kappa(i) \right) + |q'(x) - q'(y)| \leq D + C \quad (5.10)$$

and $(x, y) \in V \cap \psi(i)$.

If $(x, y) \in V \cap (Y_i \times Y_i)$ for all $i \in I$, then

$$|sq'(x) - tq'(y)| \leq D \leq C + D$$

by the second bound claimed in Lemma 5.27.

The estimates thus show the inclusion $(g \times g)(U_h) \subseteq \tilde{V}_h$. This completes the proof of the Homotopy Theorem 5.26.

Proof of Lemma 5.27 It is here where we use that $I = \mathbb{N}$. In a first step we define the monotonously decreasing function

$$\lambda : \mathbb{N} \to (0, \infty), \quad \lambda(i) := \sup_{i \leq j} \kappa(j).$$

It satisfies $\kappa \leq \lambda$ and $\lim_{i \to \infty} \lambda(i) = 0$.

We now choose by induction a new function $\mu : \mathbb{N} \to (0, \infty)$ and an auxiliary function $a : \mathbb{N} \to \mathbb{N}$ as follows.

We set $\mu(0) := \lambda(0)$ and $a(0) := 0$.

Assume now that a is defined on all numbers $\leq n$ and μ is defined for all numbers $\leq a(n)$. Then we choose $a(n + 1)$ in $\mathbb{N}$ such that $a(n + 1) > a(n)$ and $V[Y_{a(n)}] \subseteq Y_{a(n+1)}$. We then define for all k in $(a(n), a(n + 1)]$

$$\mu(k) := \max \left\{ \frac{\mu(a(n))}{\mu(a(n)) + 1}, \lambda(a(n)) \right\}.$$

This has the effect that μ is constant on the intervals $(a(n), a(n + 1)]$ for all n in $\mathbb{N}$, and that μ^{-1} increases by at most one if one increases the argument from $a(n)$ to $a(n) + 1$. We furthermore have the estimate $\lambda(k) \leq \mu(k)$ for all k in $\mathbb{N}$, μ is monotonously decreasing, and it satisfies $\lim_{i \to \infty} \mu(i) = 0$. We define the function

$$q : X \to [1, \infty), \quad q(x) := 1 + \frac{1}{\sup\{\mu(i) : x \in Y_i\}}.$$

Note that q is well-defined since for every x in X the index set of the supremum is non-empty and μ is positive. Let B be a bounded subset of X. Then we can find i in I such that $B \subseteq Y_i$. Then $\sup_{x \in B} q(x) \leq \mu(i)^{-1} + 1$. This shows that q is bornological. For x in $\bigcap_{i \in I} Y_i$ we have $q(x) \leq \mu(i)^{-1} + 1$ for every i. This in particular shows the second estimate claimed in the lemma.

We now argue that q is controlled. We will show that for (x, y) in V we have $|q(x) - q(y)| \leq 1$. We can choose n in $\mathbb{N}$ such that $x \in Y_{a(n)}$ and $q(x) = \mu(a(n))^{-1} + 1$, and $m \in \mathbb{N}$ such that $q(y) = \mu(a(m))^{-1} + 1$. After flipping the roles of x and y if necessary we can assume that $n \leq m$. Then $y \in V[Y_{a(n)}] \subseteq Y_{a(n+1)}$. We then have the following options:

1. If $n = m$, then $q(x) = q(y)$.
2. If $n < m$, then $m = n + 1$ (here we use that μ decreases monotonously). In this case $q(y) - q(x) \leq 1$.

We now consider x in X and i in I such that $x \notin Y_i$. We have $\kappa(i)q(x) \leq \mu(i)q(x)$. We must minimize over i. Assume that n is such that $x \in Y_{a(n+1)} \setminus Y_{a(n)}$. We see that

$$\inf_{i \in I : x \notin Y_i} \kappa(i)q(x) \leq \inf_{i \in I : x \notin Y_i} \mu(i)q(x) = \mu(a(n))(1 + \frac{1}{\mu(a(n + 1))}) \leq 2\mu(a(0)).$$

This finishes the proof.   □

Remark 5.28 The assumption that $I = \mathbb{N}$ in the statement of the homotopy theorem was used in the proof of Lemma 5.27. At the moment we do not know if this assumption is only technical or really necessary.   □

5.3.3   Uniform Homotopies and the Cone Functors

In the category of uniform spaces we have a natural notion of a uniform homotopy.
    Let X be a uniform space.

Definition 5.29 A uniform homotopy is a uniformly continuous map $[0, 1] \times X \to X$, where the product is equipped with the product uniform structure.

Let X and X' be bornological coarse spaces equipped with uniform structures $\mathcal{T}$ and $\mathcal{T}'$ and big families $\mathcal{Y} = (Y_i)_{i \in I}$ and $\mathcal{Y}' = (Y'_{i'})_{i' \in I'}$, respectively. Let $f_0, f_1 : X \to X'$ be two compatible morphisms.

Definition 5.30 We say that f_0 and f_1 are compatibly homotopic if there is a compatible morphism $h : IX \to X'$ such that $f_i = h_{|\{i\} \times X}$ for i in $\{0, 1\}$.

Note that compatibly homotopic maps are close to each other and uniformly homotopic.

Let X and X' be bornological coarse spaces equipped with uniform structures $\mathcal{T}$ and $\mathcal{T}'$ and big families $\mathcal{Y} = (Y_i)_{i \in I}$ and $\mathcal{Y}' = (Y'_{i'})_{i' \in I'}$, respectively.

Corollary 5.31 *Assume that $I = \mathbb{N}$ and that for every bounded subset B of X there exists i in I such that $B \subseteq Y_i$. If $f_0, f_1 : X \to X'$ are compatibly homotopic to each other, we have an equivalence*

$$\mathrm{Yo}^s(f_0) \simeq \mathrm{Yo}^s(f_1) : \mathrm{Yo}^s(X_h) \to \mathrm{Yo}^s(X'_h).$$

Proof Let $h : [0, 1] \times X \to X'$ be a compatible homotopy between f_0 and f_1. By the Lemma 5.15 it induces a morphism

$$(IX)_h \to X'_h$$

which yields the morphisms $f_i : X_h \to X'_h$ by precomposition. The assertion now follows from the Homotopy Theorem 5.26. $\square$

We continue Example 5.16 and the cone-at-infinity from Sect. 5.2.3. We consider two maps $f_0, f_1 : Y \to Y'$ between uniform spaces.

Corollary 5.32 *If f_0 and f_1 are uniformly homotopic, then we have an equivalence* $\mathrm{Yo}^s(\mathcal{O}(f_0)) \simeq \mathrm{Yo}^s(\mathcal{O}(f_1))$.

Therefore the cone functor $\mathrm{Yo}^s \circ \mathcal{O} : \mathbf{U} \to \mathbf{Sp}\mathcal{X}$ sends uniformly homotopy equivalent uniform spaces to equivalent motivic spectra. It shares this property with the functor $\mathcal{O}^\infty : \mathbf{U} \to \mathbf{Sp}\mathcal{X}$ introduced in (5.6) which is in addition excisive by Corollary 5.25.

Corollary 5.33 *The functor $\mathcal{O}^\infty : \mathbf{U} \to \mathbf{Sp}\mathcal{X}$ sends uniformly homotopy equivalent uniform spaces to equivalent motivic spectra.*

The following corollary to the combination of Corollaries 5.25 and 5.33 says that $\mathcal{O}^\infty$ behaves like a $\mathbf{Sp}\mathcal{X}$-valued homology theory on the subcategory $\mathbf{U}_{\mathrm{sep}}$ of $\mathbf{U}$ of uniform Hausdorff spaces (see Mitchener [Mit10, Thm. 4.9] for a related result).

Corollary 5.34 *The functor $\mathcal{O}^\infty : \mathbf{U}_{\mathrm{sep}} \to \mathbf{Sp}\mathcal{X}$ is uniformly homotopy invariant and satisfies excision for uniform decompositions.*

This observation is the starting point of a translation of the homotopy theory on $\mathbf{U}$ into motivic coarse spectra. It will be further discussed in the subsequent paper [BEb].

5.4 Flasque Hybrid Spaces

We consider a bornological coarse space $(X, \mathcal{C}, \mathcal{B})$ with a compatible uniform structure $\mathcal{T}$ and a big family $\mathcal{Y} = (Y_i)_{i \in I}$. These data determine the hybrid space X_h (Definition 5.10). In this section we analyse which additional structures on X guarantee that X_h is flasque in the generalized sense (see Definition 3.27).

We assume that the uniform structure $\mathcal{T}$ is induced from a metric d. We will denote the uniform entourage of radius r by U_r, see (2.1).

We furthermore assume that the big family is indexed by $\mathbb{N}$ and determined through a function $e \colon X \to [0, \infty)$ by $Y_i := \{e \leq i\}$ for all i in $\mathbb{N}$. Moreover, we assume that e has the following properties:

1. e is uniformly continuous.
2. e is controlled. This ensures that the family of subsets $(Y_i)_{i \in \mathbb{N}}$ as defined above is big. Indeed, let V be a coarse entourage of X. Then we can find an s in $\mathbb{N}$ such that $|e(x) - e(y)| \leq s$ for all (x, y) in V. For i in $\mathbb{N}$ we then have $V[Y_i] \subseteq Y_{i+s}$.
3. e is bornological. This implies that for every bounded subset B of X there exists i in $\mathbb{N}$ such that $B \subseteq Y_i$.

The reason why we make these restrictions on the index set of the big family and $\mathcal{T}$ is that we are going to use the simple characterization of hybrid entourages given in Remark 5.11. We do not know whether this restriction is of technical nature or essential.

The essential structure is a map $\Phi \colon [0, \infty) \ltimes X \to X$. We write $\Phi_t(x) := \Phi(t, x)$. Recall that the semi-direct product has the product coarse structure and the bounded sets are generated by $[0, \infty) \times B$ for bounded subsets B of X. We assume that Φ has the following properties:

1. Φ is a morphism in **BornCoarse**.
2. Φ is uniformly continuous.
3. $\Phi_0 = \mathrm{id}$.
4. For every bounded subset B of X there exists t in $\mathbb{R}$ such that $\Phi([t, \infty) \times X) \cap B = \emptyset$.
5. Φ is contracting in the following sense: For every i in $\mathbb{N}$, ε in $(0, \infty) > 0$ and coarse entourage V of X there exists T in $(0, \infty)$ such that we have

$$\forall x, y \in X, \ \forall t \in [T, \infty) : (x, y) \in V \cap (Y_i \times Y_i) \Rightarrow (\Phi_t(x), \Phi_t(y)) \in U_\varepsilon.$$

6. Φ is compatible with the big family in the sense that for all s, t in $[0, \infty)$ we have $\Phi_t(\{e \leq s\}) \subseteq \{e \leq s + t\}$.

Recall the Definition 3.27 of flasqueness in the generalized sense.

Theorem 5.35 *Let all the above assumptions be satisfied. Then X_h is flasque in the generalized sense.*

Proof For every S in $\mathbb{N} \setminus \{0, 1\}$ we consider the map $r_S : [0, \infty) \to [0, \log(S)]$ given by

$$r_S(t) := \begin{cases} \log(S) - \frac{t}{S} & \text{for } 0 \le t \le \frac{S}{\log(S)} \\ \log(S) - \frac{1}{\log(S)} - \left(t - \frac{S}{\log(S)}\right) & \text{for } \frac{S}{\log(S)} \le t \le \log(S) - \frac{1}{\log(S)} + \frac{S}{\log(S)} \\ 0 & \text{for } t \ge \log(S) - \frac{1}{\log(S)} + \frac{S}{\log(S)} \end{cases}$$

We further set $r_1 := 0$. The family of maps $(r_S)_{S \in \mathbb{N} \setminus \{0\}}$ is uniformly controlled and uniformly equicontinuous.

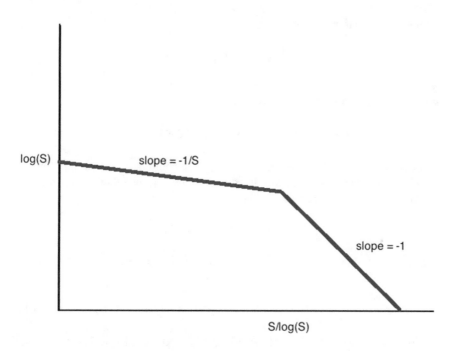

Function $r_S(t)$

We consider the composition of maps

$$\Psi_S : X \xrightarrow{(e, \mathrm{id})} [0, \infty) \times X \xrightarrow{(r_S, \mathrm{id})} [0, \log(S)] \ltimes X \xrightarrow{\Phi} X .$$

The first map is controlled since e is controlled. Furthermore it is proper since its second component is the identity. Hence it is a morphism in **BornCoarse**. The second map is also a morphism (since we use the semi-direct product it is proper). The last map is a morphism by assumption since it is a restriction of a morphism to a subset. All three maps are uniformly continuous and we have $\Psi_S(\{e \le s\}) \subseteq \{e \le s + \log(S)\}$. So the maps Ψ_S are compatible in the sense of Definition 5.14

and therefore are morphisms

$$\Psi_S : X_h \to X_h .$$

Our claim is now that these maps implement flasqueness of X_h in the generalized sense. More precisely, we will show that the family of morphisms $(\Psi_S)_{S\in\mathbb{N}\setminus\{0\}}$ satisfies the four conditions of Definition 3.27.

Condition 1
Since $r_1 \equiv 0$ we get $\Psi_1 = \mathtt{id}$.

Condition 2
We must show that the union

$$\bigcup_{S\in\mathbb{N}\setminus\{0\}} (\Psi_S \times \Psi_{S+1})(\mathrm{diag}_{X_h})$$

is an entourage of X_h. Hence we must control the pairs

$$(\Psi_S(x), \Psi_{S+1}(x)) \text{ for } x \text{ in } X \text{ and } S \text{ in } \mathbb{N}\setminus\{0\} .$$

Since the family $(r_S)_{\mathbb{N}\setminus\{0\}}$ is uniformly equicontinuous the family $(S \mapsto \Psi_S(x))_{x\in X}$ of functions is uniformly equicontinuous. Hence there exists a coarse entourage V of X such that for all x in X and S in $\mathbb{N}\setminus\{0\}$

$$(\Psi_S(x), \Psi_{S+1}(x)) \in V .$$

In view of Remark 5.11 it remains to show that for every ε in $(0, \infty)$ there exists j in $\mathbb{N}$ such that

$$\forall x \in X \; \forall S \in \mathbb{N}\setminus\{0\} \; : \; (\Psi_S(x), \Psi_{S+1}(x)) \in (Y_j \times Y_j) \cup U_\varepsilon . \tag{5.11}$$

First observe using the explicit formula for r_S that there exists a C in $(0, \infty)$ such that for all t in $[0, \infty)$ and S in $\mathbb{N}$ we have the inequality

$$|r_S(t) - r_{S+1}(t)| \leq \frac{C}{\min\{1, \log(S)\}} .$$

We now fix ε in $(0, \infty)$. Using the above estimate and that Φ is uniformly continuous we can find S_0 in $\mathbb{N}\setminus\{0, 1\}$ sufficiently large such that

$$\forall x \in X \; \forall S \in \mathbb{N} \text{ with } S \geq S_0 \; : \; (\Psi_S(x), \Psi_{S+1}(x)) \in U_\varepsilon \tag{5.12}$$

It remains to control the pairs with $1 \leq S \leq S_0$.

Set

$$i := \sup_{1 < S \le S_0} \left(\log(S) - \frac{1}{\log(S)} + \frac{S}{\log(S)} \right)$$

(note that i is finite). Then we have $r_S(e(x)) = 0$ for all x in $X \setminus Y_i$ and all S in $[1, S_0]$. Consequently, $\Psi_S(x) = x$ for all x in $X \setminus Y_i$ and all S in $[1, S_0]$.

We now choose the integer j in $\mathbb{N}$ such that $\Psi_S(Y_i) \subseteq Y_j$ for all S in $[1, S_0]$. Then we have

$$\forall x \in X \; \forall S \in \mathbb{N} \text{ with } 1 \le S \le S_0 \; : \; (\Psi_S(x), \Psi_{S+1}(x)) \in (Y_j \times Y_j) \cup \text{diag}_X .$$
(5.13)

The combination of (5.13) and (5.12) implies (5.11).

Condition 3
Let U be a coarse entourage of X_h. We have to show that the union $\bigcup_{S \in \mathbb{N} \setminus \{0\}} (\Psi_S \times \Psi_S)(U)$ is again a coarse entourage of X_h.

By definition of the hybrid structure and Remark 5.11 there is a coarse entourage V of X such that for every ε in $(0, \infty)$ we can find i in $\mathbb{N} \setminus \{0\}$ with

$$U \subseteq V \cap ((Y_i \times Y_i) \cup U_\varepsilon).$$

First of all we observe that the family $(r_S)_{S \in \mathbb{N} \setminus \{0\}}$ is uniformly controlled. Since Φ is controlled with respect to the original structures there exists an entourage V' in $\mathcal{C}$ such that

$$(\Psi_S \times \Psi_S)(V) \subseteq V'$$
(5.14)

for all S in $\mathbb{N} \setminus \{0\}$.

We now fix δ in $(0, \infty)$. Since the family $(r_S)_{S \in \mathbb{N} \setminus \{0\}}$ is uniformly equicontinuous the same is true for the family $(\Psi_S)_{S \in \mathbb{N} \setminus \{0\}}$. Hence there exists i in $\mathbb{N}$ such that

$$(\Psi_S \times \Psi_S)(U \setminus (Y_i \times Y_i)) \subseteq U_\delta$$
(5.15)

for all $S \in \mathbb{N} \setminus \{0\}$.

Assume now that is (x, y) in $(Y_i \times Y_i) \cap V$. The control of e yields a positive number R (which only depends on V) such that $|e(x) - e(y)| \le R$. Furthermore, we have $e(x), e(y) \le i$.

In the following steps we are going to choose S_0 in $\mathbb{N} \setminus \{0, 1\}$ by step increasingly bigger such that certain estimates hold true.

In the first step we choose S_0 such that

$$i \le \frac{S_0}{\log(S_0)} .$$

Then for all S in $\mathbb{N}$ with $S \geq S_0$ and (x, y) in $V \cap (Y_i \times Y_i)$ we have

$$|r_S(e(x)) - r_S(e(y))| \leq \frac{R}{S_0}. \tag{5.16}$$

We observe that

$$\lim_{S_0 \to \infty} \inf_{S \geq S_0, x \in Y_i} r_S(e(x)) = \infty. \tag{5.17}$$

We have by definition

$$\big(\Psi_S(x), \Psi_S(y)\big) = \big(\Phi_{r_S(e(x))}(x), \Phi_{r_S(e(y))}(y)\big).$$

Since Φ is uniformly continuous, in view of (5.16) we can choose S_0 so large that

$$\forall S \geq S_0, \ \forall (x, y) \in V \cap (Y_i \times Y_i) \ : \ \big(\Phi_{r_S(e(x))}(x), \Phi_{r_S(e(y))}(x)\big) \in U_{\delta/2}. \tag{5.18}$$

We employ that Φ_t is contracting and (5.17) in order to see that we can increase S_0 even more such that

$$\forall S \geq S_0, \ \forall (x, y) \in V \cap (Y_i \times Y_i) \ : \ \big(\Phi_{r_S(e(y))}(x), \Phi_{r_S(e(y))}(y)\big) \in U_{\delta/2}. \tag{5.19}$$

Equations (5.18) and (5.19) together give

$$\forall S \geq S_0, \ \forall (x, y) \in V \cap (Y_i \times Y_i) \ : \ \big(\Phi_{r_S(e(x))}(x), \Phi_{r_S(e(y))}(y)\big) \in U_\delta.$$

This implies

$$\forall S \geq S_0 \ : \ (\Psi_S \times \Psi_S)((Y_i \times Y_i) \cap V) \subseteq U_\delta. \tag{5.20}$$

We now use the compatibility of Φ with the big family and (5.14) in order to find an integer j in $\mathbb{N}$ such that for all S in $[1, S_0]$

$$(\Psi_S \times \Psi_S)((Y_i \times Y_i) \cap V) \subseteq V' \cap (Y_j \times Y_j).$$

In conclusion we have shown that there is V' in $\mathcal{C}$ and for given δ in $(0, \infty)$ we can choose j in $\mathbb{N}$ appropriately such that

$$\forall S \in \mathbb{N} \setminus \{0\} \ : \ (\Psi_S \times \Psi_S)(U) \subseteq V' \cap ((Y_j \times Y_j) \cup U_\delta).$$

In view of Remark 5.11 this proves that $\bigcup_{S \in \mathbb{N}} (\Psi_S \times \Psi_S)(U)$ is an entourage of X_h.

Condition 4
If B is a bounded subset of X, then $e_{|B}$ is bounded. We now use (5.17) and Property 4 of Φ in order to verify the condition. $\qquad\square$

5.5 Decomposition of Simplicial Complexes

The main result of the present section is Theorem 5.57. It roughly states that the motivic coarse spectrum of a simplicial complex with a hybrid structure relative to an exhaustion can be expressed as a colimit of motivic coarse spectra of discrete bornological coarse spaces.

5.5.1 Metrics on Simplicial Complexes

By a simplicial complex we understand a topological space presented as a geometric realization of an abstract simplicial complex.

We can model the n-dimensional simplex Δ^n as the intersection $S^n \cap [0, \infty)^{n+1}$ in $\mathbb{R}^{n+1}$. The restriction of the Riemannian distance on the sphere to Δ^n is called the spherical metric d_{sph}. If we model Δ^n by the subset $\{x \in \mathbb{R}^n \mid \sum_{i=0}^n x_i = 1\} \cap [0, \infty)^{n+1}$ of $\mathbb{R}^{n+1}$, then the induced metric is the Euclidean metric and will be denoted by d_{eu}. Another metric is the ℓ^1-metric d_{ℓ^1} on Δ^n given in the same model by $d(x, x') := \sum_{i=0}^n |x_i - x_i'|$.

We let X be a simplicial complex with a metric d.

Definition 5.36 The metric d is good if it has the following properties:

1. The restriction of d to every component of X a path metric.
2. There exists c in $(0, \infty)$ such that any two components of X have distance at least c.
3. For every n in $\mathbb{N}$ there exists C_n in $(0, \infty)$ such that the metric induced on every n-simplex $\Delta \subseteq X$ satisfies $C_n^{-1} \cdot d_{\mathrm{sph}} \leq d_{|\Delta} \leq C_n \cdot d_{\mathrm{sph}}$.

Note that we allow that different components of X have finite or infinite distances, i.e., our metrics are actually quasi-metrics.

Example 5.37 An example of a good metric on a simplicial complex is the path-metric induced from the spherical metrics on the simplices. In this case we can take $C_n = 1$ for all n in $\mathbb{N}$ and the components have infinite distance from each other (see Wright [Wri05, Appendix]).

In the case of finite-dimensional simplicial complexes we can also start with the Euclidean or the ℓ^1-metric to generated a good path metric. $\qquad\square$

Example 5.38 If Δ is the standard simplex, then for μ in $(0, 1]$ we let $\mu\Delta \subseteq \Delta$ be the image of the radial scaling (with respect to the barycenter) by the factor μ.

Assume that X is equipped with a good metric. For n in $\mathbb{N}$ and μ in $(0, 1)$ let Z be the subset of X given by the disjoint union of the μ-scalings of the n-simplices of X with the induced metric. After reparametrization of the simplices we can consider Z as a simplicial complex which is topologically a disjoint union of n-simplices. The restriction of the good metric of X to Z is again good.

Two simplices of Z which are scalings of simplices in the same component of X are in finite distance. $\square$

Example 5.39 In the following we describe an example of a simplicial complex X with the property that the identity of X does not induce a morphism $X_{d_e} \to X_{d_s}$ of coarse spaces, where d_e and d_s are the path metrics induced by the Euclidean and the spherical metrics on the simplices, respectively. That this can happen for infinite-dimensional simplicial complexes has essentially been observed by Wright [Wri05, Appendix].

The complex X is defined as the quotient of $\bigsqcup_{n \in \mathbb{N}} \Delta^n$ by identifying for all n in $\mathbb{N}$ the last face $\partial_n \Delta^n$ of Δ^n with the first face of the first face $\partial_0 \partial_0 \Delta^{n+1}$ of Δ^{n+1}. The entourage $\{d_e \leq 1\}$ in $\mathcal{C}_{d_e}$ does not belong to $\mathcal{C}_{d_s}$, as shown below.

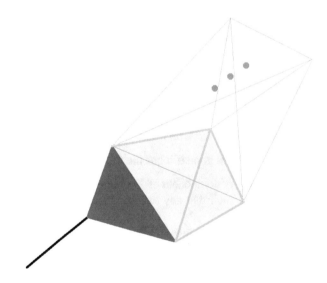

The complex X in Example 5.39

For the barycenters b_n and b_{n+1} of Δ^n and Δ^{n+1} we have

$$d_e(b_n, b_{n+1}) = \sqrt{\frac{1}{n(n+1)}} + \sqrt{\frac{2}{n(n+2)}} , \qquad (5.21)$$

while

$$d_s(b_n, b_{n+1}) = \pi/2.$$

Define $k : \mathbb{N} \to \mathbb{N}$ by

$$k(n) := \max\{m \in \mathbb{N} : d_e(b_n, b_{n+m}) \le 1\}.$$

As a consequence of (5.21) we have $\lim_{n \to \infty} k(n) = \infty$. Hence we have

$$\forall n \in \mathbb{N} \ : \ d_e(b_n, b_{n+k(n)}) \le 1$$

while

$$d_s(b_n, b_{n+k(n)}) = \frac{k(n)\pi}{2} \xrightarrow{n \to \infty} \infty.$$

Hence the identity of X does not induce a controlled map $X_{d_e} \to X_{d_s}$.   □

5.5.2   Decomposing Simplicial Complexes

Assume that X is equipped with a good metric d. We let $\mathcal{T}_d$ be the uniform structure on X determined by the metric. We further assume that X comes with a bornological coarse structure $(\mathcal{C}, \mathcal{B})$ such that $\mathcal{C}$ and $\mathcal{T}_d$ are compatible (see Definition 5.4). For example, $\mathcal{C}$ could be (but this is not necessary) the structure $\mathcal{C}_d$ induced by the metric. Note that the compatibility condition and Condition 3 of Definition 5.36 imply that the size of the simplices of a fixed dimension is uniformly bounded with respect to $\mathcal{C}$.

Finally we assume that X comes with a big family $\mathcal{Y} = (Y_i)_{i \in I}$ of subcomplexes such that every bounded subset of X is contained in Y_i for some i in I. In this situation we can consider the hybrid structure $\mathcal{C}_h$ (see Definition 5.10). We write X_h for the corresponding bornological coarse space.

We now assume that X is n-dimensional. For k in $\mathbb{N}$ we let X^k denote the k-skeleton of X. We let the subset Z of X be obtained as in Example 5.38 by scaling all n-simplices by the factor $2/3$. Furthermore we let Y be the subspace of X obtained from X by removing the interiors of the unions of the $1/3$-scaled n-simplices. Then $Y \cup Z = X$ is a decomposition of X into two closed subsets.

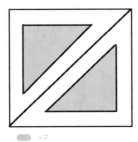

■■■ = Y   ⬤⬤ = Z   ⬤⬤

Decomposition $X = Y \cup Z$

Conditions 1 and 2 of Definition 5.36 ensure by Example 5.21 that the pair (Y, Z) is a uniform decomposition (Definition 5.19) of X. If $I = \mathbb{N}$, then the Decomposition Theorem 5.22 asserts that we have a push-out square in $\mathbf{Sp}\mathcal{X}$

$$
\begin{array}{ccc}
((Y \cap Z)_h, Y \cap Z \cap \mathcal{Y}) & \longrightarrow & (Y_h, Y \cap \mathcal{Y}) \\
\downarrow & & \downarrow \\
(Z_h, Z \cap \mathcal{Y}) & \longrightarrow & (X_h, \mathcal{Y})
\end{array}
\tag{5.22}
$$

In the following we analyse the corners of (5.22). Let C be the set of the barycenters of the n-simplices of X with the induced bornological coarse structure (see Example 2.17). The restriction of the good metric to C induces the discrete uniform structure $\mathcal{T}_{\mathrm{disc}}$. For i in I we let C_i denote the subset of C of the barycenters of the n-simplices contained in Y_i. Then $(C_i)_{i \in I}$ is a big family in C. We can consider the associated hybrid space C_h.

Lemma 5.40 *If $I = \mathbb{N}$, then in $\mathbf{Sp}\mathcal{X}$ we have an equivalence*

$$
(C_h, (C_i)_{i \in I}) \simeq (Z_h, Z \cap \mathcal{Y}).
$$

Proof The inclusion $g : C \to Z$ and the projection $p : Z \to C$ which sends each component to the center are compatible morphisms. For properness of p we use the fact noticed above that the size of the simplices of X is uniformly bounded with respect to C. Furthermore the compositions $g \circ p$ and $p \circ g$ are compatibly homotopic to the respective identities. By Lemma 5.15 we obtain morphisms $g : C_h \to Z_h$ and $p : Z_h \to C_h$. By Corollary 5.31 g induces an equivalence $\mathrm{Yo}^s(C_h) \simeq \mathrm{Yo}^s(Z_h)$. In general, g is not a coarse equivalence, but one could look at Example 4.21 in order to understand the mechanism giving this equivalence. The restriction $g_{|C_i} : C_i \to Z_h \cap Y_i$ is already an equivalence in **BornCoarse** for every i in I, where C_i has the bornological coarse structure induced from C_h. In fact, the bornological coarse structures on C_i and $Z_h \cap Y_i$ induced from the hybrid structures coincide with the bornological coarse structures induced from the original bornological coarse structures C or Z, respectively. $\qquad\square$

Recall the Definition 2.14 of the notion of a discrete bornological coarse space.

Lemma 5.41 *There exists a filtered family* $(W_j, \mathcal{W}_j)_{j \in J}$ *of pairs of discrete bornological coarse spaces and big families such that in* $\mathbf{Sp}\mathcal{X}$ *we have an equivalence*

$$(C_h, (C_i)_{i \in I}) \simeq \operatorname*{colim}_{j \in J}(W_j, \mathcal{W}_j).$$

Proof Using Point 5 of Corollary 4.5 we have the equivalence of coarse motivic spectra

$$(C_h, (C_i)_{i \in I}) \simeq \operatorname*{colim}_{U \in \mathcal{C}_h}(C_U, (C_i)_{i \in I}).$$

Let U in $\mathcal{C}_h$ be an entourage of $\mathcal{C}_h$ of the form $V \cap U_\phi$ for a coarse entourage V of C and a cofinal function $\phi : I \to \mathcal{T}_{\mathrm{disc}}$. This cofinality implies that there exists i in I such that $\phi(i) = \mathrm{diag}_C$. We let $T := C_U \setminus C_i$. Using excision (Point 2 of Corollary 4.5) with respect to the complementary pair $(T, (C_i)_{i \in I})$ we have the equivalence

$$(C_U, (C_i)_{i \in I}) \simeq (T, T \cap (C_i)_{i \in I}),$$

where T has the structures induced from C_U. The crucial observation is now that the coarse structure induced from C_U on T is discrete.   □

We now study the pair $(Y_h, Y \cap \mathcal{Y})$.

Lemma 5.42 *If $I = \mathbb{N}$, then in* $\mathbf{Sp}\mathcal{X}$ *we have an equivalence*

$$(X_h^{n-1}, X^{n-1} \cap \mathcal{Y}) \simeq (Y_h, Y \cap \mathcal{Y}).$$

Proof We consider the retraction $r : Y \to X^{n-1}$ which is locally modeled by the radial retraction of $\Delta^n \setminus \mathrm{int}(1/3\Delta^n)$ to the boundary $\partial \Delta^n$. This is a morphism in **BornCoarse** which is compatible. It provides a compatible homotopy inverse to the inclusion $X^{n-1} \to Y$. By Corollary 5.31 this inclusion induces an equivalence $\mathrm{Yo}^s(X_h^{n-1}) \simeq \mathrm{Yo}^s(Y_h)$. We again observe that for every i in I the original bornological coarse structure and the hybrid structure induce the same bornological coarse structure on $Y \cap Y_i$ and $X^{n-1} \cap Y_i$. Moreover, the restrictions of r and the inclusion are inverse to each other equivalences of bornological coarse spaces. This implies the assertion.   □

We let $D \subseteq X$ be the subset which locally is modeled by $\partial(1/2\Delta^n) \subseteq \Delta^n$. After reparametrization we can consider D as an $(n - 1)$-dimensional simplicial complex equipped with the induced bornological coarse structure and the induced good metric from X. It furthermore comes with a big family $D \cap \mathcal{Y}$.

Lemma 5.43 *If $I = \mathbb{N}$, then in $\mathbf{Sp}\mathcal{X}$ we have an equivalence*

$$(D_h, D \cap \mathcal{Y}) \simeq ((Z \cap Y)_h, (Z \cap Y \cap \mathcal{Y})).$$

Proof We have a natural inclusion $D \to Z \cap Y$ and we have also a (locally radial) retraction $r : Z \cap Y \to D$. This retraction is compatible and a compatible homotopy inverse to the inclusion. The assertion of the lemma now follows from Corollary 5.31. $\qquad\square$

5.6   Flasqueness of the Coarsening Space

In this section we will introduce the coarsening space of Wright [Wri05, Sec. 4]. The main result is that the coarsening space is flasque in the generalized sense if it is equipped with a suitable hybrid structure. This fact has first been shown by Wright [Wri05, Sec. 4]. In the present section we will adapt the arguments of Wright to our setting.

5.6.1   Construction of the Coarsening Space

Let $(X, \mathcal{C}, \mathcal{B})$ be a bornological coarse space, and let $\mathcal{U} = (U_i)_{i \in I}$ be a cover of X, i.e., a family of subsets of X such that $\bigcup_{i \in I} U_i = X$.

Definition 5.44

1. The cover $\mathcal{U}$ is bounded by V in $\mathcal{C}$ if for every i in I the subset U_i is V-bounded (see Definition 2.15). We say that $\mathcal{U}$ is bounded if it is V-bounded for some V in $\mathcal{C}$.
2. An entourage V in $\mathcal{C}$ is a Lebesgue entourage[2] of $\mathcal{U}$ if for every V-bounded subset B of X there exists i in I such that $B \subseteq U_i$.

   The nerve $N(\mathcal{U})$ of a cover $\mathcal{U}$ is a simplicial set whose (non-degenerate) n-simplices are the $(n + 1)$-tuples $(i_0, \ldots, i_n)$ in I^{n+1} (without repetitions) such that $\bigcap_{j=0}^n U_{i_j} \neq \emptyset$. The geometric realization of the nerve is a simplicial complex which will be denoted by $\|\mathcal{U}\|$. It will be equipped with the path-metric induced from the spherical metric on the simplices. The choice of the spherical metric is required in order to have Point 3 of Lemma 5.45 which would not be true in general for other metrics, e.g., the Euclidean metric (see Example 5.39) or the ℓ^1-metric, if $\|\mathcal{U}\|$ is infinite-dimensional. If $\|\mathcal{U}\|$ is not connected, then we define the metric component-

---

[2] This name should resemble the notion of Lebesgue number in the case of metric spaces.

wise and let different components have infinite distance. As in Example 2.18 the metric defines a coarse structure C_U on $\|U\|$.

We consider a cover $U = (U_i)_{i \in I}$ of X. Note that we can identify the index set I with the zero skeleton $\|U\|^0$ of the simplicial space $\|U\|$. For every x in X we can choose i in I such that $x \in U_i$. This choice defines a map

$$f : X \to \|U\|$$

which actually maps X to the zero skeleton. In the other direction we define a map

$$g : \|U\| \to X$$

such that it maps a point in the interior of a simplex in $\|U\|$ corresponding to a non-degenerate simplex $(i_0, \ldots, i_n)$ in $N(U)$ to a point in the intersection $\bigcap_{j=0}^{n} U_{i_j}$. We use the map $g : \|U\| \to X$ in order to induce a bornology $B_U := g^* B$ on $\|U\|$ (Example 2.17). Recall the Definition 2.9 of a controlled map between sets equipped with coarse structures.

The following lemma is an easy exercise.

Lemma 5.45

1. The composition $f \circ g$ is $\pi/4$-close to the identity of $\|U\|$.
2. If V in C is a Lebesgue entourage of U, then $f : (X, C\langle V \rangle) \to (\|U\|, C_U)$ is controlled.
3. If U is bounded by W in C, then $g : (\|U\|, C_U) \to (X, C\langle W \rangle)$ is controlled. Moreover $g \circ f$ is W-close to the identity of X.
4. If U is bounded, then the bornology B_U is compatible with the coarse structure C_U.

The above lemma immediately implies the following corollary.

Corollary 5.46

1. If U is a bounded cover, then $(\|U\|, C_U, B_U)$ is a bornological coarse space and $g : (\|U\|, C_U, B_U) \to (X, C, B)$ is a morphism.
2. If V is a Lebesgue entourage and W is a bound of the cover U such that $C\langle V \rangle = C\langle W \rangle$, then f and g are inverse to each other equivalences in **BornCoarse** between X_V (see Example 2.31) and $(\|U\|, C_U, B_U)$.

Let $\hat{V} = (V_i)_{i \in I}$ be a family of covers of X indexed by a partially ordered set I.

Definition 5.47 We call $\hat{V}$ an anti-Čech system if the following two conditions are satisfied:

1. For all i, j in I with $j < i$ exists a Lebesgue entourage of V_i which is a bound of V_j.
2. Every entourage of X is a Lebesgue entourage for some member of the family $\hat{V}$.

Example 5.48 Note that the notion of an anti-Čech system only depends on the coarse structure. Let $(X, \mathcal{C})$ be a coarse space. For every entourage U in $\mathcal{C}$ we consider the cover $\mathcal{V}_U$ consisting of all U-bounded subsets of X. Then $(\mathcal{V}_U)_{U \in \mathcal{C}}$ is an anti-Čech system. In this case $\mathcal{V}_U$ is U-bounded and has U as a Lebesgue entourage.   □

We will now construct the coarsening space. We consider a bornological coarse space $(X, \mathcal{C}, \mathcal{B})$ and assume that $\mathcal{C} = \mathcal{C}\langle U \rangle$ for an entourage U in $\mathcal{C}$. We consider an anti-Čech system $\hat{\mathcal{V}} = (\mathcal{V}_n)_{n \in \mathbb{N}}$ indexed by the natural numbers and assume that U is a Lebesgue entourage for $\mathcal{V}_n$ for all n in $\mathbb{N}$. Our assumption on X implies that $\mathcal{C}\langle U \rangle = \mathcal{C}\langle V \rangle$ for every coarse entourage V of X containing U. Hence $(X, \mathcal{C}, \mathcal{B})$ and $(\|\mathcal{V}_n\|, \mathcal{C}_{\mathcal{V}_n}, \mathcal{B}_{\mathcal{V}_n})$ are equivalent bornological coarse spaces by Corollary 5.46.

For every n in $\mathbb{N}$ we can choose a refinement $\mathcal{V}_n \to \mathcal{V}_{n+1}$. In detail, let $\mathcal{V}_n = (V_i)_{i \in I_n}$. Then a refinement is a map $\kappa_n : I_n \to I_{n+1}$ such that $V_i \subseteq V_{\kappa(i)}$ for all i in I_n. Such a refinement determines a morphism of simplicial sets $N(\mathcal{V}_n) \to N(\mathcal{V}_{n+1})$ and finally a map

$$\phi_n : \|\mathcal{V}_n\| \to \|\mathcal{V}_{n+1}\| \tag{5.23}$$

between the geometric realizations of the nerves. Note that these maps are contracting since we are using the spherical metric on the simplices (they would be also contracting for the ℓ^1-metric on the simplices, but in general not for the Euclidean metric).

We construct the coarsening space $(\|\hat{\mathcal{V}}\|, \mathcal{C}_{\hat{\mathcal{V}}}, \mathcal{B}_{\hat{\mathcal{V}}})$ following Wright [Wri05, Def. 4.1]. It will furthermore be equipped with a uniform structure $\mathcal{T}_{\hat{\mathcal{V}}}$. The underlying topological space of the coarsening space is given by

$$\|\hat{\mathcal{V}}\| := \big([0, 1] \times \|\mathcal{V}_0\|\big) \cup_{\phi_0} \big([1, 2] \times \|\mathcal{V}_1\|\big) \cup_{\phi_1} \big([2, 3] \times \|\mathcal{V}_2\|\big) \cup_{\phi_2} \cdots . \tag{5.24}$$

We triangulate the products $[n, n+1] \times \|\mathcal{V}_n\|$ for every $n \in \mathbb{N}$ in the standard way. The structure of a simplicial complex provides a path-metric. These path-metrics induce a path-metric on $\|\hat{\mathcal{V}}\|$, which in turn induces a coarse structure $\mathcal{C}_{\hat{\mathcal{V}}}$ and a uniform structure $\mathcal{T}_{\hat{\mathcal{V}}}$. The bornology $\mathcal{B}_{\hat{\mathcal{V}}}$ is generated by sets of the form $[n, n+1] \times B$ for all n in $\mathbb{N}$ and B in $\mathcal{B}_{\mathcal{V}_n}$. One checks that this bornology is compatible with the coarse structure. We thus get a bornological coarse space $(\|\hat{\mathcal{V}}\|, \mathcal{C}_{\hat{\mathcal{V}}}, \mathcal{B}_{\hat{\mathcal{V}}})$. Furthermore, the uniform structure $\mathcal{T}_{\hat{\mathcal{V}}}$ is compatible, see Example 5.5. Since we are using the spherical metric on the simplices, we get that for every n in $\mathbb{N}$ the subspace $\{n\} \times \|\mathcal{V}_n\|$ of $\hat{\mathcal{V}}$ with the coarse structure induced from $\mathcal{C}_{\hat{\mathcal{V}}}$ is equivalent to $\|\mathcal{V}_n\|$ with the coarse structure $\mathcal{C}_{\mathcal{V}_n}$ [Wri05, Lem. A.5]. It is not known to the authors whether this property also holds if we equip our simplicial complexes with the ℓ^1-metric or with the Euclidean metric.

If we choose another family of refinements $(\kappa_n')_{n \in \mathbb{N}}$, then we get a different simplicial complex $\|\hat{\mathcal{V}}\|'$. It consists of the same pieces $[i, i+1) \times \|\mathcal{V}_i\|$ which are glued differently. One can check that the obvious (in general non-continuous) bijection which identifies these pieces is an equivalence of bornological coarse spaces between $\|\hat{\mathcal{V}}\|$ and $\|\hat{\mathcal{V}}\|'$.

Remark 5.49 Philosophically, the choice of the family of refinements $(\kappa_n)_{n \in \mathbb{N}}$ should be part of the data of an anti-Čech system. Then the notation $\|\hat{\mathcal{V}}\|$ would be unambiguous. But we prefer to use the definition as stated since in arguments below we need the freedom to change the family of refinements. Also Example 5.48 would need some adjustment.   □

We show below flasqueness of the coarsening space for two different coarse structures, which we define now.

We consider the projection to the first coordinate

$$\pi : \|\hat{\mathcal{V}}\| \to [0, \infty) \tag{5.25}$$

and define the subsets $\|\hat{\mathcal{V}}\|_{\leq n} := \pi^{-1}([0, n])$. We observe that $(\|\hat{\mathcal{V}}\|_{\leq n})_{n \in \mathbb{N}}$ is a big family. Following Wright [Wri05] we introduce the following notation:

- By $\|\hat{\mathcal{V}}\|_0$ we denote $\|\hat{\mathcal{V}}\|$ equipped with the C_0-structure (see Example 5.12).
- By $\|\hat{\mathcal{V}}\|_h$ we denote $\|\hat{\mathcal{V}}\|$ with the hybrid structure (see Definition 5.10) associated to the big family $(\|\hat{\mathcal{V}}\|_{\leq n})_{n \in \mathbb{N}}$.

5.6.2 Flasqueness for the C_0-Structure

We consider a bornological coarse space $(X, \mathcal{C}, \mathcal{B})$, an entourage U of X, and an anti-Čech system $\hat{\mathcal{V}}$ for X.

Theorem 5.50 ([Wri05, Thm. 4.5]) *We assume that*

1. $\mathcal{C} = \mathcal{C}\langle U \rangle$,
2. $\mathcal{B}$ *is generated by* $\mathcal{C}$ *(see Example 2.16), and*
3. *for every n in* $\mathbb{N}$ *there exists a bound* U_n *in* $\mathcal{C}$ *of* $\mathcal{V}_n$ *such that* U_n^4 *is a Lebesgue entourage of* $\mathcal{V}_{n+1}$.

Then $\|\hat{\mathcal{V}}\|_0$ *is flasque in the generalized sense.*

Proof We will apply Theorem 5.35. Every connected component of $\|\hat{\mathcal{V}}\|$ meets the front face $\{0\} \times \|\mathcal{V}_0\|$. For every component we choose a base point in this front face. We then define the function

$$e : \|\hat{\mathcal{V}}\| \to [0, \infty), \quad e(x) := d(*_x, x),$$

where $*_x$ denotes the chosen base point $*_x$ in the component of x. This function is Lipschitz continuous with Lipschitz constant 1. It is therefore uniformly continuous and controlled. By Corollary 5.46 the bornology on the pieces $\|\mathcal{V}_n\|$ is the bornology of metrically bounded subsets. This implies that e is also bornological.

Consequently, e satisfies the three conditions listed in Sect. 5.4. Furthermore note that the family $((\{e \le n\}))_{n \in \mathbb{N}}$ of subsets is a cofinal sequence in the bornology $\mathcal{B}_{\hat{\mathcal{V}}}$. Hence the hybrid structure determined by e is the $\mathcal{C}_0$-structure.

It remains to construct the morphism $\Phi: [0, \infty) \ltimes \|\hat{\mathcal{V}}\| \to \|\hat{\mathcal{V}}\|$ with the properties listed in Sect. 5.4. First note that we are free to choose the family of refinements $(\kappa_n: \mathcal{V}_n \to \mathcal{V}_{n+1})_{n \in \mathbb{N}}$ since being flasque in the generalized sense is invariant under equivalences of bornological coarse spaces.

We will choose the refinements later after we have explained which properties they should have. For the moment we adopt some choice. Then we define [Wri05, Def. 4.3] the map $\Phi: [0, \infty) \ltimes \|\hat{\mathcal{V}}\| \to \|\hat{\mathcal{V}}\|$ by

$$\Phi(t, (s, x)) := \begin{cases} \left(t, (\phi_{\lfloor t \rfloor - 1} \circ \cdots \circ \phi_{\lfloor s \rfloor})(x)\right) & \text{for } \lfloor s \rfloor < \lfloor t \rfloor \\ \left(t, \phi_{\lfloor s \rfloor}(x)\right) & \text{for } s \le t, \lfloor s \rfloor = \lfloor t \rfloor \qquad (5.26) \\ (s, x) & \text{for } s \ge t \end{cases}$$

(see (5.23) for ϕ_n). Note that

$$\Phi(t, \Phi(s, (u, x))) = \Phi(\max\{t, s\}, (u, x)) . \qquad (5.27)$$

We must check that Φ has the properties required for Theorem 5.35. By definition of Φ we have $\Phi_0 = \mathrm{id}$ and Conditions 4 and 6 are satisfied. Using that the maps ϕ_n are contracting we see that Φ is uniformly continuous. Since the maps ϕ_n are coarse equivalences one furthermore checks that Φ is a morphism in **BornCoarse**.

The only non-trivial one is that Φ is contracting. Below we show that we can choose the refinements for the anti-Čech system such that Φ is contracting. So by Theorem 5.35 we conclude that $\|\hat{\mathcal{V}}\|_0$ is flasque in the generalized sense.

We now reprove [Wri05, Lem. 4.4] in our somewhat more general setting, which will imply that Φ is contracting (we will give the argument for this at the end of this proof). The statement is that the refinements $\kappa_n: I_n \to I_{n+1}$ can be chosen with the following property: for every bounded subset B of X which is contained in a single coarse component (see Definition 2.30) there exists an integer N such that for all k in $\mathbb{N}$ with $k \le N$.

$$(\kappa_N \circ \kappa_{N-1} \circ \cdots \circ \kappa_k)(\{i \in I_k : U_i \cap B \ne \emptyset\}) \subseteq I_N \quad \text{consists of a single point.} \qquad (5.28)$$

As a first step we observe that we can assume that X is coarsely connected. Indeed we can do the construction below for every coarse component separately.

We now assume that X is coarsely connected and we fix a base point x_0 in X. For every n in $\mathbb{N}$ we let J_n be the index subset of I_n of members of the covering $\mathcal{V}_n$ which intersect $U_n[x_0]$ non-trivially. We now define the connecting maps with the property that $\kappa_n(J_n)$ consists of a single point: Since U_n is a bound of $\mathcal{V}_n$ the union $\bigcup_{j \in J_n} V_j$ is bounded by U_n^4. Since this is by assumption a Lebesgue entourage of

$\mathcal{V}_{n+1}$ there exists i in I_{n+1} such that $V_j \subseteq V_i$ for all j in J_n. We define $\kappa_n(j) := i$ for all j in J_n. Then we extend κ_n to $I_n \setminus J_n$.

Assume now that B is bounded. Since we assume that the bornology is generated by the coarse structure there exists an entourage U_B in $\mathcal{C}$ such that $B \subseteq U_B[x_0]$. If we choose N in $\mathbb{N}$ so large that $U_B \subseteq U_N$, then (5.28) holds.

We now show now that (5.28) implies that Φ is contracting. Given i in $\mathbb{N}$, ε in $(0, \infty)$ and a coarse entourage V of $\|\hat{\mathcal{V}}\|$ we must provide a number T in $(0, \infty)$ with the property

$$\forall x, y \in \|\hat{\mathcal{V}}\|, \ \forall t \in [T, \infty) \ : \ (x, y) \in V \cap (Y_i \times Y_i) \Rightarrow (\Phi_t(x), \Phi_t(y)) \in U_\varepsilon \, .$$

Here $Y_i = \{e \le i\}$ is a member of the big family which is used to define the $\mathcal{C}_0$-structure. Now Y_i is a bounded subset of $\|\hat{\mathcal{V}}\|$. In particular, it is contained in a finite union of generating bounded subsets which are of the form $[n, n + 1] \times B$ with B in $\mathcal{B}_{\mathcal{V}_n}$ for n in $\mathbb{N}$.

Let x and y be points in Y_i. There are an integer j, natural numbers $n_1, \ldots, n_j$ and bounded subsets B_l in $\mathcal{B}_{\mathcal{V}_{n_l}}$ for $1 \le l \le j$ such that $Y_i \subseteq \bigcup_{l=1}^{j} [n_l, n_l + 1] \times B_{n_l}$. Corollary 5.46 provides an equivalence of bornological coarse spaces $\omega_0 : X \to \|\mathcal{V}_0\|$, and we get equivalences $\omega_r : X \to \|\mathcal{V}_r\|$ by composing ω with the equivalence $\phi_{r-1} \circ \cdots \circ \phi_0$, see (5.23). We get a bounded subset

$$B_X := \omega_{n_0}^{-1}(B_{n_0}) \cup \ldots \cup \omega_{n_l}^{-1}(B_{n_l})$$

of X. By (5.28) we get a corresponding integer N, and then $T := \max\{N + 1, i + 1\}$ does the job. □

Remark 5.51 If $\hat{\mathcal{V}}$ is an anti-Čech system on a bornological coarse space X whose coarse structure has a single generating entourage and whose bornology is the underlying one, then by passing to a subsystem we can satisfy the third condition of Theorem 5.50. □

5.6.3 Flasqueness for the Hybrid Structure

We now discuss flasqueness of $\|\hat{\mathcal{V}}\|_h$.

Let $(X, \mathcal{C})$ be a coarse space.

Definition 5.52 ([DG07, Def. 2.7]) $(X, \mathcal{C})$ is exact, if for all U in $\mathcal{C}$ and all ε in $(0, \infty)$ there exists a bounded cover $\mathcal{W} = (W_i)_{i \in I}$ and a partition of unity $(\varphi_i)_{i \in I}$ subordinate to $\mathcal{W}$ such that

$$\forall (x, y) \in U \ : \ \sum_{i \in I} |\varphi_i(x) - \varphi_i(y)| \le \varepsilon \, . \tag{5.29}$$

Remark 5.53 We assume that (X, d) is a metric space and consider the coarse structure $\mathcal{C} := \mathcal{C}_d$ on X defined by the metric. In this situation Dadarlat and Guentner [DG07, Prop. 2.10] proved that we have the chain of implications

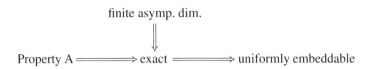

Here uniformly embeddable abbreviates uniformly embeddable into a Hilbert space [Yu00].

Property A has been introduced by Yu [Yu00, Sec. 2], where he also showed that finitely generated, amenable groups have Property A [Yu00, Ex. 2.3].

A metric space (X, d) is uniformly discrete if there exists an ε in $(0, \infty)$ such that $d(x, y) > \varepsilon$ for all x, y in X with $x \neq y$. Such a space has bounded geometry if for all r in $(0, \infty)$ we have

$$\sup_{x \in X} |B_d(x, r)| < \infty,$$

where $|A|$ denotes the number of points in the set A. Equivalently, a uniformly discrete metric space X has bounded geometry, if the bornological coarse space X_d has strongly bounded geometry in the sense of Definition 7.75. If (X, d) is uniformly discrete and has bounded geometry, then exactness is equivalent to Property A [DG07, Prop. 2.10(b)]. Tu proved that under the bounded geometry assumptions there are many more equivalent reformulations of Property A [Tu01, Prop. 3.2].

Higson and Roe [HR00a, Lem. 4.3] proved that bounded geometry metric spaces with finite asymptotic dimension (Definition 5.54) have Property A. Hence such spaces are exact.

Willett [Wil09, Cor. 2.2.11] has shown that a metric space of finite asymptotic dimension is exact. Note that the cited corollary states that spaces of finite asymptotic dimension have Property A under the assumption of bounded geometry. A close examination of the proof reveals that it is first shown that spaces of finite asymptotic dimension are exact, and then bounded geometry is used to verify Property A.

Exactness has the flavor of being a version of coarse para-compactness. The relationship of Property A to versions of coarse para-compactness was further investigated in [CDV14] and [Dyd16].   □

Let $(X, \mathcal{C})$ be a coarse space.

Definition 5.54 ([Gro93, Sec. 1.E]) $(X, \mathcal{C})$ has finite asymptotic dimension if it admits an anti-Čech system $\hat{\mathcal{V}} = (\mathcal{V}_n)_{n \in \mathbb{N}}$ with $\sup_{n \in \mathbb{N}} \dim \|\mathcal{V}_n\| < \infty$.

We consider a bornological coarse space X with coarse structure $\mathcal{C}$ and bornology $\mathcal{B}$. Furthermore, let $\hat{\mathcal{V}} = (\mathcal{V}_n)_{n \in \mathbb{N}}$ be an anti-Čech system on X.

Theorem 5.55 (Wright [Wri05, Thm. 5.9]) *Assume:*

1. *There exists U in C such that $C = C\langle U \rangle$.*
2. *$\sup_{n \in \mathbb{N}} \dim \|\mathcal{V}_n\| < \infty$.*

Then the bornological coarse space $\|\hat{\mathcal{V}}\|_h$ is flasque in the generalized sense.

Proof Since X admits an anti-Čech system $\hat{\mathcal{V}} = (\mathcal{V}_n)_{n \in \mathbb{N}}$ with $\sup_{n \in \mathbb{N}} \dim \|\mathcal{V}_n\| < \infty$ it has finite asymptotic dimension.

We will apply Theorem 5.35. We let $e := \pi$ be the projection defined in (5.25). It is straightforward to check that e is uniformly continuous, controlled and bornological, i.e., e satisfies all three conditions listed in Sect. 5.4.

We must provide the morphism $\Phi_h : [0, \infty) \ltimes \|\hat{\mathcal{V}}\| \to \|\hat{\mathcal{V}}\|$ (we use the subscript h in order to distinguish it from the morphism Φ in Eq. (5.26)).

Corollary 5.46 provides an equivalence of bornological coarse spaces $\omega_0 : X \to \|\mathcal{V}_0\|$, and we get equivalences $\omega_r : X \to \|\mathcal{V}_r\|$ by composing ω with the equivalence $\phi_{r-1} \circ \cdots \circ \phi_0$, see (5.23). Since asymptotic dimension is a coarsely invariant notion it follows that each $\|\mathcal{V}_r\|$ also has finite asymptotic dimension. Furthermore, the zero skeleton $\|\mathcal{V}_r\|^0$ with the induced bornological coarse structure is coarsely equivalent to $\|\mathcal{V}_r\|$ and therefore also has finite asymptotic dimension. By Willett [Wil09, Cor. 2.2.11] a quasi-metric space of finite asymptotic dimension is exact, hence $\|\mathcal{V}_r\|^0$ is exact. For each r we fix an inverse equivalence ω_r^0 to the inclusion $\|\mathcal{V}_r\|^0 \to \|\mathcal{V}_r\|$.

The argument below will involve the following construction. It depends on an integer T in $\mathbb{N}$, a number ε in $(0, \infty)$ and an entourage V of $\|\mathcal{V}_T\|^0$. The construction then produces an integer S with $T \leq S$ and a map $\Psi_{T,S} : \|\mathcal{V}_T\| \to \|\mathcal{V}_S\|$ such that

$$\forall (x, y) \in V : \operatorname{dist}(\Psi_{T,S}(x), \Psi_{T,S}(y)) \leq C \cdot \varepsilon, \tag{5.30}$$

where $C > 0$ is a constant only depending on the dimension of $\|\hat{\mathcal{V}}\|$. At this point we use Assumption 2. Note that (5.29) gives an estimate with respect to the ℓ^1-metric and the constant arrises since we work with the spherical metric.

We now describe the construction. Let T in $\mathbb{N}$, an entourage V of $\|\mathcal{V}_T\|^0$, and ε in $(0, \infty)$ be given. By exactness of $\|\mathcal{V}_T\|^0$ we can find a bounded cover $\mathcal{W}_T$ of $\|\mathcal{V}_T\|^0$ and a subordinate partition of unity $(\varphi_W)_{W \in \mathcal{W}_T}$ with Property (5.29) for the given ε and entourage V (in place of U). We define a map $\Psi_T : \|\mathcal{V}_T\|^0 \to \|\mathcal{W}_T\|$ by

$$\Psi_T(x) := \sum_{W \in \mathcal{W}_T} \varphi_W(x)[W].$$

Note that we can identify $\|\mathcal{V}_T\|^0 = I_T$. Since $\hat{\mathcal{V}}$ is an anti-Čech system and $\mathcal{W}_T$ is bounded, there is an S in $\mathbb{N}$ with $S > T$ such that for every member W of $\mathcal{W}_T$ there exists an element denoted by $\kappa(W)$ in I_S such that $V_i \subseteq V_{\kappa(W)}$ for all i in W. The map κ from the index set of $\mathcal{W}_T$ to I_S determines a map $\phi_{T,S} : \|\mathcal{W}_T\| \to \|\mathcal{V}_S\|$. The composition $\phi_{T,S} \circ \Psi_T : \|\mathcal{V}_T\|^0 \to \|\mathcal{V}_S\|$ can be extended linearly to a map $\Psi_{T,S} : \|\mathcal{V}_T\| \to \|\mathcal{V}_S\|$.

We now repeatedly apply the construction above. We start with the data $T_0 := 0$, $\varepsilon_0 := 1$ and $V := ((\omega_0^0 \circ \omega_0) \times (\omega_0^0 \circ \omega_0))(U)$, where the entourage U of X is as in the statement of this theorem. We get the integer T_1 and a map $\Psi_{0,T_1} : \|\mathcal{V}_0\| \to \|\mathcal{V}_{T_1}\|$. In the n'th step we use the data T_n, $\varepsilon_n := 1/n$ and $V := ((\omega_{T_n}^0 \circ \omega_{T_n}) \times (\omega_{T_n}^0 \circ \omega_{T_n}))(U^n)$, and get a integer T_{n+1} and a map $\Psi_{T_n,T_{n+1}} : \|\mathcal{V}_{T_n}\| \to \|\mathcal{V}_{T_{n+1}}\|$.

Let i in I be given and assume that (s,x) is a point in $[T_{i-1}, T_{i-1}+1) \times \|\mathcal{V}_{T_{i-1}}\|$. If Φ is the map from Eq. (5.26), then we define $\Theta_{i-1}(s,x) \in \{T_i\} \times \|\mathcal{V}_{T_i}\|$ by linear interpolation (writing $s = T_{i-1} + \bar{s}$)

$$\Theta_{i-1}(s,x) := (1-\bar{s})(T_i, \Psi_{T_{i-1},T_i}(x)) + \bar{s}\Phi(T_i, (s,x)),$$

where Φ is the map from Eq. (5.26).

We now define the map $\Phi_h : [0, \infty) \times \|\hat{\mathcal{V}}\| \to \|\hat{\mathcal{V}}\|$ by

$$\Phi_h(t, (s,x)) := \begin{cases} \Phi(t, \Psi_{T_{i(t)-1}, T_{i(t)}} \circ \Phi_{T_{i(t)-1}}(s,x)) & \text{for } s \leq T_{i(t)-1} \\ \Phi(t, \Theta_{i(t)-1}(s,x)) & \text{for } T_{i(t)-1} \leq s < T_{i(t)-1} + 1 \\ \Phi(t, (s,x)) & \text{for } T_{i(t)-1} + 1 < s \leq t \\ (s,x) & \text{for } s > t \end{cases}$$

where we have used the notation

$$i(t) := \sup_{T_i \leq t} i. \tag{5.31}$$

In order to check the continuity at $s = T_{i(t)-1} + 1$ we use the relation (5.27).

We have to check that Φ_h has the required properties listed above Theorem 5.35. The functions Φ and Ψ are 1-Lipschitz. This implies that Φ_h is uniformly continuous. The only remaining non-obvious property is contractivity.

It remains to show that Φ_h is contractive. Given j in $\mathbb{N}$, ε in $(0, \infty)$, and a coarse entourage V of $\|\hat{\mathcal{V}}\|$ we have to provide a number T in $(0, \infty)$ with the property

$$\forall (s,x), (r,y) \in \|\hat{\mathcal{V}}\|, \ \forall t \in [T, \infty) :$$

$$((s,x),(r,y)) \in V \cap (Y_j \times Y_j) \Rightarrow (\Phi_h(t,(s,x)), \Phi_h(t,(r,y))) \in U_\varepsilon.$$

We define a map (which is not a morphism in **BornCoarse**) $\Omega : \|\hat{\mathcal{V}}\| \to X$ which sends the point (s,x) in $\|\hat{\mathcal{V}}\|$ to $\omega_m'(x)$ for s in $[m, m+1)$, where ω_m' is a choice of inverse equivalence to ω_m. Let n in $\mathbb{N}$ be such that $(\Omega \times \Omega)(V \cap (Y_j \times Y_j)) \subseteq U^n$. We choose T in $\mathbb{R}$ with

1. $i(T) - 1 > \max\{\lceil C/\varepsilon \rceil, n\}$ (see (5.31)), and
2. $T_{i(T)-1} > j$.

The second inequality implies that the application of Φ_h to $(t, (s,x))$ and to $(t, (r,y))$ will fall into the first case of the definition of Φ_h, i.e., the function

$\Psi_{T_{i(T)-1}, T_{i(T)}}$ will be applied. But the latter function, by the first inequality, maps pairs of points lying in the entourage $(\omega_{T_{i(T)-1}} \times \omega_{T_{i(T)-1}})(U^n)$ to pairs of points lying a distance at most ε from each other due to (5.30) and the choice of $\varepsilon_n = 1/n$ in the construction above. Since Φ is 1-Lipschitz we get the desired statement for Φ_h. $\square$

Remark 5.56   We do not know whether the assumption of finite asymptotic dimension in Theorem 5.55 is necessary. It would be interesting to construct a space X such that its coarsening space equipped with the hybrid coarse structure is not flasque. One could try to adapt the constructions of surjectivity counter-examples to the coarse Baum–Connes conjecture [Hig99] in order to show that there exists a non-trivial element in the coarse K-homology $K\mathcal{X}_*(\|\hat{\mathcal{V}}\|_h)$ in the case of an expander X. But we were not able to carry out this idea. $\square$

5.7   The Motivic Coarse Spectra of Simplicial Complexes and Coarsening Spaces

Recall Definition 4.11. Let $\mathcal{A}_{\text{disc}}$ denote the set of discrete bornological coarse spaces.

We consider a simplicial complex X equipped with a bornological coarse structure $(\mathcal{C}, \mathcal{B})$, a metric d, and an ordered family of subcomplexes $\mathcal{Y} := (Y_n)_{n \in \mathbb{N}}$. Let $\mathcal{T}_d$ denote the uniform structure associated to the metric. In the statement of the following theorem X_h denotes the set X with the hybrid structure defined in Definition 5.10. We furthermore use the notation (4.5).

Theorem 5.57   *We assume:*

1. *d is good (Definition 5.36).*
2. *$\mathcal{C}$ and $\mathcal{T}_d$ are compatible (Definition 5.4).*
3. *The family $(Y_n)_{n \in \mathbb{N}}$ is big (Definition 3.2).*
4. *For every B in $\mathcal{B}$ there exists i in $\mathbb{N}$ such that $B \subseteq Y_i$.*
5. *X is finite-dimensional.*

Then we have $(X_h, \mathcal{Y}) \in \mathbf{Sp}\mathcal{X}\langle \mathcal{A}_{\text{disc}} \rangle$

Proof   We argue by induction on the dimension. We fix n in $\mathbb{N}$ and assume that the theorem has been shown for all dimensions k in $\mathbb{N}$ with $k < n$. We now assume that $\dim(X) = n$. We consider the decomposition (5.22). Combining Lemmas 5.40 and 5.41 we see that the lower left corner of (5.22) belongs to $\mathbf{Sp}\mathcal{X}\langle \mathcal{A}_{\text{disc}} \rangle$. In view of the Lemmas 5.42 and 5.43 the induction hypothesis implies that the two upper corners of the push-out square belong to $\mathbf{Sp}\mathcal{X}\langle \mathcal{A}_{\text{disc}} \rangle$. We conclude that $(X_h, \mathcal{Y}) \in \mathbf{Sp}\mathcal{X}\langle \mathcal{A}_{\text{disc}} \rangle$. $\square$

Let $(X, \mathcal{C})$ be a coarse space.

Definition 5.58 X has weakly finite asymptotic dimension if here exists a cofinal subset $\mathcal{C}'$ of $\mathcal{C}$ such that for every U in $\mathcal{C}'$ the coarse space $(X, \mathcal{C}\langle U \rangle)$ has finite asymptotic dimension (Definition 5.54).

If X is a bornological coarse space, then we apply this definition to its underlying coarse space.

We consider a bornological coarse space X.

Theorem 5.59 *If X has weakly finite asymptotic dimension, then we have* $\mathrm{Yo}^s(X) \in \mathbf{Sp}\mathcal{X}\langle \mathcal{A}_{\mathrm{disc}}\rangle$.

Proof Let $\mathcal{C}$ be the coarse structure of X, and let $\mathcal{C}'$ be as in Definition 5.58. By u-continuity we have

$$\mathrm{Yo}^s(X) \simeq \underset{U \in \mathcal{C}'}{\mathrm{colim}}\, \mathrm{Yo}^s(X_U)\,.$$

It therefore suffices to show that $\mathrm{Yo}^s(X_U) \in \mathbf{Sp}\mathcal{X}\langle \mathcal{A}_{\mathrm{disc}}\rangle$ for all U in $\mathcal{C}'$.

We now consider U in $\mathcal{C}'$. The bornological coarse space X_U admits an anti-Čech system $\hat{\mathcal{V}}$ satisfying the condition formulated in Definition 5.54.

We consider the coarsening space $\|\hat{\mathcal{V}}\|_h$ with the big family $(\|\hat{\mathcal{V}}\|_{\leq n})_{n \in \mathbb{N}}$. By Theorem 5.55 and Lemma 4.6 we have $\mathrm{Yo}^s(\|\hat{\mathcal{V}}\|_h) \simeq 0$. By Point 1 of Corollary 4.5 this implies

$$\Sigma\,\mathrm{Yo}^s((\|\hat{\mathcal{V}}\|_{\leq n})_{n \in \mathbb{N}}) \simeq (\|\hat{\mathcal{V}}\|_h, (\|\hat{\mathcal{V}}\|_{\leq n})_{n \in \mathbb{N}})\,.$$

We apply Theorem 5.57 to the pair $(\|\hat{\mathcal{V}}\|_h, (\|\hat{\mathcal{V}}\|_{\leq n})_{n \in \mathbb{N}})$ in order to conclude that

$$\Sigma\,\mathrm{Yo}^s((\|\hat{\mathcal{V}}\|_{\leq n})_{n \in \mathbb{N}}) \in \mathbf{Sp}\mathcal{X}\langle \mathcal{A}_{\mathrm{disc}}\rangle\,.$$

We use that $X_U \to \|\hat{\mathcal{V}}\|_0 \to \|\hat{\mathcal{V}}\|_{\leq n}$ is an equivalence in **BornCoarse** for every n in $\mathbb{N}$ and consequently $\mathrm{Yo}^s(X_U) \simeq \mathrm{Yo}^s((\|\hat{\mathcal{V}}\|_{\leq n})_{n \in \mathbb{N}})$ (see (4.4) for notation) in $\mathbf{Sp}\mathcal{X}$. It follows that $\mathrm{Yo}^s(X_U) \in \mathbf{Sp}\mathcal{X}\langle \mathcal{A}_{\mathrm{disc}}\rangle$ as required.   □

Part II
Coarse and Locally Finite Homology Theories

Chapter 6
First Examples and Comparison of Coarse Homology Theories

The notion of a coarse homology theory was introduced in Definition 4.22. We first show that the condition of u-continuity can be enforced. This result may be helpful for the construction of coarse homology theories. We furthermore discuss some additional additivity properties for coarse homology theories.

We then construct first examples of coarse homology theories, namely coarse ordinary homology and the coarsification of stable homotopy.

Before we discuss further examples of coarse homology theories in subsequent sections we state and prove the comparison theorems which motivated this book.

6.1   Forcing u-Continuity

Let $\mathbf{C}$ be a cocomplete stable ∞-category. In the following we describe a construction which turns a functor $E \colon \mathbf{BornCoarse} \to \mathbf{C}$ into a u-continuous one $E_u \colon \mathbf{BornCoarse} \to \mathbf{C}$.

We will have a natural transformation $E_u \to E$ which is an equivalence on most bornological coarse spaces of interest (e.g., path metric spaces like Riemannian manifolds, or finitely generated groups with a word metric). Furthermore, if E satisfies one of the conditions coarse invariance, excision or vanishes on flasque spaces, then E_u retains these properties.

The obvious idea to define $E_u \colon \mathbf{BornCoarse} \to \mathbf{C}$ is to set

$$E_u(X) := \operatorname*{colim}_{U \in \mathcal{C}} E(X_U), \tag{6.1}$$

where $\mathcal{C}$ denotes the coarse structure of X and we use the notation introduced in Example 2.31. This formula defines the value of E_u on objects, only. In order to define E_u as a functor we first introduce the category $\mathbf{BornCoarse}^{\mathcal{C}}$. Its objects are

© The Editor(s) (if applicable) and The Author(s), under exclusive licence
to Springer Nature Switzerland AG 2020
U. Bunke, A. Engel, *Homotopy Theory with Bornological Coarse Spaces*,
Lecture Notes in Mathematics 2269, https://doi.org/10.1007/978-3-030-51335-1_6

pairs (X, U) of a bornological coarse space X and a coarse entourage U of X. A morphism $f : (X, U) \to (X', U')$ in **BornCoarse**$^{\mathcal{C}}$ is a morphism $f : X \to X'$ in **BornCoarse** such that $(f \times f)(U) \subseteq U'$. We have a forgetful functors

$$\textbf{BornCoarse} \xleftarrow{q} \textbf{BornCoarse}^{\mathcal{C}} \xrightarrow{p} \textbf{BornCoarse}, \quad X_U \leftarrow (X, U) \to X.$$

We now consider the diagram

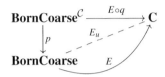

The functor E_u is then obtained from $E \circ q$ by a left Kan extension along p.

There is a natural transformation $q \to p$ of functors **BornCoarse**$^{\mathcal{C}} \to$ **BornCoarse** given on (X, U) in **BornCoarse**$^{\mathcal{C}}$ by the morphism $X_U \to X$. Applying E we get a transformation $E \circ q \to E \circ p$. This now induces the natural transformation $E_u \to E$ via the universal property of the left Kan extension.

The pointwise formula for left Kan extensions gives

$$E_u(X) \simeq \operatorname*{colim}_{\textbf{BornCoarse}^{\mathcal{C}}/X} E \circ q,$$

where **BornCoarse**$^{\mathcal{C}}/(X, \mathcal{C}, \mathcal{B})$ is the slice category of objects from **BornCoarse**$^{\mathcal{C}}$ over the bornological coarse space X. We observe that the subcategory of objects $((X, U), \mathrm{id}_X)$ for U in $\mathcal{C}$ is cofinal in **BornCoarse**$^{\mathcal{C}}/X$. Hence we can restrict the colimit to this subcategory. We then get the formula (6.1) as desired.

Proposition 6.1 *For a bornological coarse space X the transformation $E_u(X) \to E(X)$ is an equivalence if the coarse structure of X is generated by a single entourage.*

The following properties of E are inherited by E_u:

1. *coarse invariance,*
2. *excision,*
3. *vanishing on flasque bornological coarse spaces.*

Proof The first claim is clear from (6.1).

We now show that if E is coarsely invariant, then E_u is coarsely invariant. It suffices to show that for every bornological coarse space X the projection $\{0, 1\} \times X \to X$ induces an equivalence

$$E_u(\{0, 1\} \otimes X) \to E_u(X), \tag{6.2}$$

where $\{0, 1\}$ has the maximal bornology and coarse structure.

For every entourage U of X we have an isomorphism of bornological coarse spaces

$$(\{0, 1\} \otimes X)_{\tilde{U}} \cong \{0, 1\} \otimes X_U ,$$

where $\tilde{U} := \{0, 1\}^2 \times U$. Moreover, the entourages of $\{0, 1\} \otimes X$ of the form $\tilde{U}$ are cofinal in all entourages of this bornological coarse space. We now have an equivalence

$$E((\{0, 1\} \otimes X)_{\tilde{U}}) \simeq E(\{0, 1\} \otimes X_U) \simeq E(X_U),$$

where for the second we use that E is coarsely invariant. Forming the colimit of these equivalences over all entourages U of X and using the description (6.1) of the evaluation of E_u we get the desired equivalence (6.2).

Let E be coarsely invariant. Let $f : X \to X'$ be an equivalence with inverse g, let $g \circ f$ be W-close to id_X and $f \circ g$ be W'-close to $\mathrm{id}_{X'}$. For an entourage U of X we will then have an equivalence $f : X_V \to X'_{V'}$, where we have set $V := U \circ W \circ (g \times g)(W')$ and $V' := (f \times f)(U) \circ (f \times f)(W) \circ W'$. Note that to any entourage occurring we also have to add the diagonal. This shows that E_u is also coarsely invariant.

If E is coarsely excessive, then E_u is also coarsely excessive. This follows from the fact that complementary pairs on X are also complementary pairs on X_U.

Assume that E vanishes on flasque spaces. Let X be a flasque space. In order to show that $E_u(X) \simeq 0$ it suffices to show that $E(X_W) \simeq 0$ for a cofinal set of entourages W of X. Let $f : X \to X$ be a morphism implementing flasqueness. Then there exists an entourage V of X such that f is V-close to id_X. For any entourage U of X consider the entourage

$$W := \bigcup_{k \in \mathbb{N}} (f^k \times f^k)(U \cup V).$$

Then X_W is flasque with flasqueness implemented by the morphism f and $U \subseteq W$. Hence we have $E(X_W) \simeq 0$.   □

Let $\mathbf{C}$ be a cocomplete stable ∞-category, and let $E : \mathbf{BornCoarse} \to \mathbf{C}$ be a functor.

Corollary 6.2 *Assume:*

1. *E is coarsely invariant.*
2. *E vanishes on flasques.*
3. *E satisfies excision.*

Then:

1. *E_u is a coarse homology theory.*
2. *If E is coarse homology theory, then $E_u \to E$ is an equivalence.*

*3. If the coarse structure of X is generated by a single entourage, then $E_u(X) \to$
$E(X)$ is an equivalence.*

Note that Assertion 3. applies, e.g. if X is a path metric space.

6.2 Additivity and Coproducts

Note that the axioms for a coarse homology theory do not include an additivity
property for, say, infinite coproducts (finite coproducts are covered by excision). We
refer to the discussion in Example 4.10 and Lemma 4.12. But most of the known
coarse homology theories are in one way or the other compatible with forming
infinite coproducts and / or infinite free unions. In this section we are going to
formalize this.

6.2.1 Additivity

Let $E : \mathbf{BornCoarse} \to \mathbf{C}$ be a coarse homology theory and assume that $\mathbf{C}$ is in
addition complete.

Definition 6.3 E is strongly additive if for every family $(X_i)_{i \in I}$ of bornological
coarse spaces the morphism

$$E\left(\bigsqcup_{i \in I}^{\mathrm{free}} X_i\right) \to \prod_{i \in I} E(X_i)$$

induced by the collection of morphisms (4.9) is an equivalence.

Definition 6.4 We call E additive if for every set I the morphism

$$E\left(\bigsqcup_{I}^{\mathrm{free}} *\right) \to \prod_{I} E(*)$$

is an equivalence.

Clearly, a strongly additive coarse homology theory is additive. The notion of
additivity features in Theorem 6.44.

Example 6.5 Coarse ordinary homology (Proposition 6.24), the coarsification of
stable homotopy (Proposition 6.38) and coarsifications of locally finite homology
theories (Proposition 7.46) are strongly additive. In [BEe] we show that coarse K-
homology $K\mathcal{X}$ is strongly additive. We refer to [BEb, Ex. 10.5] for an extended

list of strongly additive coarse homology theories which have been constructed in sequels to the present book.

We were not able to show that the quasi-local version $K\mathcal{X}_{\mathrm{ql}}$ of coarse K-homology is strongly additive. But both $K\mathcal{X}$ and $K\mathcal{X}_{\mathrm{ql}}$ are additive (Corollary 8.115).

In Example 6.40 we will provide a coarse homology theory which is not additive. □

Let $E\colon \mathbf{BornCoarse} \to \mathbf{C}$ be a coarse homology theory, and let $(X_i)_{i\in I}$ be a family of bornological coarse spaces.

Lemma 6.6 *If E is strongly additive, then we have a fibre sequence*

$$\bigoplus_{i\in I} E(X_i) \to E\Big(\overset{\text{mixed}}{\bigsqcup_{i\in I}} X_i\Big) \to \operatorname*{colim}_{J\subseteq I \text{ finite}} \prod_{i\in I\setminus J} E(X_{i,\mathrm{disc}}) . \qquad (6.3)$$

Proof This follows from Lemma 4.12 and the fact that we can rewrite the remainder term (4.12) using strong additivity of E (note that for discrete bornological coarse spaces the mixed and the free union coincide). □

Example 6.7 Let E be an additive coarse homology theory with $E(*) \not\simeq 0$. Such a coarse homology theory exists, e.g., $H\mathcal{X}$ defined in Sect. 6.3. Applying (6.3) to E in the case that $X_i = *$ for all i in I (in this case it suffices to assume that E is just additive in order to have the fibre sequence (6.3)), we get the fibre sequence

$$\bigoplus_I E(*) \to \prod_I E(*) \to \operatorname*{colim}_{J\subseteq I \text{ finite}} \prod_{i\in I\setminus J} E(*) ,$$

which shows that the remainder term is non-trivial provided I is infinite. Here we have again used that for discrete bornological spaces the mixed and free union coincide. □

Example 6.8 We refer to Theorem 6.13 for further examples of additive coarse homology theories. □

6.2.2 Coproducts

Let $E\colon \mathbf{BornCoarse} \to \mathbf{C}$ be a coarse homology theory.

Definition 6.9 E preserves coproducts if for every family $(X_i)_{i\in I}$ of bornological coarse spaces the canonical morphism

$$\bigoplus_{i\in I} E(X_i) \to E\Big(\coprod_{i\in I} X_i\Big)$$

is an equivalence.

Remark 6.10 Due to the nature of the remainder term (4.13) derived in the proof of Lemma 4.12 in order to show that a coarse homology theory preserves coproducts it suffices to check this only on discrete bornological coarse spaces.   □

Example 6.11 We show in this paper that coarse ordinary homology and both versions of coarse K-homology preserve coproducts. Coarsifications of locally finite homology theories preserve coproducts if the locally finite theory does so.
                                                                                            □

Example 6.12 Let E: **BornCoarse** → **C** be a coarse homology theory. In [BEb] we construct a new coarse homology theory denoted by $E\mathcal{O}^\infty \mathbf{P}$: **BornCoarse** → **C**. It serves as the domain of the coarse assembly map μ_E: $E\mathcal{O}^\infty \mathbf{P}$ → ΣE which is the main topic of that paper. Using information on this assembly map and Lemma 4.12 in [BEb] we will observe the following theorem.

Theorem 6.13 *Assume that E is strong (see [BEKW20, [Def. 4.19]).*

1. *If E is additive, then so is $E\mathcal{O}^\infty \mathbf{P}$.*
2. *If E preserves coproducts, then so does $E\mathcal{O}^\infty \mathbf{P}$.*

In particular, we get more examples of coarse homology theories which preserve coproducts.   □

In [BEKW20] we introduce the notion of continuity. If X is a bornological coarse space, then by $\mathrm{LF}(X)$ we denote its poset of locally finite subsets, see Definition 2.3. The elements of $\mathrm{LF}(X)$ will be equipped with the bornological coarse structure induced from X. Note that by definition the induced bornology is the minimal one.

Let **C** be a cocomplete stable ∞-category, and let E : **BornCoarse** → **C** be a functor.

Definition 6.14 E is continuous if for every X in **BornCoarse** the canonical map

$$\operatorname*{colim}_{L \in \mathrm{LF}(X)} E(L) \to E(X)$$

is an equivalence.

Remark 6.15 It is straightforward to check that a continuous coarse homology theory preserves coproducts [BEKW20, Lem. 5.17].

The coarse ordinary homology (Definition 6.18) and the two versions $K\mathcal{X}$ and $K\mathcal{X}_{\mathrm{ql}}$ of coarse K-homology (Definition 8.78) are continuous.

In general, given a coarse homology theory, using a method similar to the one in Sect. 6.1 one can construct a new coarse homology which is continuous [BEKW20, Lem. 5.26]. So this construction produces further examples of coarse homology theories which preserve coproducts.

Continuity of coarse homology theories is relevant for their application in the study of assembly maps. In the present book we will not discuss continuity further and refer to [BEKW20] and [BEKWa] for more information and applications.   □

6.3 Coarse Ordinary Homology

In this section we introduce a coarse version $H\mathcal{X}$ of ordinary homology. After defining the functor our main task is the verification of the properties listed in Definition 4.22. The fact that the coarse homology $H\mathcal{X}$ exists and is non-trivial is a first indication that $\mathbf{Sp}\mathcal{X}$ is non-trivial.

Since the following constructions and arguments are in principle well-known we will be sketchy at various points in order to keep this section short. Our task here will be mainly to put the well-known construction into the setup developed in this paper. The structure of this section is very much the model for the analogous Sects. 6.4 and 8.6 dealing with coarsification and coarse K-homology.

It is hard to trace back the first definition of coarse ordinary homology. One of the first definitions was the uniform version (i.e., a so-called rough homology theory) by Block and Weinberger [BW92].

Let $\mathbf{Ch}$ denote the small category of very small chain complexes. Its objects are chain complexes of very small abelian groups. The morphisms in $\mathbf{Ch}$ are chain maps. The target category of the coarse ordinary homology will be the stable ∞-category $\mathbf{Ch}_\infty$ obtained by a Dwyer–Kan localization

$$\ell : \mathbf{Ch} \to \mathbf{Ch}[W^{-1}] =: \mathbf{Ch}_\infty\,, \tag{6.4}$$

where W is the set of quasi-isomorphisms in $\mathbf{Ch}$. The homotopy category of $\mathbf{Ch}_\infty$ is the usual unbounded derived category of abelian groups. It is well known that $\mathbf{Ch}_\infty$ is stable, very small complete and cocomplete. We refer to [Lur17, Sec. 1.3] for details. We will use the following properties of the functor ℓ:

1. ℓ sends chain homotopic maps to equivalent morphisms.
2. ℓ preserves filtered colimits.
3. ℓ sends short exact sequences of chain complexes to fibre sequences of spectra.
4. ℓ preserves sums and products.

We start by defining a functor

$$C\mathcal{X} : \mathbf{BornCoarse} \to \mathbf{Ch}$$

which associates to a bornological coarse space X the chain complex of locally finite controlled chains.

In order to talk about properties of chains on a bornological coarse space X we introduce the following language elements. Let n be in $\mathbb{N}$ and B be a subset of X. We say that a point $(x_0, \ldots, x_n)$ in X^{n+1} meets B if there exists an index i in $\{0, \ldots, n\}$ such that $x_i \in B$.

Let U be an entourage of X. We say that $(x_0, \ldots, x_n)$ is U-controlled if for every two indices i, j in $\{0, \ldots, n\}$ we have $(x_i, x_j) \in U$.

An n-chain on X is by definition just a function $c : X^{n+1} \to \mathbb{Z}$. Its support is the set

$$\mathrm{supp}(c) := \{x \in X^{n+1} \mid c(x) \neq 0\}\,.$$

Let X be a bornological coarse space with an entourage U. Let furthermore c be an n-chain on X.

Definition 6.16

1. c is locally finite, if for every bounded B the set of points in $\mathrm{supp}(c)$ which meet B is finite.
2. c is U-controlled if every point in $\mathrm{supp}(c)$ is U-controlled.

We say that an n-chain is controlled if it is U-controlled for some entourage U of X.

Let X be a bornological coarse space.

Definition 6.17 We define $C\mathcal{X}_n(X)$ to be the abelian group of locally finite and controlled n-chains.

It is often useful to represent n-chains on X as (formal, potentially infinite) sums

$$\sum_{x \in X^{n+1}} c(x)x\,.$$

For i in $\{0, \ldots, n\}$ we define $\partial_i : X^{n+1} \to X^n$ by

$$\partial_i(x_0, \ldots, x_n) := (x_0, \ldots \hat{x}_i, \ldots, x_n)\,.$$

So ∂_i is the projection leaving out the i'th coordinate.

If x is U-controlled, then so is $\partial_i x$.

Using that the elements of $C\mathcal{X}_n(X)$ are locally finite and controlled one checks that ∂_i extends linearly (also for infinite sums) in a canonical way to a map

$$\partial_i : C\mathcal{X}_n(X) \to C\mathcal{X}_{n-1}(X)\,.$$

We consider the sum $\partial := \sum_{i=0}^{n}(-1)^i \partial_i$. One easily verifies that $\partial \circ \partial = 0$. The family $(C\mathcal{X}_n(X))_{n \in \mathbb{N}}$ of abelian groups together with the differential ∂ is the desired chain complex $C\mathcal{X}(X)$ in **Ch**.

Let now $f : X \to X'$ be a morphism of bornological coarse spaces. The map

$$(x_0, \ldots, x_n) \mapsto (f(x_0), \ldots, f(x_n))$$

extends linearly (also for infinite sums) in a canonical way to a map

$$CX(f) : CX_n(X) \to CX_n(X') .$$

This map involves sums over fibres of f which become finite since f is proper and controlled and since we apply it to locally finite chains. Using that f is controlled we conclude that the induced map sends controlled chains to controlled chains. It is easy to see that $CX(f)$ is a chain map. This finishes the construction of the functor

$$CX : \mathbf{BornCoarse} \to \mathbf{Ch} .$$

Definition 6.18 We define the functor $HX : \mathbf{BornCoarse} \to \mathbf{Ch}_\infty$ by

$$HX := \ell \circ CX .$$

Theorem 6.19 *HX is a **Sp**-valued coarse homology theory.*

Proof In the following four propositions we will verify the properties listed in Definition 4.22, which will complete the proof. □

Proposition 6.20 *HX is coarsely invariant.*

Proof Let $f, g : X \to X'$ be two morphisms which are close to each other (Definition 3.13). We must show that the morphisms $HX(f)$ and $HX(g)$ from $HX(X)$ to $HX(X')$ are equivalent. We will construct a chain homotopy from $CX(g)$ to $CX(f)$ and then apply Property 1. of the functor ℓ.

For i in $\{0, \ldots, n\}$ we consider the map

$$h_i : X^{n+1} \to X'^{n+2} , \quad (x_0, \ldots, x_n) \mapsto (f(x_0) \ldots, f(x_i), g(x_i), \ldots, g(x_n)) .$$

We observe that it extends linearly (again for possibly infinite sums) to

$$h_i : CX_n(X) \to CX_{n+1}(X')$$

in a canonical way. Here we need the fact that f and g are close to each other in order to control the pairs $(f(x_i), g(x_j))$. We now define

$$h := \sum_{i=0}^{n} (-1)^i h_i .$$

One checks in a straightforward manner that

$$\partial \circ h + h \circ \partial = CX(g) - CX(f) .$$

Hence h is a chain homotopy between $CX(g)$ and $CX(f)$ as desired. □

Proposition 6.21 *$H\mathcal{X}$ satisfies excision.*

Proof Let X be a bornological coarse space, and let $(Z, \mathcal{Y})$ a complementary pair (Definition 3.5) with the big family $\mathcal{Y} = (Y_i)_{i \in I}$. By Property 2. of the functor ℓ we have an equivalence

$$H\mathcal{X}(\mathcal{Y}) \simeq \ell(\underset{i \in I}{\mathrm{colim}}\, C\mathcal{X}(Y_i))\,.$$

Since for an injective map f between bornological coarse spaces the induced map $C\mathcal{X}(f)$ is obviously injective we can interpret the colimit on the right-hand side as a union of subcomplexes in $C\mathcal{X}(X)$. Similarly we can consider $C\mathcal{X}(Z)$ as a subcomplex of $C\mathcal{X}(X)$ naturally.

We now consider the following diagram

$$
\begin{array}{ccccccccc}
0 & \longrightarrow & C\mathcal{X}(Z \cap \mathcal{Y}) & \longrightarrow & C\mathcal{X}(Z) & \longrightarrow & C\mathcal{X}(Z)/C\mathcal{X}(Z \cap \mathcal{Y}) & \longrightarrow & 0 \\
& & \downarrow & & \downarrow & & \downarrow & & \\
0 & \longrightarrow & C\mathcal{X}(\mathcal{Y}) & \longrightarrow & C\mathcal{X}(X) & \longrightarrow & C\mathcal{X}(X)/C\mathcal{X}(\mathcal{Y}) & \longrightarrow & 0
\end{array}
$$

We want to show that the left square gives a push-out square after applying ℓ. Using Property 3. of the functor ℓ we see that it suffices to show that the right vertical map is a quasi-isomorphism. We will actually show that it is an isomorphism of chain complexes.

For injectivity we consider a chain c in $C\mathcal{X}(Z)$ whose class $[c]$ in $C\mathcal{X}(Z)/C\mathcal{X}(Z \cap \mathcal{Y})$ is sent to zero. Then $c \in C\mathcal{X}(\mathcal{Y}) \cap C\mathcal{X}(Z) = C\mathcal{X}(Z \cap \mathcal{Y})$. Hence $[c] = 0$.

For surjectivity consider c in $C\mathcal{X}(X)$. We fix i in I such that $Z \cup Y_i = X$. We can find a coarse entourage U of X such that c is U-controlled. Let c_Z be the restriction of c to Z^{n+1}, where n is the degree of c. Using that $\mathcal{Y}$ is big we choose j in I such that $U[Y_i] \subseteq Y_j$. Then one can check that $c - c_Z \in C\mathcal{X}(Y_j)$. Consequently, the class $[c_Z]$ in $C\mathcal{X}(Z)/C\mathcal{X}(Z \cap \mathcal{Y})$ is mapped to the class of $[c]$. $\qquad\square$

Proposition 6.22 *If X is a flasque bornological coarse space, then $H\mathcal{X}(X) \simeq 0$.*

Proof Let $f : X \to X$ implement flasqueness of X (Definition 3.21). For every n in $\mathbb{N}$ we define

$$S \colon X^{n+1} \to C\mathcal{X}_n(X)\,, \quad S(x) := \sum_{k \in \mathbb{N}} C\mathcal{X}(f^k)(x)\,.$$

Using the properties of f listed in Definition 3.21 one checks that S extends linearly to a chain map

$$S \colon C\mathcal{X}(X) \to C\mathcal{X}(X)\,.$$

For example, we use Point 3 of Definition 3.21 in order to see that S takes values in locally finite chains. We further use Point 2 of Definition 3.21 in order to see that S maps controlled chains to controlled chains.

From the construction of S we immediately conclude that

$$\mathrm{id}_{C\mathcal{X}(X)} + C\mathcal{X}(f) \circ S = S$$

and applying ℓ we get

$$\mathrm{id}_{H\mathcal{X}(X)} + H\mathcal{X}(f) \circ \ell(S) \simeq \ell(S).$$

We now use that f is close to id_X (3.21.1) and Proposition 6.20 to conclude that

$$\mathrm{id}_{H\mathcal{X}(X)} + \ell(S) = \ell(S).$$

This implies that $\mathrm{id}_{H\mathcal{X}(X)} \simeq 0$ and hence $H\mathcal{X}(X) \simeq 0$.   □

Proposition 6.23 *$H\mathcal{X}$ is u-continuous.*

Proof Let X be a bornological coarse space with coarse structure $\mathcal{C}$. Then $C\mathcal{X}(X)$ is the union of its subcomplexes of U-controlled locally finite chains over all U in $\mathcal{C}$. Furthermore, for fixed U in $\mathcal{C}$ the union of the U^n-controlled chains, where n runs over n in $\mathbb{N}$, is the image of $C\mathcal{X}(X_U)$. Using Property 2. of the functor ℓ we immediately get the desired equivalence

$$H\mathcal{X}(X) \simeq \ell(\operatorname*{colim}_{U \in \mathcal{C}} C\mathcal{X}(X_U)) \simeq \operatorname*{colim}_{U \in \mathcal{C}} \ell(C\mathcal{X}(X_U)) \simeq \operatorname*{colim}_{U \in \mathcal{C}} H\mathcal{X}(X_U).$$

□

Proposition 6.24 *$H\mathcal{X}$ is strongly additive.*

Proof Let $(X_i)_{i \in I}$ be a family of bornological coarse spaces. An inspection of the definitions shows that controlled simplices on $\bigsqcup_{i \in I}^{\mathrm{free}} X_i$ can not mix the components. It follows that

$$C\mathcal{X}\Big(\bigsqcup_{i \in I}^{\mathrm{free}} X_i\Big) \cong \prod_{i \in I} C\mathcal{X}(X_i).$$

We now use Property 4. of the functor ℓ in order to conclude the equivalence

$$H\mathcal{X}\Big(\bigsqcup_{i \in I}^{\mathrm{free}} X_i\Big) \simeq \prod_{i \in I} H\mathcal{X}(X_i).$$

One easily checks that this equivalence is induced by the correct morphism.   □

Lemma 6.25 *$H\mathcal{X}$ preserves coproducts.*

Proof Let $(X_i)_{i \in I}$ be a family of bornological coarse spaces. Since coarse chains are locally finite, we conclude that every coarse chain on $\coprod_{i \in I} X_i$ is supported on finitely many components X_i. Furthermore, a controlled simplex is supported on a single component. So we get an isomorphism of complexes

$$\bigoplus_{i \in I} C\mathcal{X}(X_i) \cong C\mathcal{X}\left(\coprod_{i \in I} X_i \right)$$

and the assertion of the lemma follows with Property 4. of the functor ℓ.   □

In the following we consider abelian groups as chain complexes concentrated in degree 0.

Lemma 6.26 *If X is a set considered as a bornological coarse space with the maximal structures, then $H\mathcal{X}(X) \simeq \ell(\mathbb{Z})$.*

Proof We have $H\mathcal{X}(*) \simeq \ell(\mathbb{Z})$. Indeed,

$$C\mathcal{X}(*) \cong \ldots 0 \leftarrow \mathbb{Z} \xleftarrow{0} \mathbb{Z} \xleftarrow{1} \mathbb{Z} \xleftarrow{0} \mathbb{Z} \xleftarrow{1} \mathbb{Z} \xleftarrow{0} \mathbb{Z} \ldots$$

and the inclusion of the left-most $\mathbb{Z}$ sitting in degree 0 induces an equivalence after applying ℓ. The inclusion $* \to X$ is an equivalence. Consequently,

$$\ell(\mathbb{Z}) \simeq H\mathcal{X}(*) \simeq H\mathcal{X}(X),$$

as claimed.   □

6.4   Coarsification of Stable Homotopy

Let **Sp** denote the small stable ∞-category of very small spectra. In usual homotopy theory the sphere spectrum S in **Sp** represents the homology theory

$$\Sigma_+^{\infty,\mathrm{top}} : \mathbf{Top} \to \mathbf{Sp}$$

called the stable homotopy theory. The goal of the present section is to construct a coarse homology theory

$$Q : \mathbf{BornCoarse} \to \mathbf{Sp}$$

which is a coarse version of stable homotopy theory. The construction uses the process of coarsification of a locally finite homology theory which was introduced by Roe [Roe03, Def. 5.37] using anti-Čech systems. We employ spaces of probability measures as Emerson and Meyer [EM06, Sec. 4]. In the present section we go directly to the construction of Q. In the subsequent Sect. 7 we will develop the theory of locally finite homology theories and their coarsification in general, generalizing the ideas of the present subsection.

The coarsification of stable homotopy theory Q gives rise to may further coarse homology theories. If $\mathbf{C}$ is a cocomplete stable ∞-category, then it is tensorized over $\mathbf{Sp}$, i.e, we have a bi-functor

$$\mathbf{C} \times \mathbf{Sp} \to \mathbf{C}, \quad (C, R) \mapsto C \wedge R \tag{6.5}$$

which preserves very small colimits in both arguments. Every object C in $\mathbf{C}$ gives rise to a $\mathbf{C}$-valued homology theory

$$C \wedge \Sigma_+^{\infty,\mathrm{top}} : \mathbf{Top} \to \mathbf{C}, \quad X \mapsto C \wedge \Sigma_+^{\infty,\mathrm{top}}(X)$$

whose value on a point is C. Similarly, we can define a $\mathbf{C}$-valued coarse homology theory

$$C \wedge Q : \mathbf{BornCoarse} \to \mathbf{C}, \quad X \mapsto C \wedge Q(X),$$

see Corollary 6.39.

Let $\mathbf{C}$ be a complete and cocomplete ∞-category. In some arguments below we need the following property of $\mathbf{C}$.

Let $(J_i)_{i \in I}$ be a very small family of very small filtered categories. Then $J := \prod_{i \in I} J_i$ is again a very small filtered category. If $(C_i : J_i \to \mathbf{C})_{i \in I}$ is a family of diagrams, then we can form a diagram $\prod_{i \in I} C_i : J \to \mathbf{C}$.

Definition 6.27 ([ALR03, Sec. 1.2.(iv)]) $\mathbf{C}$ has the property that products distribute over filtered colimits, if for every family of diagrams $(C_i : J_i \to \mathbf{C})_{i \in I}$ as above the canonical morphism

$$\operatorname*{colim}_{J} \prod_{i \in I} C_i \to \prod_{i \in I} \operatorname*{colim}_{J_i} C_i$$

is an equivalence.

Remark 6.28 It is easy to see that the category $\mathbf{Set}$ has this property. Using the fact that the functors $\pi_i : \mathbf{Spc} \to \mathbf{Set}$ (the group structure is not relevant here) for i in $\mathbb{N}$ preserves filtered colimits and products, and that they jointly detect equivalences we can conclude that $\mathbf{Spc}$ has the property, too. One can then conclude that every compactly generated ∞-category $\mathbf{C}$ has the property defined in Definition 6.27. To this end one uses the functors $\operatorname{Map}_{\mathbf{C}}(K, -) : \mathbf{C} \to \mathbf{Spc}$ for compact objects K in $\mathbf{C}$ in order to reduce to the case of $\mathbf{Spc}$.

In particular, in the ∞-categories $\mathbf{Sp}$ and $\mathbf{Ch}_\infty$ products distribute over filtered colimits.

We will also need the dual property: sums distribute over cofiltered limits in $\mathbf{C}$ if products distribute over filtered colimits, in $\mathbf{C}^{\mathrm{op}}$. $\square$

6.4.1   Rips Complexes and a Coarsification of Stable Homotopy

A set X can be considered as a discrete topological space. It is locally compact and Hausdorff. Its compact subsets are the finite subsets. We let $C_0(X)$ be the Banach space closure of the space of compactly supported, continuous (this a void condition) functions on X with respect to the sup-norm. For every point x in X we define the function

$$e_x \in C_0(X), \quad e_x(y) := \begin{cases} 1 & y = x \\ 0 & \text{else} \end{cases}$$

We can consider $(X, \mathcal{P}(X))$ as the Borel measurable space associated to the topological space $(X, \mathcal{P}(X))$. By $P(X)$ we denote the space of regular probability measures on X. Note that a regular probability measure is in this case just a positive function on the set X with ℓ^1-norm 1. The topology on $P(X)$ is induced from the inclusion of $P(X)$ into the unit sphere of the Banach dual of $C_0(X)$ which is compact and Hausdorff in the weak $*$-topology. A map of sets $f : X \to X'$ functorially induces a continuous map

$$f_* : P(X) \to P(X'), \quad \mu \mapsto f_* \mu .$$

The support of a probability measure μ in $P(X)$ is defined by

$$\mathrm{supp}(\mu) := \{x \in X \mid \mu(e_x) \neq 0\} .$$

We assume that X is a bornological coarse space with coarse structure $\mathcal{C}$ and bornology $\mathcal{B}$. For an entourage U in $\mathcal{C}$ we let

$$P_U(X) := \{\mu \in P(X) : \mathrm{supp}(\mu) \text{ is } U\text{-bounded}\} \tag{6.6}$$

(see Definition 2.15 for the notion of U-boundedness).

We claim that $P_U(X)$ is a closed subspace of $P(X)$. We will show that $P(X) \setminus P_U(X)$ is open. Let μ be in the complement $P(X) \setminus P_U(X)$. Then there exist two points x, y in X such that $(x, y) \notin U$ and $\mu(e_x) \neq 0$ and $\mu(e_y) \neq 0$. Then $\{\nu \in P(X) \mid \nu(e_x) \neq 0 \,\&\, \nu(e_y) \neq 0\}$ is an open neighbourhood of μ contained in $P(X) \setminus P_U(X)$. This shows the claim.

Let X' be a second bornological coarse space with structures C' and B', and let U' be in C'. If $f : X \to X'$ is map of sets and $(f \times f)(U) \subseteq U'$, then f_* restricts to a map

$$f_* : P_U(X) \to P_{U'}(X').$$

If Y is a subset of X, then we get a map

$$P_{U_Y}(Y) \to P_U(X),$$

where $U_Y := U \cap (Y \times Y)$. In the following we will simplify the notation and write $P_U(Y)$ instead of $P_{U_Y}(Y)$.

In the following we use the functor

$$\Sigma_+^{\infty,\mathrm{top}} : \mathbf{Top} \to \mathbf{Sp}. \tag{6.7}$$

It is defined as the composition

$$\mathbf{Top} \xrightarrow{\mathrm{sing}} \mathbf{sSet} \xrightarrow{\ell} \mathbf{sSet}[W^{-1}] \xrightarrow{!} \mathbf{Spc} \xrightarrow{\Sigma_+^{\infty}} \mathbf{Sp}. \tag{6.8}$$

Here sing sends a topological space to the very small simplicial set of singular simplices. The functor ℓ is the Dwyer–Kan localization at the set W of weak homotopy equivalences. The equivalence marked by ! depends on the choice of a model for $\mathbf{Spc}$ and could be taken as a definition of $\mathbf{Spc}$. Finally, the functor Σ_+^{∞} is the small version of the left-adjoint in (4.1). In order to distinguish (6.7) from this functor we added the superscript "top".

In the arguments below we will use the following well-known properties of this functor:

1. $\Sigma_+^{\infty,\mathrm{top}}$ sends homotopic maps to equivalent maps.
2. If $(T_i)_{i \in I}$ is a filtered family of subspaces of some space such that for every i in I there exists j in J such that T_j contains an open neighbourhood of T_i, then

$$\operatorname*{colim}_{i \in I} \Sigma_+^{\infty,\mathrm{top}}(T_i) \simeq \Sigma_+^{\infty,\mathrm{top}}\left(\bigcup_{i \in I} T_i\right).$$

3. If (U, V) is an open covering of a topological space X, then we have a push-out square in spectra

$$\begin{array}{ccc}
\Sigma_+^{\infty,\mathrm{top}}(U \cap V) & \longrightarrow & \Sigma_+^{\infty,\mathrm{top}} V \\
\downarrow & & \downarrow \\
\Sigma_+^{\infty,\mathrm{top}} U & \longrightarrow & \Sigma_+^{\infty,\mathrm{top}} X
\end{array}$$

Definition 6.29 We define the functor

$$Q : \mathbf{BornCoarse} \to \mathbf{Sp}$$

on objects by

$$Q(X) := \operatorname*{colim}_{U \in \mathcal{C}} \lim_{B \in \mathcal{B}} \operatorname{Cofib}\big(\Sigma_+^{\infty,\mathrm{top}} P_U(X \setminus B) \to \Sigma_+^{\infty,\mathrm{top}} P_U(X)\big), \qquad (6.9)$$

where $\mathcal{C}$ and $\mathcal{B}$ denote the coarse structure and the bornology of X. For the complete definition also on morphisms see Remark 6.31 below.

Example 6.30 We have $Q(*) \simeq S$, where S in **Sp** is the sphere spectrum. $\qquad\qquad\square$

Remark 6.31 The formula (6.9) defines Q on objects. In the following we recall the standard procedure to interpret this formula as a definition of a functor. Let **BornCoarse**$^{\mathcal{C},\mathcal{B}}$ be the category of triples (X, U, B) of a bornological coarse space X, a coarse entourage U of X, and a bounded subset B of X. A morphism

$$f : (X, U, B) \to (X', U', B')$$

in **BornCoarse**$^{\mathcal{C},\mathcal{B}}$ is a morphism $f : X \to X'$ in **BornCoarse** such that $(f \times f)(U) \subseteq U'$ and $f^{-1}(B') \subseteq B$.

A similar category **BornCoarse**$^{\mathcal{C}}$ of pairs (X, U) has been considered in Sect. 6.1.

We have a chain of forgetful functors

$$\mathbf{BornCoarse}^{\mathcal{C},\mathcal{B}} \to \mathbf{BornCoarse}^{\mathcal{C}} \to \mathbf{BornCoarse}$$

where the first forgets the bounded subset, and the second forgets the entourage. We now consider the diagram

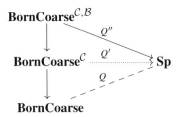

The functor Q'' is defined as the composition

$$\mathbf{BornCoarse}^{\mathcal{C},\mathcal{B}} \to \mathrm{Mor}(\mathbf{Sp}^{sm}) \overset{\mathtt{Cofib}}{\to} \mathbf{Sp}^{sm},$$

where the first functor sends (X, U, B) to the morphism of spectra

$$\Sigma_+^{\infty,\text{top}} P_U(X \setminus B) \to \Sigma_+^{\infty,\text{top}} P_U(X),$$

and the second takes the cofibre. Therefore, Q'' is the functor

$$(X, U, B) \mapsto \text{Cofib}\big(\Sigma_+^{\infty,\text{top}} P_U(X \setminus B) \to \Sigma_+^{\infty,\text{top}} P_U(X)\big).$$

We now define Q' as the right Kan extension of Q'' along the forgetful functor

$$\mathbf{BornCoarse}^{\mathcal{C},\mathcal{B}} \to \mathbf{BornCoarse}^{\mathcal{C}}.$$

By the pointwise formula for the evaluation of the Kan extension on objects we have

$$Q'(X, U) := \lim_{(X,U)/\mathbf{BornCoarse}^{\mathcal{C},\mathcal{B}}} Q'',$$

where $(X, U)/\mathbf{BornCoarse}^{\mathcal{C},\mathcal{B}}$ is the category of objects from $\mathbf{BornCoarse}^{\mathcal{C},\mathcal{B}}$ under (X, U). One now observes that the subcategory of objects $\big((X, U, B), \text{id}_{(X,U)}\big)$ with B in $\mathcal{B}$ is final in $(X, U)/\mathbf{BornCoarse}^{\mathcal{C},\mathcal{B}}$. Hence we can restrict the limit to this final subcategory and get

$$Q'(X, U) \simeq \lim_{B \in \mathcal{B}} \text{Cofib}\big(\Sigma_+^{\infty,\text{top}} P_U(X \setminus B) \to \Sigma_+^{\infty,\text{top}} P_U(X)\big).$$

The functor Q is then obtained from Q' by a left Kan extension along the forgetful functor

$$\mathbf{BornCoarse}^{\mathcal{C}} \to \mathbf{BornCoarse}.$$

The pointwise formula gives

$$Q(X) \simeq \underset{\mathbf{BornCoarse}^{\mathcal{C}}/X}{\text{colim}}\ Q',$$

where $\mathbf{BornCoarse}^{\mathcal{C}}/X$ is the category of objects from $\mathbf{BornCoarse}^{\mathcal{C}}$ over the bornological coarse space X. As in Sect. 6.1 we observe that the subcategory of objects $\big((X, U), \text{id}_X\big)$ for U in $\mathcal{C}$ is cofinal in $\mathbf{BornCoarse}^{\mathcal{C}}/X$. Hence we can restrict the colimit to this subcategory. We then get the formula (6.9) as desired. $\square$

Theorem 6.32 *The functor Q is an* **Sp***-valued coarse homology theory.*

The proof of Theorem 6.32 will be given in Sect. 6.4.2 below.

Remark 6.33 In the literature one often works with the subspace $P_U^{\text{fin}}(X) \subseteq P_U(X)$ of probability measures whose support is U-bounded and in addition finite (also we do in [BEb, BEKW20]). The constructions above go through with $P_U(X)$ replaced by the finite version $P_U^{fin}(X)$. We have no particular reason to prefer the larger space.

Note that if X has the property that its U-bounded subsets are finite, then we have an equality $P_U^{fin}(X) = P_U(X)$. This is the case, for example, if X has strongly locally bounded geometry (Definition 8.137). □

6.4.2   Proof of Theorem 6.32

In view of Definition 4.22 the Theorem 6.32 follows from the following four lemmas which verify the required four properties.

Lemma 6.34 Q *is coarsely invariant.*

Proof Let X be a bornological coarse space. We consider the two inclusions

$$i_0, i_1 : X \to \{0, 1\} \otimes X .$$

We must show that $Q(i_0)$ and $Q(i_1)$ are equivalent.

For every entourage U of X we have a continuous map

$$[0, 1] \times P_U(X) \to P_{\tilde{U}}(\{0, 1\} \times X) , \quad (t, \mu) \mapsto t i_{0,*}\mu + (1 - t) i_{1,*}\mu ,$$

where $\tilde{U} := [0, 1]^2 \times U$. This shows that $P_U(i_0)$ and $P_U(i_1)$ are naturally homotopic. Using Property 1. of the functor $\Sigma_+^{\infty, top}$ we see that $\Sigma_+^{\infty, top} P_U(i_0)$ and $\Sigma_+^{\infty, top} P_U(i_1)$ are naturally equivalent. Using the naturality of the homotopies with respect to X we see that this equivalence survives the limit over $\mathcal{B}$ and the colimit over $\mathcal{C}$ as well as the cofibre functor. We conclude that $Q(i_0) \simeq Q(i_1)$ are equivalent. □

Lemma 6.35 Q *is excessive.*

Proof Let X be a bornological coarse space, U an entourage of X, and let $\mathcal{Y} = (Y_i)_{i \in I}$ be a big family on X. For every i in I the inclusion $Y_i \to X$ induces an inclusion of a closed subset $P_U(Y_i) \to P_U(X)$. Since $\mathcal{Y}$ is big we can find j in I be such that $U[Y_i] \subseteq Y_j$. We claim that $P_U(Y_j)$ contains an open neighbourhood of $P_U(Y_i)$. Let μ be in $P_U(Y_i)$, and let x be in $\mathrm{supp}(\mu)$. Then we consider the open neighbourhood

$$W := \{v \in P_U(X) : |\mu(e_x) - v(e_x)| < \mu(e_x)/2\}$$

of μ. If v belongs to this neighbourhood, then $x \in \mathrm{supp}(v)$ and hence $\mathrm{supp}(v) \subseteq U[x] \subseteq Y_j$. This shows that $W \subseteq P_U(Y_j)$. Consequently, the union

$$P_U(\mathcal{Y}) := \bigcup_{i \in I} P_U(Y_i)$$

is an open subset of $P_U(X)$. Furthermore, it implies by Property 2. of the functor $\Sigma_+^{\infty,\mathrm{top}}$ that

$$\underset{i\in I}{\mathrm{colim}}\, \Sigma_+^{\infty,\mathrm{top}} P_U(Y_i) \simeq \Sigma_+^{\infty,\mathrm{top}} P_U(\mathcal{Y}).$$

If $(Z, \mathcal{Y})$ is a complementary pair on X, then $(P_U(\{Z\}), P_U(\mathcal{Y}))$ is an open decomposition of $P_U(X)$. By Property 3. of the functor $\Sigma_+^{\infty,\mathrm{top}}$ this gives a push-out square

$$\begin{array}{ccc}
\Sigma_+^{\infty,\mathrm{top}} P_U(\{Z\}\cap\mathcal{Y}) & \longrightarrow & \Sigma_+^{\infty,\mathrm{top}} P_U(\mathcal{Y}) \\
\downarrow & & \downarrow \\
\Sigma_+^{\infty,\mathrm{top}} P_U(\{Z\}) & \longrightarrow & \Sigma_+^{\infty,\mathrm{top}} P_U(X)
\end{array}$$

The same can be applied to the complementary pair $(Z \setminus B, \mathcal{Y} \setminus B)$ on $X \setminus B$ for every B in $\mathcal{B}$. We now form the limit of the corresponding cofibres of the map of push-out squares induced by $X \setminus B \to X$ over B in $\mathcal{B}$ and the colimit over U in $\mathcal{C}$. Using stability (and hence that push-outs are pull-backs and cofibres are fibres up to shift) we get the push-out square

$$\begin{array}{ccc}
Q(\{Z\}\cap\mathcal{Y}) & \longrightarrow & Q(\mathcal{Y}) \\
\downarrow & & \downarrow \\
Q(\{Z\}) & \longrightarrow & Q(X)
\end{array}$$

By Lemma 6.34 we have an equivalence $Q(Z) \simeq Q(\{Z\})$. Similarly we have an equivalence $Q(Z \cap \mathcal{Y}) \simeq Q(\{Z\}\cap\mathcal{Y})$. We conclude that

$$\begin{array}{ccc}
Q(Z\cap\mathcal{Y}) & \longrightarrow & Q(\mathcal{Y}) \\
\downarrow & & \downarrow \\
Q(Z) & \longrightarrow & Q(X)
\end{array}$$

is a push-out square. $\qquad\square$

Lemma 6.36 *Q vanishes on flasque bornological coarse spaces.*

Proof Let X be a bornological coarse space, and let U be an entourage of X. For a subset Y of X with the induced structures we write

$$P_U(X, Y) := \mathrm{Cofib}\big(\Sigma_+^{\infty,\mathrm{top}} P_U(Y) \to \Sigma_+^{\infty,\mathrm{top}} P_U(X)\big).$$

Let us now assume that flasquenss of X is implemented by the endomorphism $f :$ $X \to X$. We consider the diagram

$$
\begin{array}{ccc}
\mathrm{colim}_{U \in \mathcal{C}} \lim_{B \in \mathcal{B}} P_U(X, X \setminus B) & \xrightarrow{\mathrm{id}_{X,*}} & \mathrm{colim}_{U \in \mathcal{C}} P_U(X, X) \\
\| & & \uparrow \\
\mathrm{colim}_{U \in \mathcal{C}} \lim_{B \in \mathcal{B}} P_U(X, X \setminus B) & \xrightarrow{\mathrm{id}_{X,*}+f_*} & \mathrm{colim}_{U \in \mathcal{C}} P_U(X, f(X)) \\
\| & & \uparrow \\
\mathrm{colim}_{U \in \mathcal{C}} \lim_{B \in \mathcal{B}} P_U(X, X \setminus B) & \xrightarrow{\mathrm{id}_{X,*}+f_*+f_*^2} & \mathrm{colim}_{U \in \mathcal{C}} P_U(X, f^2(X)) \\
\| & & \uparrow \\
\vdots & & \vdots \\
\| & & \uparrow \\
\mathrm{colim}_{U \in \mathcal{C}} \lim_{B \in \mathcal{B}} P_U(X, X \setminus B) & \xrightarrow{\sum_k f_*^k} & \mathrm{colim}_{U \in \mathcal{C}} \lim_{n \in \mathbb{N}} P_U(X, f^k(X)) \\
\| & & \downarrow \\
Q(X) & \xrightarrow{\quad F \quad} & Q(X)
\end{array}
$$

All the small cells in the upper part commute. For example the filler of the upper square is an equivalence $f_* \simeq 0$. It is obtained from the factorization of f as $X \to$ $f(X) \to X$. Indeed, we have the diagram

$$
\begin{array}{ccc}
\Sigma_+^{\infty,\mathrm{top}} P_U(X \setminus B) \longrightarrow \Sigma_+^{\infty,\mathrm{top}} P_U(f(X)) \longrightarrow \Sigma_+^{\infty,\mathrm{top}} P_U(f(X)) \\
\downarrow \qquad\qquad\qquad \downarrow \qquad\qquad\qquad \downarrow \\
\Sigma_+^{\infty,\mathrm{top}} P_U(X) \longrightarrow \Sigma_+^{\infty,\mathrm{top}} P_U(f(X)) \longrightarrow \Sigma_+^{\infty,\mathrm{top}} P_U(X) \\
\downarrow \qquad\qquad\qquad \downarrow \qquad\qquad\qquad \downarrow \\
P_U(X, X \setminus B) \longrightarrow 0 \longrightarrow P_U(X, f(X))
\end{array}
$$

where the columns are pieces of fibre sequences and the lower map is the definition of the induced map f_*. The fillers provide the desired equivalence $f_* \simeq 0$.

The upper part of the diagram defines the map denoted by the suggestive symbol $\sum_k f_*^k$. This map further induces F.

We now observe that

$$
Q(f) + F \simeq Q(f) + Q(f) \circ F \simeq F .
$$

This implies $Q(f) \simeq 0$. Since Q maps close maps to equivalent morphisms and f is close to id_X we get the first equivalence in $\mathrm{id}_{Q(X)} \simeq Q(f) \simeq 0$. $\qquad\square$

Lemma 6.37 *For every X in* **BornCoarse** *with coarse structure $\mathcal{C}$ we have*

$$Q(X) = \operatorname*{colim}_{U \in \mathcal{C}} Q(X_U).$$

Proof This is clear from the definition. $\square$

6.4.3 Further Properties of the Functor Q and Generalizations

In this section we show that Q is strongly additive. We furthermore show how Q can be used to construct more examples of coarse homology theories.

Proposition 6.38 *The coarse homology theory Q is strongly additive.*

Proof Let $(X_i)_{i \in I}$ be a family of bornological coarse spaces and set $X := \bigsqcup_{i \in I}^{\text{free}} X_i$. For an entourage U of X every point of $P_U(X)$ is a probability measure supported on one of the components. We conclude that

$$P_U(X) \cong \coprod_{i \in I} P_{U_i}(X_i), \tag{6.10}$$

where we set $U_i := U \cap (X_i \times X_i)$.

We now analyze

$$\mathrm{Cofib}(\Sigma_+^{\infty,\mathrm{top}} P_U(X \setminus B) \to \Sigma_+^{\infty,\mathrm{top}} P_U(X))$$

for a bounded subset B of X. We set $B_i := B \cap X_i$ and define the set $J_B := \{i \in I : B_i \neq \emptyset\}$. Then we have an equivalence

$$\mathrm{Cofib}(\Sigma_+^{\infty,\mathrm{top}} P_U(X \setminus B) \to \Sigma_+^{\infty,\mathrm{top}} P_U(X)) \tag{6.11}$$

$$\simeq \mathrm{Cofib}\Big(\Sigma_+^{\infty,\mathrm{top}}\big(\coprod_{i \in J_B} P_{U_i}(X_i \setminus B_i)\big) \to \Sigma_+^{\infty,\mathrm{top}}\big(\coprod_{i \in J_B} P_{U_i}(X_i)\big)\Big).$$

By the definition of the bornology of a free union, J_B is a finite set. We can use Property 3. of the functor $\Sigma_+^{\infty,\mathrm{top}}$ to identify (6.11) with

$$\bigoplus_{i \in J_B} \mathrm{Cofib}\big(\Sigma_+^{\infty,\mathrm{top}} P_{U_i}(X_i \setminus B_i) \to \Sigma_+^{\infty,\mathrm{top}} P_{U_i}(X_i)\big). \tag{6.12}$$

In order to get $Q(X)$ we now take the limit over B in $\mathcal{B}$ and then the colimit over U in $\mathcal{C}$. We can restructure the index set of the limit replacing $\lim_{B \in \mathcal{B}}$ by the two limits $\lim_{J \subseteq I, J \text{finite}} \lim_{(B_i)_{i \in J} \in \prod_{i \in J} \mathcal{B}_i}$ and arrive at

$$Q(X) \simeq \operatorname*{colim}_{U \in \mathcal{C}} \lim_{\substack{J \subseteq I \\ J\text{finite}}} \lim_{(B_i)_{i \in J} \in \prod_{i \in J} \mathcal{B}_i} \bigoplus_{i \in J} \mathrm{Cofib}\big(\Sigma^{\infty,\text{top}}_+ P_{U_i}(X_i \setminus B_i) \to \Sigma^{\infty,\text{top}}_+ P_{U_i}(X_i)\big)$$

$$\simeq \operatorname*{colim}_{U \in \mathcal{C}} \lim_{\substack{J \subseteq I \\ J\text{finite}}} \bigoplus_{i \in J} \lim_{B_i \in \mathcal{B}_i} \mathrm{Cofib}\big(\Sigma^{\infty,\text{top}}_+ P_{U_i}(X_i \setminus B_i) \to \Sigma^{\infty,\text{top}}_+ P_{U_i}(X_i)\big)$$

$$\simeq \operatorname*{colim}_{U \in \mathcal{C}} \prod_{i \in I} \lim_{B_i \in \mathcal{B}_i} \mathrm{Cofib}\big(\Sigma^{\infty,\text{top}}_+ P_{U_i}(X_i \setminus B_i) \to \Sigma^{\infty,\text{top}}_+ P_{U_i}(X_i)\big).$$

In the last step we want to commute the colimit of U in $\mathcal{C}$ with the product. To this end we note that in view of the definition of the coarse structure of a free union we have a bijection $\prod_{i \in I} \mathcal{C}_i \cong \mathcal{C}$ sending the family $(U_i)_{i \in I}$ of entourages to the entourage $\bigsqcup_{i \in I} U_i$. The inverse sends U in $\mathcal{C}$ to the family $(U \cap (X_i \times X_i))_{i \in I}$. Using that in **Sp** products distribute over filtered colimits (see Remark 6.28) we get an equivalence

$$\operatorname*{colim}_{U \in \mathcal{C}} \prod_{i \in I} \lim_{B_i \in \mathcal{B}_i} \mathrm{Cofib}\big(\Sigma^{\infty,\text{top}}_+ P_{U_i}(X_i \setminus B) \to \Sigma^{\infty,\text{top}}_+ P_{U_i}(X_i)\big)$$

$$\simeq \prod_{i \in I} \operatorname*{colim}_{U_i \in \mathcal{C}_i} \lim_{B_i \in \mathcal{B}_i} \mathrm{Cofib}\big(\Sigma^{\infty,\text{top}}_+ P_{U_i}(X_i \setminus B) \to \Sigma^{\infty,\text{top}}_+ P_{U_i}(X_i)\big)$$

$$\simeq \prod_{i \in I} Q(X_i)$$

which proves the assertion of the lemma.   □

Let **C** be a complete and cocomplete stable ∞-category. We now use the tensor structure (6.5).

Let C be an object of **C**.

Corollary 6.39

$$C \wedge Q : \mathbf{BornCoarse} \to \mathbf{Sp}, \qquad X \mapsto (C \wedge Q(X)) \tag{6.13}$$

is a **C**-*valued coarse homology theory.*

Proof The proof follows from Theorem 6.32 and the fact that the functor $C \wedge -$ preserves colimits.   □

Example 6.40 We can use the coarsification in order to produce an example of a non-additive coarse homology theory. We choose a spectrum C in **Sp** such that the functor $C \wedge - : \mathbf{Sp} \to \mathbf{Sp}$ does not send $\prod_{i \in I} S$ to $\prod_{i \in I} C$. Here S is the sphere spectrum and I is some suitable infinite set. For example we could take $E := \bigoplus_p H\mathbb{Z}/p\mathbb{Z}$, where p runs over all primes. Then $C \wedge Q$ is not additive. $\square$

6.5 Comparison of Coarse Homology Theories

Let $\mathbf{C}$ be a cocomplete stable ∞-category. In the present subsection we discuss the problem to which extend a coarse homology theory $E : \mathbf{BornCoarse} \to \mathbf{C}$ is determined by the value $E(*)$, or at least by the values $E(X)$ on discrete bornological coarse spaces X.

Example 6.41 We consider the category $\mathbf{Sp}$ and the complex K-theory spectrum KU in $\mathbf{Sp}$. In the present paper we construct various examples of coarse homology theories which all assume the value KU on the one-point space:

1. $KU \wedge Q$ (Corollary 6.39)
2. $Q(KU \wedge \Sigma_+^\infty)^{\mathrm{lf}}$ (Definition 7.44)
3. $QK^{\mathrm{an,lf}}$ (Definition 7.44 and Definition 7.66)
4. $K\mathcal{X}$ and $K\mathcal{X}_{\mathrm{ql}}$ (Definition 8.78)

A further example is constructed in [BEb]. One could ask whether these homology theories are really different.

Note that $Q(KU \wedge \Sigma_+^\infty)^{\mathrm{lf}}$ and $QK^{\mathrm{an,lf}}$ are additive. But we do not expect that $KU \wedge Q$ is additive for a similar reason as in Example 6.40.

If X is a bornological coarse spaces of locally bounded geometry (see Definition 8.138), then we have the coarse assembly map (see Definition 8.139)

$$\mu \colon QK_*^{\mathrm{an,lf}}(X) \to K\mathcal{X}_*(X).$$

The known counter-examples to the coarse Baum–Connes conjecture show that this map is not always surjective. But this leaves open the possibility of the existence of a different transformation which is an equivalence.

In general we have a natural transformation

$$K\mathcal{X} \to K\mathcal{X}_{\mathrm{ql}}.$$

The goal of the present subsection is to provide positive results asserting that such a transformation is an equivalence on certain spaces. $\square$

In the following we will assume that $\mathbf{C}$ is a cocomplete stable ∞-category. We consider two coarse homology theories $E, F : \mathbf{BornCoarse} \to \mathbf{C}$ and a natural transformation

$$T : E \to F.$$

Let $\mathcal{A}$ be a set of objects in $\mathbf{BornCoarse}$ and recall Definition 4.11 of $\mathbf{Sp}\mathcal{X}\langle\mathcal{A}\rangle$.

Let X be a bornological coarse space.

Lemma 6.42 *Assume:*

1. $T : E(Y) \to F(Y)$ *is an equivalence for all Y in $\mathcal{A}$.*
2. $\mathrm{Yo}^s(X) \in \mathbf{Sp}\mathcal{X}\langle\mathcal{A}\rangle$.

Then $E(X) \to F(X)$ is an equivalence.

Proof We use Corollary 4.25. The assertion follows from the fact that the extension of E and F to functors $E^{\mathrm{la}} : \mathbf{Sp}\mathcal{X} \to \mathbf{C}^{\mathrm{la}}$ and $F^{\mathrm{la}} : \mathbf{Sp}\mathcal{X} \to \mathbf{C}^{\mathrm{la}}$ preserve colimits.
□

Combining Theorem 5.59 and Lemma 6.42 we get the following consequence. Let X be a bornological coarse space.

Corollary 6.43 *Assume:*

1. $T : E(Y) \to F(Y)$ *is an equivalence for all discrete bornological coarse spaces Y.*
2. *X has weakly finite asymptotic dimension (Definition 5.58).*

Then $T : E(X) \to F(X)$ is an equivalence.

For a transformation between additive coarse homology theories we can simplify the assumption on the transformation considerably.

We assume that $\mathbf{C}$ is a complete and cocomplete stable ∞-category. Let $T : E \to F$ be as above, and let X be a bornological coarse space.

Theorem 6.44 *Assume:*

1. $T : E(*) \to F(*)$ *is an equivalence.*
2. *E and F are additive.*
3. *X has weakly finite asymptotic dimension.*
4. *X has the minimal compatible bornology.*

Then $T : E(X) \to F(X)$ is an equivalence.

Proof Let $\mathcal{A}_*$ be the class consisting of the motivic coarse spectra of the form $\mathrm{Yo}^s(\bigsqcup_I^{\mathrm{free}} *)$ for a set I. An inspection of the intermediate steps of Theorem 5.59 shows that $\mathrm{Yo}^s(X) \in \mathbf{Sp}\mathcal{X}\langle\mathcal{A}_*\rangle$. We now conclude using Lemma 6.42. □

Chapter 7
Locally Finite Homology Theories and Coarsification

7.1 Locally Finite Homology Theories

In this section we review locally finite homology theories in the context of **TopBorn** of topological spaces with an additional bornology. It gives a more systematic explanation of the appearance of the limit over bounded subsets in (6.9). We will see that the constructions of Sect. 6.4 can be extended to a construction of coarse homology theories from locally finite homology theories.

Note that our notion of a locally finite homology theory differs from the one studied by Weiss and Williams [WW95], but our set-up contains a version of classical Borel–Moore homology of locally finite chains. There is also related work by Carlsson and Pedersen [CP98].

One motivation to develop this theory in the present paper is to capture the example $K^{an,lf}$ of analytic locally finite K-homology. In order to relate this with the homotopy theoretic version of the locally finite homology theory (as discussed in Example 7.40) associated to spectrum KU we derive the analogs of the classification results of Weiss and Williams [WW95] in the context of topological bornological spaces.

We think that the notions and results introduced in this section are interesting in itself, independent of their application to coarse homology theories.

The coarsification of the analytic locally finite K-homology is the domain of the assembly map discussed in Sect. 8.10. In [BEb] we will introduce the closely related concept of a local homology theory which is better suited as a domain of the coarse assembly map in general. On nice spaces it behaves like a locally finite homology theory.

© The Editor(s) (if applicable) and The Author(s), under exclusive licence
to Springer Nature Switzerland AG 2020
U. Bunke, A. Engel, *Homotopy Theory with Bornological Coarse Spaces*,
Lecture Notes in Mathematics 2269, https://doi.org/10.1007/978-3-030-51335-1_7

7.1.1 Topological Bornological Spaces

In order to talk about locally finite homology theories we introduce the category
TopBorn of topological bornological spaces.

Let X be a set with a topology $\mathcal{S}$ and a bornology $\mathcal{B}$.

Definition 7.1 $\mathcal{B}$ is compatible with $\mathcal{S}$ if $\mathcal{S} \cap \mathcal{B}$ is cofinal in $\mathcal{B}$ and $\mathcal{B}$ is closed under taking closures.

Example 7.2 In words, compatibility means that every bounded subset has a bounded closure and a bounded open neighbourhood.

A compatible bornology on a topological space X contains all relatively compact subsets. Indeed, let B be a relatively compact subset of X. By compatibility, for every point b in X there is a bounded open neighbourhood U_b of b. Since B is relatively compact, there exists a finite subset I of $\bar{B}$ such that $B \subseteq \bigcup_{b \in B} U_b$. But then B is bounded since a finite union of bounded subsets is bounded and a bornology is closed under taking subsets.   □

Definition 7.3 A topological bornological space is a triple $(X, \mathcal{S}, \mathcal{B})$ consisting of a set X with a topology $\mathcal{S}$ and a bornology $\mathcal{B}$ which are compatible.

In general we will just use the symbol X to denote topological bornological spaces.

Let X, X' be topological bornological spaces, and let $f : X \to X'$ be a map between the underlying sets.

Definition 7.4 f is a morphism between topological bornological spaces if f is continuous and proper.

Note that the word *proper* refers to the bornologies and means that $f^{-1}(\mathcal{B}') \subseteq \mathcal{B}$. We let **TopBorn** denote the small category of very small topological bornological spaces and morphisms.

Example 7.5 A topological space is compact if it is Hausdorff and every open covering admits a finite subcovering. A subspace of a topological spaces is relatively compact if its closure is compact. By convention, a locally compact topological space is a topological space which is Hausdorff and such that every point admits a compact neighbourhood.

If X is a locally compact topological space, then the family of relatively compact subsets of X turns X into a topological bornological space. We get an inclusion

$$\mathbf{Top}^{\mathrm{lc}} \to \mathbf{TopBorn}$$

of the category of locally compact spaces and proper maps as a full subcategory.

Assume that X is locally compact and U is an open subset of X. If we equip U with the induced topology and bornology, then $U \to X$ is a morphism in **TopBorn**. But in general, U equipped with these induced structures is not contained in the image of the functor **Top**$^{lc} \to$ **TopBorn** since the bornology on U induced from X is too large.

Consider, e.g., the open subset $(0, 1)$ of $\mathbb{R}$. The induced bornology of $(0, 1)$ is the maximal bornology, while e.g. $(0, 1)$ is not relatively compact in $(0, 1)$.

In order to define the notion of descent, on the one hand we want to work with open coverings. On the other hand the inclusions of the open subsets should be morphisms in the category. In **Top**^{lc} this is impossible and motivates to work in **TopBorn**. This category just provides to minimal amount of structure for the consideration of locally finite homology theories. □

Lemma 7.6 *The category* **TopBorn** *has all very small products.*

Proof Let $(X_i)_{i \in I}$ be a family of topological bornological spaces. Then the product of the family is represented by the topological bornological space $\prod_{i \in I} X_i$ whose underlying topological space is the product in of the underlying topological spaces, and whose bornology is the minimal one such that the projections to the factors are proper (compare with the bornology of a product of bornological coarse spaces in see Lemma 2.23). □

Example 7.7 The category **TopBorn** is not complete. It does not have a final objects. In order to get a complete and cocomplete category one could remove the condition on the bornology that all points are bounded, see [Hei]. □

Example 7.8 The category **TopBorn** has a symmetric monoidal structure

$$(X, X') \mapsto X \otimes X'.$$

The underlying topological space of $X \times X'$ is the product of the underlying topological spaces of X and X'. The bornology of $X \times X'$ is generated by the sets $B \times B'$ for all bounded subsets B of X and B' of X'. The tensor unit is the one-point space. As in the case of bornological coarse spaces (Example 2.32) the tensor product in general differs from the cartesian product. □

Lemma 7.9 *The category* **TopBorn** *has all very small coproducts.*

Proof Let $(X_i)_{i \in I}$ be a family of topological bornological spaces. Then the coproduct of the family is represented by the topological bornological space $\coprod_{i \in I} X_i$ whose underlying topological space is the coproduct of the underlying topological spaces, and whose bornology is given by

$$\mathcal{B} := \{B \subseteq X \mid (\forall i \in I : B \cap X_i \in \mathcal{B}_i)\},$$

where $\mathcal{B}_i$ is the bornology of X_i. □

Note that the bornology of the coproduct is given by the same formula as in the case of bornological coarse spaces; see Definition 2.25.

Let $(X_i)_{i \in I}$ be a family of topological bornological spaces.

Definition 7.10 The free union $\bigsqcup_{i \in I}^{\text{free}} X_i$ of the family is the following topological bornological space:

1. The underlying set of the free union is the disjoint union $\bigsqcup_{i \in I} X_i$.
2. The topology of the free union is the one of the coproduct of topological spaces.
3. The bornology of the free union is given by $\mathcal{B}\left(\bigcup_{i \in I} \mathcal{B}_i\right)$, where $\mathcal{B}_i$ is the bornology of X_i.

Remark 7.11 The free union should not be confused with the coproduct. The topology of the free union is the same, but the bornology is smaller. The free union plays a role in the discussion of additivity of locally finite theories.   □

7.1.2 Definition of Locally Finite Homology Theories

We will use the following notation. Let E be any functor from some category to a stable ∞-category. If $Y \to X$ is a morphism in the domain of E, then we write

$$E(X, Y) := \text{Cofib}\big(E(Y) \to E(X)\big).$$

We now introduce the notion of local finiteness. It is this property of a functor from topological bornological spaces to spectra which involves the bornology and distinguishes locally finite homology theories amongst all homotopy invariant and excisive functors.

Let **C** be a complete stable ∞-category. We consider a functor

$$E : \textbf{TopBorn} \to \textbf{C}.$$

Definition 7.12 E is locally finite if the natural morphism

$$E(X) \to \lim_{B \in \mathcal{B}} E(X, X \setminus B)$$

is an equivalence for all X in **TopBorn**.

Remark 7.13 E is locally finite if and only if

$$\lim_{B \in \mathcal{B}} E(X \setminus B) \simeq 0$$

for all X in **TopBorn**.   □

Example 7.14 A typical feature which is captured by the notion of local finiteness is the following. Let X be in **TopBorn** and assume that its bornology $\mathcal{B}$ has a countable cofinal subfamily $(B_n)_{n \in \mathbb{N}}$. Then we have the equivalence

$$E(X) \simeq \lim_{n \in \mathbb{N}} E(X, X \setminus B_n).$$

Assume that E takes values in spectra **Sp**. Then in homotopy groups this is reflected by the presence of Milnor $\lim^1$-sequences

$$0 \to \lim_{n \in \mathbb{N}}{}^1 \pi_{k+1}(E(X, X \setminus B_n)) \to \pi_k(E(X)) \to \lim_{n \in \mathbb{N}} \pi_k(E(X, X \setminus B_n)) \to 0$$

for all k in $\mathbb{Z}$. The presence of the $\lim^1$-term shows that one can not construct graded abelian group valued locally finite homology as the limit $\lim_{n \in \mathbb{N}} \pi_k(E(X, X \setminus B_n))$. In general the latter would not satisfy descent in the sense that we have Mayer–Vietoris sequences for appropriate covers. □

Let **C** be a complete stable ∞-category, and let $E : \textbf{TopBorn} \to \textbf{C}$ be any functor.

Definition 7.15 We define the locally finite evaluation

$$E^{\mathrm{lf}} : \textbf{TopBorn} \to \textbf{C}$$

by

$$E^{\mathrm{lf}}(X) := \lim_{B \in \mathcal{B}} E(X, X \setminus B).$$

Remark 7.16 In order to turn this description of the locally finite evaluation on objects into a definition of a functor we use right Kan extensions as described in Remark 6.31. We consider the category $\textbf{TopBorn}^{\mathcal{B}}$ of pairs (X, B), where X is a topological bornological space and B is a bounded subset of X. A morphism $f : (X, B) \to (X', B')$ is a continuous map such that $f(B) \subseteq B'$. We have a forgetful functor

$$p : \textbf{TopBorn}^{\mathcal{B}} \to \textbf{TopBorn}, \quad (X, B) \mapsto X.$$

The locally finite evaluation is then defined as the right Kan extension of the functor

$$\tilde{E} : \textbf{TopBorn}^{\mathcal{B}} \to \textbf{C}, \quad (X, B) \mapsto E(X, X \setminus B)$$

as indicated in the following diagram

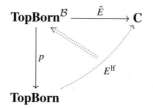

□

We have a canonical natural transformation

$$E \to E^{\mathrm{lf}}. \tag{7.1}$$

By Definition 7.12 the functor E is locally finite if and only if the natural transformation (7.1) is an equivalence.

Let **C** be a complete stable ∞-category, and let $E : \mathbf{TopBorn} \to \mathbf{C}$ be any functor.

Lemma 7.17 E^{lf} *is locally finite.*

Proof Let X be a topological bornological space. We have

$$
\begin{aligned}
\lim_{B \in \mathcal{B}} E^{\mathrm{lf}}(X \setminus B) &\simeq \lim_{B \in \mathcal{B}} \lim_{B' \in \mathcal{B}} \mathrm{Cofib}\big(E(X \setminus (B \cup B')) \to E(X \setminus B)\big) \\
&\simeq \lim_{B' \in \mathcal{B}} \lim_{B \in \mathcal{B}} \mathrm{Cofib}\big(E(X \setminus (B \cup B')) \to E(X \setminus B)\big) \\
&\simeq \lim_{B' \in \mathcal{B}} \lim_{B \in \mathcal{B}, B' \subseteq B} \mathrm{Cofib}\big(E(X \setminus (B \cup B')) \to E(X \setminus B)\big) \\
&\simeq \lim_{B' \in \mathcal{B}} \lim_{B \in \mathcal{B}, B' \subseteq B} \mathrm{Cofib}\big(E(X \setminus B) \to E(X \setminus B)\big) \\
&\simeq 0.
\end{aligned}
$$

By Remark 7.13 this equivalence implies that E^{lf} is locally finite.   □

Note that Lemma 7.17 implies that the natural transformation (7.1) induces an equivalence

$$E^{\mathrm{lf}} \simeq (E^{\mathrm{lf}})^{\mathrm{lf}}.$$

We now introduce the notion of homotopy invariance of functors defined on **TopBorn**. To this end we equip the unit interval $[0, 1]$ with its maximal bornology. Then for every topological bornological space X the projection $[0, 1] \otimes X \to X$ (see Example 7.8 for $- \otimes -$) is a morphism in **TopBorn**. Moreover, the inclusions $X \to [0, 1] \otimes X$ are morphisms. This provides a notion of homotopy in the category **TopBorn**. In the literature one often talks about proper homotopy. We will not add this word *proper* here since it is clear from the context **TopBorn** that all morphisms are proper, hence also the homotopies.

Let $E : \mathbf{TopBorn} \to \mathbf{C}$ be a functor.

Definition 7.18 E is homotopy invariant if the projection $[0, 1] \otimes X \to X$ induces an equivalence $E([0, 1] \otimes X) \to E(X)$.

Next we shall see that the combination of homotopy invariance and local finiteness of a functors allows Eilenberg swindle arguments. Similar as in the bornological coarse case we encode this in the property that the functor vanishes on certain flasque spaces.

Definition 7.19 A topological bornological space X is flasque if it admits a morphism $f : X \to X$ such that:

1. f is homotopic to id.
2. For every bounded B in X there exists k in $\mathbb{N}$ such that $f^k(X) \cap B = \emptyset$.

Example 7.20 Let X be a topological bornological space, and let $[0, \infty)$ have the bornology given by relatively compact subsets. Then $[0, \infty) \otimes X$ is flasque. We can define the morphism $f : [0, \infty) \otimes X \to [0, \infty) \otimes X$ by $f(t, x) := (t + 1, x)$. □

Let $\mathbf{C}$ be a complete stable ∞-category, and let $E : \mathbf{TopBorn} \to \mathbf{C}$ be a functor.

Lemma 7.21 *Assume:*

1. *E is locally finite.*
2. *E is homotopy invariant.*
3. *X is flasque.*

Then $E(X) \simeq 0$.

Proof The argument is very similar to the proof of Lemma 6.36. Let X be a topological bornological space with bornology $\mathcal{B}$, and let $f : X \to X$ implement flasqueness. We consider the diagram

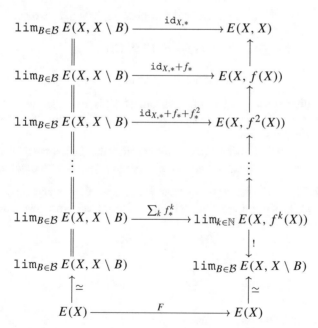

In order to define the morphism marked by ! we note that for every B in $\mathcal{B}$ we can choose k in $\mathbb{N}$ such that $f^k(X) \cap B = \emptyset$. Then we have a map of pairs $(X, f^k(X)) \to (X, X \setminus B)$.

All the small cells in the upper part commute. For example the filler of the upper square is an equivalence $f_* \simeq 0$. It is obtained from the factorization of f as $X \to f(X) \to X$. Indeed, we have the diagram

$$
\begin{array}{ccccc}
E(X \setminus B) & \longrightarrow & E(f(X)) & \longrightarrow & E(f(X)) \\
\downarrow & & \downarrow & & \downarrow \\
E(X) & \longrightarrow & E(f(X)) & \longrightarrow & E(X) \\
\downarrow & & \downarrow & & \downarrow \\
E(X, X \setminus B) & \longrightarrow & 0 & \longrightarrow & E(X, f(X))
\end{array}
$$

where the columns are pieces of fibre sequences and the lower map is the definition of the induced map f_*. The fillers provide the desired equivalence $f_* \simeq 0$.

The upper part of the diagram defines the map denoted by the suggestive symbol $\sum_k f_*^k$. This map further induces F.

The construction of F implies that

$$
E(\mathrm{id}_X) + E(f) \circ E(F) \simeq E(F).
$$

Using that E is homotopy invariant and f is homotopic to id_X we get

$$
E(\mathrm{id}_X) + E(F) \simeq E(F)
$$

and hence $E(\mathrm{id}_X) \simeq 0$. This implies $E(X) \simeq 0$.   □

We now discuss the notion of excision. If Y is a subset of a topological bornological space X, then we consider Y as a topological bornological space with the induced structures.

In order to capture all examples we will consider three versions of excision.

Let **C** be a stable ∞-category, and let $E : \mathbf{TopBorn} \to \mathbf{C}$ be a functor.

Definition 7.22 E satisfies (open or closed) excision if for every (open or closed) decomposition (Y, Z) of a topological bornological space X we have a push-out

$$
\begin{array}{ccc}
E(Y \cap Z) & \longrightarrow & E(Y) \\
\downarrow & & \downarrow \\
E(Z) & \longrightarrow & E(X)
\end{array}
$$

Example 7.23 We consider the functor

$$
\ell \circ C^{\mathrm{sing}} \circ \mathcal{F}_{\mathcal{B}} : \mathbf{TopBorn} \to \mathbf{Ch}_\infty ,
$$

where ℓ is as in (6.4), $\mathcal{F}_\mathcal{B} : \mathbf{TopBorn} \to \mathbf{Top}$ forgets the bornology, and C^{sing} : $\mathbf{Top} \to \mathbf{Ch}$ is the singular chain complex functor. This composition is homotopy invariant and well-known to satisfy open excision. Closed excision fails.

The functor

$$\Sigma_+^{\infty,\mathrm{top}} \circ \mathcal{F}_\mathcal{B} : \mathbf{TopBorn} \to \mathbf{Sp}$$

(see (6.7)) is homotopy invariant and satisfies open excision (Property 3. of $\Sigma_+^{\infty,\mathrm{top}}$). Again, closed excision fails for $\Sigma_+^{\infty,\mathrm{top}} \circ \mathcal{F}_\mathcal{B}$.

We will see below (combine Assertion 1 and 7.54) that analytic locally finite K-homology $K^{\mathrm{an,lf}}$ satisfies closed excision. We do not know if it satisfies open excision. $\square$

The following notion of weak excision is an attempt for a concept which comprises both open and closed excision. Let X be a topological space, and let $\mathcal{Y} := (Y_i)_{i \in I}$ be a filtered family of subsets.

Definition 7.24 $\mathcal{Y}$ is called a big family if for every i in I there exists i' in I such that $Y_{i'}$ contains an open neighbourhood of $\bar{Y}_i$.

Example 7.25 The bornology of a topological bornological space is a big family. The condition introduced in Definition 7.24 is satisfied by compatibility between the topology and the bornology, see Definition 7.1. $\square$

Let $\mathbf{C}$ be a cocomplete stable ∞-category. If $\mathcal{Y}$ is a big family on X and $E : \mathbf{TopBorn} \to \mathbf{C}$ is a functor, then we write

$$E(\mathcal{Y}) := \operatorname*{colim}_{i \in I} E(Y_i) .$$

We say that a pair $(\mathcal{Y}, \mathcal{Z})$ of two big families $\mathcal{Z}$ and $\mathcal{Y}$ is a decomposition of X if there exist members Z and Y of $\mathcal{Z}$ and $\mathcal{Y}$, respectively, such that $X = Z \cup Y$.

Let $\mathbf{C}$ be a cocomplete stable ∞-category, and let $E : \mathbf{TopBorn} \to \mathbf{C}$ be a functor.

Definition 7.26 E satisfies weak excision if for every decomposition $(\mathcal{Y}, \mathcal{Z})$ of a topological bornological space X into two big families we have a push-out

$$
\begin{array}{ccc}
E(\mathcal{Y} \cap \mathcal{Z}) & \longrightarrow & E(\mathcal{Y}) \\
\downarrow & & \downarrow \\
E(\mathcal{Z}) & \longrightarrow & E(X)
\end{array}
$$

By cofinality arguments it is clear that open or closed excision implies weak excision.

Let $\mathbf{C}$ be a complete and cocomplete stable ∞-category, and let $E : \mathbf{TopBorn} \to \mathbf{C}$ be a functor.

Definition 7.27 E is a locally finite homology theory if it has the following properties:

1. E is locally finite.
2. E is homotopy invariant.
3. E satisfies weak excision.

Remark 7.28 A locally finite homology theory E gives rise to wrong way maps for bornological open inclusions. We consider a topological bornological space X with bornology $\mathcal{B}_X$ and an open subset U in X. Let $\mathcal{B}_U$ be some compatible bornology on U such that the inclusion $U \to X$ is bornological. So $\mathcal{B}_U$ may be smaller than the induced bornology $\mathcal{B}_X \cap U$. We further assume that every B in $\mathcal{B}_U$ has an open neighbourhood V such that $\mathcal{B}_U \cap V = \mathcal{B}_X \cap V$.

We let $\tilde{U}$ denote the topological bornological space with the induced topology and the bornology $\mathcal{B}_U$. We use the tilde symbol in order to distinguish that space from the topological bornological space U which has by definition the induced bornology $U \cap \mathcal{B}_X$.

In contrast to $U \to X$ the inclusion $\tilde{U} \to X$ is in general not a morphism in **TopBorn**. The observation in this remark is that we have a wrong-way morphism

$$E(X) \to E(\tilde{U})$$

defined as follows: For a bounded closed subset B in $\mathcal{B}_U$ we get open coverings $(X \setminus B, U)$ of X and $(U \setminus B, V)$ of U. Using open descent for E twice we get the excision equivalence

$$E(X, X \setminus B) \simeq E(V, V \setminus B) \simeq E(\tilde{U}, \tilde{U} \setminus B) . \tag{7.2}$$

The desired wrong-way map is now given by

$$E(X) \simeq \lim_{B \in \mathcal{B}_X} E(X, X \setminus B) \to \lim_{B \in \mathcal{B}_U} E(X, X \setminus B) \overset{(7.2)}{\to} \lim_{B \in \mathcal{B}_U} E(\tilde{U}, \tilde{U} \setminus B) \simeq E(\tilde{U}),$$

where the first morphism is induced by the restriction of index sets along $\mathcal{B}_U \to \mathcal{B}_X$ which is defined since the inclusion $\tilde{U} \to X$ was assumed to be bornological.

A typical instance of this is the inclusion of an open subset U into a locally compact space X. In general, the induced bornology $\mathcal{B}_X \cap U$ is larger than the bornology $\mathcal{B}_U$ of relatively compact subsets. If B is relatively compact in U, then it has a relatively compact neighbourhood. So in this case the wrong-way map $E(X) \to E(\tilde{U})$ is defined. These wrong-way maps are an important construction in index theory, but since they do not play any role in the present paper we will not discuss them further.

Note that in [WW95, Sec. 2] the wrong way maps are encoded in a completely different manner by defining the functors themselves on the larger category $\mathcal{E}^\bullet$ instead of $\mathcal{E}$ (notation in [WW95]).   □

7.1.3 Additivity

In the following we discuss additivity. Let $E : \mathbf{TopBorn} \to \mathbf{C}$ be a functor which satisfies weak excision. We consider a family $(X_i)_{i \in I}$ of topological bornological spaces. Recall the Definition 7.10 of the free union $\bigsqcup_{i \in I}^{\text{free}} X_i$. For j in I we can form the two big one-member families of $\bigsqcup_{i \in I}^{\text{free}} X_i$ consisting of X_j and $\bigsqcup_{i \in I \setminus \{j\}}^{\text{free}} X_i$. Excision provides the first equivalence in the following definition of projection morphisms

$$E\Big(\bigsqcup_{i \in I}^{\text{free}} X_i\Big) \simeq E(X_j) \oplus E\Big(\bigsqcup_{i \in I \setminus \{j\}}^{\text{free}} X_i\Big) \to E(X_j).$$

These projections for all j in I together induce a morphism

$$E\Big(\bigsqcup_{i \in I}^{\text{free}} X_i\Big) \to \prod_{i \in I} E(X_i). \tag{7.3}$$

Let $\mathbf{C}$ be a complete stable ∞-category, and let $E : \mathbf{TopBorn} \to \mathbf{C}$ be a weakly excisive functor.

Definition 7.29 E is additive if (7.3) is an equivalence for all families $(X_i)_{i \in I}$ of topological bornological spaces.

Lemma 7.30 *If E is locally finite and satisfies weak excision, then E is additive.*

Proof We consider a family $(X_i)_{i \in I}$ of topological bornological spaces and their free union $X := \bigsqcup_{i \in I}^{\text{free}} X_i$. If B is a bounded subset of X, then the subset $J(B) := \{i \in I \mid B \cap X_i \neq \emptyset\}$ of I is finite. By excision we have an equivalence

$$E(X, X \setminus B) \simeq \bigoplus_{j \in J(B)} E(X_j, X_j \setminus B).$$

We now form the limit over B in $\mathcal{B}$ in two stages

$$\lim_{B \in \mathcal{B}} \cdots \simeq \lim_{J \subseteq I} \lim_{B \in \mathcal{B}, J(B)=J} \cdots,$$

where J runs over the finite subsets of I. The inner limit gives by local finiteness of E

$$\lim_{B \in \mathcal{B}, J(B)=J} \bigoplus_{j \in J} E(X_j, X_j \setminus B) \simeq \bigoplus_{j \in J} E(X_j).$$

Taking now the limit over the finite subsets J of I, and using

$$\lim_{\substack{J \subseteq I}} \bigoplus_{j \in J(B)} E(X_j) \simeq \prod_{i \in I} E(X_i)$$

and the local finiteness of E again we get the equivalence

$$E(X) \simeq \prod_{i \in I} E(X_i)$$

as claimed.   □

Let **C** be a complete and cocomplete stable ∞-category, and let $E : \mathbf{TopBorn} \to$ **C** be a functor.

Corollary 7.31 *If E is a locally finite homology theory, then E is additive.*

In general it is notoriously difficult to check that a given functor is locally finite if it is not already given as E^{lf}. In the following we show a result which can be used to deduce local finiteness from weak excision and additivity, see Remark 7.13.

Let **C** be a complete stable ∞-category, and let $E : \mathbf{TopBorn} \to$ **C** be a weakly excisive functor.

Definition 7.32 E is countably additive if (7.3) is an equivalence for all countable families $(X_i)_{i \in I}$ of topological bornological spaces.

An additive functor is countably additive.

We consider an increasing family $(Y_k)_{k \in \mathbb{N}}$ of subsets of a topological bornological space X.

Lemma 7.33 *Assume:*

1. *For every bounded subset B of X there exists k in $\mathbb{N}$ such that $B \subseteq Y_k$.*
2. *E is weakly excisive.*
3. *E is countably additive.*

Then $\lim_{n \in \mathbb{N}} E(X \setminus Y_n) \simeq 0$.

Proof For k, ℓ in $\mathbb{N}$ and $k \geq \ell$ we let

$$f_{k,\ell} : X \setminus Y_k \to X \setminus Y_\ell$$

be the inclusion. We have a fibre sequence

$$\to \lim_{n \in \mathbb{N}} E(X \setminus Y_n) \to \prod_{n \in \mathbb{N}} E(X \setminus Y_n) \xrightarrow{d} \prod_{n \in \mathbb{N}} E(X \setminus Y_n) \to$$

where d is described in the language of elements by

$$d((\phi_n)_n) := (\phi_n - f_{n+1,n}\phi_{n+1})_n .$$

We must show that d is an equivalence.

For every m in $\mathbb{N}$ we have a morphism of topological bornological spaces

$$g_m := \sqcup_{n \geq m} f_{n,m} : \overset{\text{free}}{\underset{\substack{n \in \mathbb{N} \\ n \geq m}}{\bigsqcup}} X \setminus Y_n \to X \setminus Y_m .$$

The condition 1 ensures that g_m is proper.

Using the additivity of E, for every m in $\mathbb{N}$ we can define the morphism

$$s_m : \prod_{n \in \mathbb{N}} E(X \setminus Y_n) \overset{proj}{\to} \prod_{\substack{n \in \mathbb{N} \\ n \geq m}} E(X \setminus Y_n) \simeq E\Big(\overset{\text{free}}{\underset{\substack{n \in \mathbb{N} \\ n \geq m}}{\bigsqcup}} X \setminus Y_n\Big) \overset{g_m}{\to} E(X \setminus Y_m) .$$

Let $\mathrm{pr}_m : \prod_{n \in \mathbb{N}} E(X \setminus Y_n) \to E(X \setminus Y_m)$ denote the projection. By excision we have the relation

$$\mathrm{pr}_m + f_{m+1,m}s_{m+1} \simeq s_m . \tag{7.4}$$

We further have the relation

$$s_m((f_{n+1,n}\phi_{n+1})_n) \simeq f_{m+1,m}(s_{m+1}((\phi_n)_n)) \tag{7.5}$$

We can now define

$$h : \prod_{n \in \mathbb{N}} E(X \setminus Y_n) \to \prod_{n \in \mathbb{N}} E(X \setminus Y_n)$$

by

$$h((\phi_n)_n) = (s_m((\phi_n)_n))_m .$$

We calculate

$$\begin{aligned}
(d \circ h)((\phi_n)_n) &\simeq d((s_m((\phi_n)_n))_m) \\
&\simeq (s_m((\phi_n)_n) - f_{m+1,m}(s_{m+1}((\phi_n)_n)))_m \\
&\overset{(7.4)}{\simeq} (\phi_m)_m
\end{aligned}$$

and

$$
\begin{aligned}
(h \circ d)((\phi_n)_n) &\simeq h((\phi_n - f_{n+1,n}\phi_{n+1})_n) \\
&\simeq (s_m((\phi_n)_n))_m - (s_m((f_{n+1,n}\phi_{n+1})_n))_m \\
&\overset{(7.5)}{\simeq} (s_m((\phi_n)_n))_m - (f_{m+1,m}(s_{m+1}((\phi_n)_n)))_m \\
&\overset{(7.4)}{\simeq} (\phi_m)_m \ .
\end{aligned}
$$

This calculation shows that d is an equivalence. Consequently,

$$
\lim_{n \in \mathbb{N}} E(X \setminus Y_n) \simeq 0 \ ,
$$

finishing the proof.   □

Remark 7.34 If a functor $E : \mathbf{TopBorn} \to \mathbf{C}$ is countably additive, homotopy invariant, and satisfies weak excision, then it vanishes on flasque spaces. In fact, the weaker version of local finiteness shown in Lemma 7.33 suffices to make a modification of the proof of Lemma 7.21 work.   □

7.1.4 Construction of Locally Finite Homology Theories

Let $\mathbf{C}$ be a complete stable ∞-category, and let $E : \mathbf{TopBorn} \to \mathbf{C}$ be a functor. Recall the Definition 7.15 of the locally finite evaluation.

Lemma 7.35 *If E is homotopy invariant, then E^{lf} is homotopy invariant.*

Proof We consider the projection $[0,1] \otimes X \to X$. Then the subsets $[0,1] \times B$ for bounded subsets B of X are cofinal in the bounded subsets of $[0,1] \otimes X$. So we can conclude that $E^{\mathrm{lf}}([0,1] \otimes X) \to E^{\mathrm{lf}}(X)$ is the limit of equivalences (by homotopy invariance of E)

$$
E([0,1] \otimes X, [0,1] \times (X \setminus B)) \to E(X, X \setminus B)
$$

and consequently an equivalence, too.   □

Let $\mathbf{C}$ be a complete and cocomplete stable ∞-category, and let $E : \mathbf{TopBorn} \to \mathbf{C}$ be a functor.

Lemma 7.36 *If E satisfies (open, closed or weak) excision, then E^{lf} also satisfies (open, closed or weak) excision.*

Proof We discuss open excision. The other two cases are similar. Let (Y, Z) be an open decomposition of X. For every bounded B we get an open decomposition

$(Y \setminus B, Z \setminus B)$ of $X \setminus B$. The cofibre of the corresponding maps of push-out diagrams (here we use that E satisfies open excision) is the push-out diagram

$$
\begin{array}{ccc}
E(Y \cap Z, (Y \cap Z) \setminus B) & \longrightarrow & E(Y, Y \setminus B) \\
\downarrow & & \downarrow \\
E(Z, Z \setminus B) & \longrightarrow & E(X, X \setminus B)
\end{array}
$$

Using stability of the category $\mathbf{C}$ the limit of these push-out diagrams over B in $\mathcal{B}$ is again a push-out diagram. $\qquad\square$

Let $\mathbf{C}$ be a complete and cocomplete stable ∞-category, and let $E \colon \mathbf{TopBorn} \to \mathbf{C}$ be a functor.

Proposition 7.37 *If E is homotopy invariant and satisfies weak excision, then its locally finite evaluation E^{lf} is a locally finite homology theory.*

Proof By Lemma 7.17 the functor E^{lf} is locally finite. Furthermore, by Lemma 7.35 shown below it is homotopy invariant. Finally, by Lemma 7.36 it satisfies excision. This shows that E is a locally finite homology theory. $\qquad\square$

Let $\mathbf{C}$ be a complete and cocomplete stable ∞-category, and let $E \colon \mathbf{TopBorn} \to \mathbf{C}$ be a functor.

Lemma 7.38 *Assume:*

1. *$\mathbf{C}$ has the property that filtered limits distribute over sums (Remark 6.28).*
2. *E preserves coproducts.*

Then E^{lf} also preserves coproducts.

Proof Let $(X_i)_{i \in I}$ be a family of topological bornological spaces. We must show that the canonical morphism

$$
\bigoplus_{i \in I} E^{\mathrm{lf}}(X_i) \to E^{\mathrm{lf}}\Big(\coprod_{i \in I} X_i\Big)
$$

is an equivalence. We have the chain of equivalences

$$
E^{\mathrm{lf}}\Big(\coprod_{i \in I} X_i\Big) := \lim_{B \in \mathcal{B}} E\Big(\coprod_{i \in I} X_i, \coprod_{i \in I} X_i \setminus B_i\Big)
$$

$$
\stackrel{!!}{\simeq} \lim_{B \in \mathcal{B}} \bigoplus_{i \in I} E(X_i, X_i \setminus B_i)
$$

$$
\stackrel{!}{\simeq} \bigoplus_{i \in I} \lim_{B_i \in \mathcal{B}_i} E(X_i, X_i \setminus B_i)
$$

$$
\simeq \bigoplus_{i \in I} E^{\mathrm{lf}}(X_i) .
$$

For the equivalence marked by !! we use that E preserves coproducts. Furthermore, for the equivalence marked by ! we use that the bornology of $\mathcal{B}$ of the coproduct is can be identified with $\prod_{i \in I} \mathcal{B}_i$, and Assumption 1. □

Remark 7.39 Note that the condition on **C** that filtered limits distribute over sums seems to be quite exotic. It is satisfied, e.g. for $\mathbf{C} = \mathbf{Sp}^{\mathrm{op}}$ since in **Sp** filtered colimits distribute over products. □

Example 7.40 The functor $\Sigma_+^{\infty,\mathrm{top}} : \mathbf{TopBorn} \to \mathbf{Sp}$ (we omit to write the forgetful functor $\mathcal{F}_{\mathcal{B}}$) is homotopy invariant and satisfies open excision (see Example 7.23 and the list of properties of this functor in Sect. 6.4.1). Consequently, its locally finite evaluation

$$\Sigma_+^{\infty,\mathrm{lf}} := (\Sigma_+^{\infty,\mathrm{top}})^{\mathrm{lf}} : \mathbf{TopBorn} \to \mathbf{Sp}$$

is an additive locally finite homology theory.

Let **C** be a complete and cocomplete stable ∞-category, and let C be an object of **C**. Then

$$(C \wedge \Sigma_+^{\infty,\mathrm{top}})^{\mathrm{lf}} : \mathbf{TopBorn} \to \mathbf{C}$$

is a locally finite homology theory.

The order of applying $C \wedge -$ and forming the local finite evaluation in general matters. If the functor $C \wedge - : \mathbf{Sp} \to \mathbf{C}$ commutes with limits, then we have an equivalence

$$C \wedge \Sigma_+^{\infty,\mathrm{lf}} \simeq (C \wedge \Sigma_+^{\infty,\mathrm{top}})^{\mathrm{lf}}.$$

But if $C \wedge -$ does not preserve limits, then applying this functor will destroy local finiteness.

For example, $\mathbf{C} = \mathbf{Sp}$ and C is a dualizable spectrum, then $C \wedge -$ preserves limits. In Example 6.40 we gave an example of a spectrum C where $C \wedge -$ does not commute with limits. □

Example 7.41 The functor

$$\ell \circ C^{\mathrm{sing}} : \mathbf{TopBorn} \to \mathbf{Ch}_\infty$$

is homotopy invariant and satisfies open excision (see Example 7.23). Its locally finite evaluation

$$(\ell \circ C^{\mathrm{sing}})^{\mathrm{lf}} : \mathbf{TopBorn} \to \mathbf{Ch}_\infty$$

is a locally finite homology theory. It is the **TopBorn**-version of Borel–Moore homology for locally compact spaces. □

7.1.5 Classification of Locally Finite Homology Theories

In this subsection we provide a partial classification of locally finite homology theories. The result Proposition 7.43, and also our approach are very similar to the classification result of Weiss and Williams [WW95].

Let X be a set, and let $\mathcal{F}$ be a subset of $\mathcal{P}_X$. Then we can consider $\mathcal{F}$ as a poset with respect to the inclusion relation. We define $X \setminus \mathcal{F} := \{X \setminus Y \mid Y \in \mathcal{F}\}$.

We consider the full subcategory **F** of **TopBorn** of topological bornological spaces which are homotopy equivalent to spaces X which admit a cofinal subset $\mathcal{B}'$ of the bornology with the following properties:

1. For every B in $\mathcal{B}'$ there exists B' in $\mathcal{B}'$ and a subset $\{B\}$ of $\mathcal{P}_X$ satisfying:

   (a) $\{B\}$ is filtered and a big family (Definition 7.24).
   (b) Every member $\tilde{B}$ of $\{B\}$ satisfies $B \subseteq \tilde{B} \subset B'$.
   (c) For every member $\tilde{B}$ of $\{B\}$ the inclusion $B \to \tilde{B}$ is a homotopy equivalence.
   (d) The family $X \setminus \{B\} := \{X \setminus B'' : B'' \in \{B\}\}$ is filtered and a big family.

2. Every B in $\mathcal{B}'$ is homotopy equivalent to a finite CW-complex.

The Condition 1 on the members of $\mathcal{B}'$ is a sort of cofibration condition. It is satisfied, e.g., if the inclusion $B \to X$ has a normal unit disc bundle D. Then we can take $B' = D_1$ and $\{B\} := \{D_r \mid r \in (0, 1)\}$, where D_r is the total space of the subbundle of discs of radius r.

Example 7.42 A locally finite simplicial complex with the topological and bornological structures induced by the spherical path metric belongs to **F**. For $\mathcal{B}'$ we can take the family of finite subcomplexes.

A typical example of such a complex is the coarsening space $\|\hat{\mathcal{Y}}\|$, see (5.24), provided the complex $\|\mathcal{Y}_n\|$ is finite-dimensional for every n in $\mathbb{N}$. The space $\|\hat{\mathcal{Y}}\|$ itself need not be finite-dimensional. □

The following is an adaptation of a result of Weiss and Williams [WW95].

Let **C** be a cocomplete stable ∞-category. In the following we use the tensor structure (6.5) of **C** over **Sp**.

Let $E : \textbf{TopBorn} \to \textbf{C}$ be a functor.

Proposition 7.43 *If E is homotopy invariant, then there exists a natural transformation*

$$E(*) \wedge \Sigma_+^{\infty,\mathrm{top}} \to E.$$

If E is a locally finite homology theory, then the induced transformation $(\Sigma_+^{\infty} \wedge E(*))^{\mathrm{lf}} \to E$ induces an equivalence on all objects of **F**.

Proof Let Δ denote the usual category of the finite posets $[n]$ for n in $\mathbb{N}$. We have a functor

$$\Delta : \mathbf{\Delta} \to \mathbf{TopBorn}, \quad [n] \mapsto \Delta^n$$

which sends the poset $[n]$ to the n-dimensional topological simplex. The simplex has the maximal bornology since it is compact, see Example 7.2.

We consider the category $\mathbf{TopBorn}^{\mathrm{simp}}$ whose objects are pairs (X, σ) of a topological bornological space X and a singular simplex $\sigma : \Delta^n \to X$. A morphism $(X, \sigma) \to (X', \sigma')$ is a commutative diagram

$$\begin{array}{ccc} \Delta^n & \xrightarrow{\phi} & \Delta^{n'} \\ \downarrow{\sigma} & & \downarrow{\sigma} \\ X & \xrightarrow{f'} & X' \end{array}$$

where f is a morphism in $\mathbf{TopBorn}$ and ϕ is induced by a morphism $[n] \to [n']$ in $\mathbf{\Delta}$.

We have two forgetful functors

$$\mathbf{TopBorn} \xleftarrow{p} \mathbf{TopBorn}^{\mathrm{simp}} \xrightarrow{q} \mathbf{TopBorn}, \quad p(X, \sigma) := X, \quad q(X, \sigma) := \Delta^n .$$

We then define the functor

$$E^{\%} : \mathbf{TopBorn} \to \mathbf{C}$$

by left-Kan extension

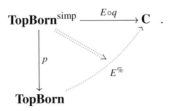

The objectwise formula for Kan extensions gives

$$E^{\%}(X) := \operatorname*{colim}_{(\Delta^n \to X)} E(\Delta^n) .$$

We have a natural transformation

$$E \circ q \to E \circ p, \quad (X, \sigma) \mapsto E(\sigma) : E(\Delta^n) \to E(X) . \tag{7.6}$$

The universal property of the Kan extension provides an equivalence of mapping spaces in functor categories

$$\mathrm{Map}(E \circ q, E \circ p) \simeq \mathrm{Map}(E^{\%}, E).$$

Consequently, the transformation (7.6) provides a natural transformation

$$E^{\%} \to E. \tag{7.7}$$

We use now that E is homotopy invariant. The projection $\Delta^n \to *$ is a homotopy equivalence in **TopBorn** for every n in $\mathbb{N}$. We get an equivalence of functors **TopBorn**$^{\mathrm{simp}} \to \mathbf{C}$

$$E \circ q \xrightarrow{\simeq} \mathrm{const}(E(*)). \tag{7.8}$$

We now use the definition (6.8) of $\Sigma_{+}^{\infty,\mathrm{top}}$ and the well-know facts that

$$\mathop{\mathrm{colim}}_{\mathbf{Top}^{simp}/X} \ell(\mathrm{sing}(*)) \simeq \ell(\mathrm{sing}(X))$$

in **Spc** (where **Top**^{simp} similar as **TopBorn**^{simp} is the category of pairs (X, σ) of topological spaces X with a singular simplex), and that Σ_{+}^{∞} preserves colimits. By the point-wise formula the left-Kan extension of $\mathrm{const}(E(*))$ can be identified with the functor

$$X \mapsto E(*) \wedge \Sigma_{+}^{\infty} \ell(\mathrm{sing}(X)) \simeq E(*) \wedge \Sigma_{+}^{\infty,\mathrm{top}} X.$$

The left Kan extension of the equivalence (7.8) therefore yields an equivalence

$$E^{\%} \xrightarrow{\simeq} E(*) \wedge \Sigma_{+}^{\infty}.$$

The composition of the inverse of this equivalence with (7.7) yields the asserted transformation

$$E(*) \wedge \Sigma_{+}^{\infty} \to E.$$

We now use that E is locally finite. We apply the $(-)^{\mathrm{lf}}$-construction and use Lemma 7.17 for the second equivalence in order to get the transformation between locally finite homology theories

$$(E(*) \wedge \Sigma_{+}^{\infty})^{\mathrm{lf}} \to E^{\mathrm{lf}} \simeq E. \tag{7.9}$$

If one evaluates this transformation on the one-point space, then it induces an equivalence. By excision and homotopy invariance we get an equivalence on all bounded spaces which are homotopy equivalent to finite CW-complexes.

Assume now that X belongs to $\mathbf{F}$. After replacing X by a homotopy equivalent space we can assume that it admits a cofinal subset $\mathcal{B}'$ of its bornology satisfying the assumptions stated above. Every member B of $\mathcal{B}'$ is homotopy equivalent to a finite CW-complex. Furthermore there exists B' in $\mathcal{B}'$ such that $(X \setminus \{B\}, \{B'\})$ is a decomposition of X into two big families. By excision we get a natural equivalence

$$(\Sigma_+^\infty \wedge E(*))^{\mathrm{lf}}(X, X \setminus \{B\}) \simeq (\Sigma_+^\infty \wedge E(*))^{\mathrm{lf}}(\{B'\}, \{B'\} \setminus \{B\})$$

$$\simeq E(\{B'\}, \{B'\} \setminus \{B\})$$

$$\simeq E(X, X \setminus \{B\}) .$$

The two middle terms involve a double colimit $\mathrm{colim}_{\{B\}} \mathrm{colim}_{\{B'\}}$.

In the middle equivalence we use that the natural transformation (7.9) induces an equivalence on spaces which are bounded and homotopy equivalent to finite CW-complexes.

If we now take the limit over B in $\mathcal{B}'$ and use the cofinality of $\mathcal{B}'$ in $\mathcal{B}$ and the local finiteness of E, then we get

$$(\Sigma_+^\infty \wedge E(*))^{\mathrm{lf}}(X) \xrightarrow{\simeq} E(X) .$$

which is the desired equivalence.   □

7.2 Coarsification of Locally Finite Theories

In this section we explain the construction of coarsifying a locally finite homology theory in the sense of Definition 7.27. This construction is a generalization of the construction of the coarse stable homotopy theory Q in Definition 6.29.

We refer to [BEb] for a similar construction whose input is a local homology theory on the category of uniform bornological coarse spaces.

Let X be a bornological coarse space. If U is an entourage of X, then we can consider the space of controlled probability measures $P_U(X)$ defined in (6.6) as a topological bornological space with the bornology generated by the subsets $P_U(B)$ for all bounded subsets B of X. The arguments given in the proof of Lemma 6.35 applied to the big family $\mathcal{B}$ show that the bornology and the topology of $P_U(X)$ are compatible in the sense of Definition 7.1. From now we will consider $P_U(X)$ as an object of **TopBorn**.

Let **BornCoarse**$^{\mathcal{C}}$ be as in Remark 6.31 the category of pairs (X, U) of a bornological coarse space X and an entourage U of X. We define the functor

$$\mathcal{P} : \textbf{BornCoarse}^{\mathcal{C}} \to \textbf{TopBorn}, \quad \mathcal{P}(X, U) := P_U(X).$$

On the level of morphisms $\mathcal{P}$ is defined in the obvious way in terms of the push-forward of measures.

Let **C** be a cocomplete stable ∞-category, and let $E : \textbf{TopBorn} \to \textbf{C}$ be a functor.

Definition 7.44 We define the functor

$$QE : \textbf{BornCoarse} \to \textbf{C}$$

by left Kan-extension

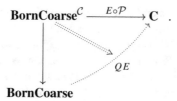

We call QE the coarsification of E.

Remark 7.45 The evaluation of QE on a bornological coarse space X with coarse structure $\mathcal{C}$ is given by

$$QE(X) \simeq \operatorname*{colim}_{U \in \mathcal{C}} E(P_U(X)). \tag{7.10}$$

This expression follows from the object-wise formula for the left Kan extension. $\square$

Let **C** be a complete and cocomplete stable ∞-category, and let $E : \textbf{TopBorn} \to$ **C** be a functor.

Proposition 7.46

1. *If E is a locally finite homology theory, then QE is a coarse homology theory.*
2. *If in* **C** *filtered colimits distribute over products (Definition 6.27), then QE is strongly additive.*
3. *If E preserves coproducts, then so does QE.*

Proof The proof of 1 is very similar to the proof of Theorem 6.32.

Assume that E is a locally finite homology theory. We must verify that QE satisfies the four conditions listed in Definition 4.22 (with QE in place of E).

In order to see that QE is coarsely invariant we use that E is homotopy invariant and then argue as in the proof of Lemma 6.34.

Let X be in **BornCoarse**, and let $(Z, \mathcal{Y})$ be a complementary pair on X. Let furthermore U be an entourage of X. We have shown in the proof of Lemma 6.35 that $(P_U(\{Z\}), P_U(\mathcal{Y}))$ is a decomposition of the topological bornological space $P_U(X)$ into two big families. We now use weak excision of E and form the colimit of the resulting push-out diagrams over the coarse entourages U of X in order to conclude that

$$
\begin{array}{ccc}
QE(\{Z\} \cap \mathcal{Y}) & \longrightarrow & QE(\{Z\}) \\
\downarrow & & \downarrow \\
QE(\mathcal{Y}) & \longrightarrow & QE(X)
\end{array}
$$

is a push-out diagram. We finally use that QE is coarsely invariant in order to replace $\{Z\}$ by Z.

Assume that X is flasque and that flasqueness of X is implemented by $f : X \to X$. As in the proof of Proposition 6.1.3. we can find a cofinal set set of entourages U of X such that $(f \times f)(U) \subseteq U$ and $(\mathrm{id}_X \times f)(\mathrm{diag}_X) \subseteq U$. For such U the map $P_U(f) : P_U(X) \to P_U(X)$ implements flasqueness of $P_U(X)$ as a topological bornological space. We now use Lemma 7.21 and that E is homotopy invariant and locally finite in order to see that $E(P_U(X)) \simeq 0$ for those U. By taking the colimit over these entourages we conclude that $QE(X) \simeq 0$.

Finally, QE is u-continuous by definition. This finishes the argument that QE is a coarse homology theory.

We now show 2. Let $(X_i)_{i \in I}$ be a family of bornological coarse spaces and set $X := \bigsqcup_{i \in I}^{\mathrm{free}} X_i$. Using (7.10) and that $P_U(X) \cong \bigsqcup_{i \in I}^{\mathrm{free}} P_{U_i}(X_i)$ we get

$$
QE(X) \simeq \operatorname*{colim}_{U \in \mathcal{C}} E\Big(\bigsqcup_{i \in I}^{\mathrm{free}} P_{U_i}(X_i)\Big),
$$

where we set $U_i := U \cap (X_i \times X_i)$. Note that contrary to (6.10), where we were working with topological spaces, here we have to write the free union in order to take the bornology into account properly (which is not the bornology of the coproduct in this case). Since E is additive by Lemma 7.31, we get

$$
\operatorname*{colim}_{U \in \mathcal{C}} E\Big(\bigsqcup_{i \in I}^{\mathrm{free}} P_{U_i}(X_i)\Big) \simeq \operatorname*{colim}_{U \in \mathcal{C}} \prod_{i \in I} E(P_{U_i}(X_i)).
$$

We finish the argument by interchanging the colimit with the product (see the end of the proof of Lemma 6.38 for a similar argument.) At this point we use the additional assumption on **C**.

We now show 3. We assume that E preserves coproducts. Let $(X_i)_{i \in I}$ be a family of bornological coarse spaces and set $X := \coprod_{i \in I} X_i$. We now have an isomorphism $P_U(X) \cong \coprod_{i \in I} P_{U_i}(X_i)$ (the coproduct is understood in topological bornological spaces) and therefore the following chain of equivalences

$$QE(X) \simeq \operatorname*{colim}_{U \in \mathcal{C}} \bigoplus_{i \in I} E(P_{U_i}(X_i)) \simeq \bigoplus_{i \in I} \operatorname*{colim}_{U \in \mathcal{C}} E(P_{U_i}(X_i)) \simeq \bigoplus_{i \in I} QE(X_i).$$

This completes the proof of Proposition 7.46.   □

Example 7.47 Let **C** be a complete and cocomplete stable ∞-category, and let C be an object of **C**. Then we can construct the locally finite homology theory $(C \wedge \Sigma_+^{\infty, \mathrm{top}})^{\mathrm{lf}}$ as in Example 7.40.

By Proposition 7.46 we obtain to obtain a coarse homology theory

$$Q(C \wedge \Sigma_+^{\infty, \mathrm{top}})^{\mathrm{lf}} : \mathbf{BornCoarse} \to \mathbf{C}.$$

If **C** has the property that filtered colimits distribute over products, then $Q(C \wedge \Sigma_+^{\infty, \mathrm{top}})^{\mathrm{lf}}$ is strongly additive.

We have

$$Q(C \wedge \Sigma_+^{\infty, \mathrm{top}})^{\mathrm{lf}}(*) \simeq C.$$

Note that $C \wedge Q$ (Corollary 6.39) is also coarse homology theory with the value C on the one-point space, but this coarse homology theory is in general not even additive.

We always have a natural transformation

$$C \wedge Q \to QC(\wedge \Sigma_+^{\infty, \mathrm{top}})^{\mathrm{lf}} \tag{7.11}$$

of coarse homology theories. On the other hand, if the functor $C \wedge -$ preserves limits, then the transformation (7.11) is an equivalence, see Example 6.40.   □

7.3 Analytic Locally Finite K-Homology

7.3.1 Extending Functors from Locally Compact Spaces to **TopBorn**

In order to deal with analytic locally finite K-homology we consider the full subcategory $\mathbf{Top}^{\mathrm{lc}}$ of $\mathbf{TopBorn}$ of separable, locally compact spaces with the bornology of relatively compact subsets. By definition, morphisms in $\mathbf{Top}^{\mathrm{lc}}$ are proper continuous maps. Analytic locally finite K-homology is initially defined on

Top^{lc}. So we must extend this functor to **TopBorn** preserving good properties. We first discuss such extensions in general and then apply the theory to K-homology.

In the following it is useful to remember the Examples 7.5 and 7.2.

Let X be a topological bornological space, and let C be a subset of X.

Definition 7.48 The subset C is X-locally compact[1] if C with the induced topology is separable and locally compact and the induced bornology is given by the relatively compact subsets of C.

We let $\mathrm{Loc}(X)$ denote the poset of locally compact subsets of X.

Let X be a topological bornological space, and let C be a subset of X.

Lemma 7.49 *Assume:*

1. *X is Hausdorff.*
2. *C is X-locally compact.*

Then C is a closed subset of X.

Proof We show that $X \setminus C$ is open. To this end we consider a point x in $X \setminus C$. Let $\mathcal{J}$ be the family of closed and bounded neighbourhoods of x in X. Since $\{x\}$ is bounded such neighbourhoods exist by the compatibility of the bornology with the topology of X (see Definition 7.1). For every W in $\mathcal{J}$ the intersection $W \cap C$ is closed in C and relatively compact in C, therefore it is compact. We claim that $\bigcap_{W \in \mathcal{J}}(W \cap C) = \emptyset$. Indeed, if $\bigcap_{W \in \mathcal{J}}(W \cap C) \neq \emptyset$, then because of $\bigcap_{W \in \mathcal{J}} W = \{x\}$ (since X is Hausdorff by assumption) we would have $x \in C$. By the claim and the compactness of the intersections $W \cap C$ there exists W in $\mathcal{J}$ such that $W \cap C = \emptyset$. Hence W is a (closed) neighbourhood of x contained in $X \setminus C$. $\qquad\square$

If C is a X-locally compact subset of X, then C is Hausdorff. A closed subset of a locally compact subset C of X is therefore again X-locally compact. By Lemma 7.49 the X-locally compact subsets of X contained in the X-locally compact subset C are precisely the closed subsets of C.

Let **C** be a stable ∞-category, and let $E : \mathbf{Top}^{lc} \to \mathbf{C}$ be a functor. Then we can say that E is countably additive, locally finite, homotopy invariant or satisfies closed descent by interpreting the corresponding definitions made for **TopBorn** in the obvious way. For the first two properties we must assume that **C** admits countable products.

Remark 7.50 Since **Top**^{lc} only has countable free unions (because of the separability assumption) it does not make sense to consider additivity for larger families.

Furthermore, by Lemma 7.49 we are forced to consider closed descent. The condition of open descent could not even be formulated since open subsets of a locally compact space in general do not belong to **Top**^{lc}, see Example 7.5.

---

[1] We omit saying "separable" in order to shorten the text.

The interpretation of the local finiteness condition is as

$$\lim_{B \in \mathcal{B} \cap \mathcal{S}} E(C \setminus B) \simeq 0.$$

We must restrict the limit to open, relatively compact subsets since then $C \setminus B$ is a closed subset of C and therefore also locally compact. Since a separable, locally compact space is σ-compact, the intersection $\mathcal{B} \cap \mathcal{S}$ contains a countable cofinal subfamily. $\square$

We now discuss the left Kan extension of functors defined on $\mathbf{Top}^{\mathrm{lc}}$ to $\mathbf{TopBorn}$. Let $\mathbf{C}$ be a cocomplete stable ∞-category, and let $E : \mathbf{Top}^{\mathrm{lc}} \to \mathbf{C}$ be a functor.

Definition 7.51 We define

$$L(E) : \mathbf{TopBorn} \to \mathbf{C}, \quad L(E)(X) := \operatorname*{colim}_{C \in \mathrm{Loc}(X)} E(C).$$

Remark 7.52 Using the Kan extension technique we can turn the above description of the functor on objects into a proper definition of a functor. We consider the category $\mathbf{TopBorn}^{\mathrm{Loc}}$ of pairs (X, C), where X is a topological bornological space and C is an element of $\mathrm{Loc}(X)$. A morphism $f : (X, C) \to (X', C')$ is a morphism $f : X \to X'$ of topological bornological spaces such that $f(C) \subseteq C'$. We have the forgetful functors

$$\mathbf{TopBorn}^{\mathrm{Loc}} \to \mathbf{TopBorn}, \ (X, C) \mapsto X, \ p : \mathbf{TopBorn}^{\mathrm{Loc}} \to \mathbf{Top}^{\mathrm{lc}}, \ (X, C) \mapsto C.$$

We then define $L(E)$ as left Kan extension

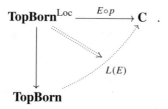

$\square$

If C is a separable locally compact space, then C is final in the poset $\mathrm{Loc}(C)$. This implies that

$$L(E)_{|\mathbf{Top}^{\mathrm{lc}}} \simeq E. \tag{7.12}$$

Lemma 7.53 *If E is homotopy invariant, then so is $L(E)$.*

Proof The subsets $[0, 1] \times C$ of $[0, 1] \times X$ for C in $\mathrm{Loc}(X)$ are cofinal in $\mathrm{Loc}([0, 1] \times X)$. $\square$

Lemma 7.54 *If E satisfies closed excision, then so does* $L(E)$.

Proof Here we use that for C in $\mathrm{Loc}(X)$ a closed decomposition of X induces a closed decomposition of C. We thus get a colimit of push-out diagrams which is again a push-out diagram.   □

Lemma 7.55 *If E satisfies closed excision, then* $L(E)$ *preserves all coproducts.*

Proof Let $X := \coprod_{i \in I} X_i$ be the coproduct of the family $(X_i)_{i \in I}$ in the category **TopBorn** and let C be in $\mathrm{Loc}(X)$. Since the induced bornology on C is the bornology of relatively compact subsets the set $I_C := \{i \in I : C \cap X_i \neq \emptyset\}$ is finite. Furthermore, for every i in I the space $C_i := C \cap X_i$ belongs to $\mathrm{Loc}(X)$. Since E satisfies closed excision it commutes with finite coproducts. Therefore we get the chain of equivalences:

$$L(E)(X) \simeq \operatorname*{colim}_{C \in \mathrm{Loc}(X)} E(C)$$

$$= \operatorname*{colim}_{C \in \mathrm{Loc}(X)} E\Big(\coprod_{i \in I_C} C_i \Big)$$

$$\simeq \operatorname*{colim}_{C \in \mathrm{Loc}(X)} \bigoplus_{i \in I_C} E(C_i)$$

$$\simeq \bigoplus_{i \in I} \operatorname*{colim}_{C_i \in \mathrm{Loc}(X_i)} E(C_i)$$

$$\simeq \bigoplus_{i \in I} L(E)(X_i)$$

which finishes the proof.   □

Let **C** be a complete and cocomplete stable ∞-category, and $E : \mathbf{Top}^{\mathrm{lc}} \to \mathbf{C}$ be a functor.

Proposition 7.56

1. *If E is homotopy invariant and satisfies closed excision, then* $L(E)^{\mathrm{lf}} : \mathbf{TopBorn} \to$ **C** *is a locally finite homology theory.*
2. *If in* **C** *filtered limits distribute over sums, then* $L(E)^{\mathrm{lf}}$ *preserves coproducts.*

Proof We show Assertion 1. By Lemmas 7.54 and 7.53 the functor $L(E)$ satisfies closed excision and is homotopy invariant. By Proposition 7.37 the functor $L(E)^{\mathrm{lf}}$ is a locally finite homology theory.

In order to show Assertion 2 we combine the Lemmas 7.55 and 7.38.   □

Note that $L(E)^{\mathrm{lf}}$ satisfies closed excision.

The following lemma ensures that $L(E)^{\mathrm{lf}}$ really extends E if it was locally finite.

Lemma 7.57 *If E is locally finite, then* $L(E)^{\mathrm{lf}}_{|\mathbf{Top}^{\mathrm{lc}}} \simeq E$.

Proof Let X be in $\mathbf{Top}^{lc}$. For an open bounded subset B of C we then have $X \setminus B \in \mathbf{Top}^{lc}$. We now use that by compatibility of the bornology and the topology we can define the locally finite evaluation as a limit over open bounded subsets of X. This gives

$$L(E)^{lf}(X) \simeq \lim_{B \in \mathcal{B} \cap \mathcal{S}} L(E)(X, X \setminus B) \overset{(7.12)}{\simeq} \lim_{B \in \mathcal{B} \cap \mathcal{S}} E(X, X \setminus B) \simeq E(X),$$

where for the last equivalence we use that E is locally finite. □

7.3.2 Cohomology for C*-Algebras

In order to construct the analytic K-homology and its locally finite version we employ a K-cohomology functor for C^*-algebras. In this subsection we recall the required properties of such a cohomology functor in an axiomatic way and provide references to the corresponding literature. Our main examples are the K-cohomology functors from Definition 7.59.

Let $C^*\mathbf{Alg}^{sep}$ denote the category of separable (very small) C^*-algebras. These algebras need not be unital, and morphisms need not preserve units in case they exist.

Let $\mathbf{C}$ be a complete stable ∞-category. We consider a functor

$$F : C^*\mathbf{Alg}^{sep,op} \to \mathbf{C}.$$

Definition 7.58 We define the following properties of F.

1. homotopy invariance: For every A in $C^*\mathbf{Alg}^{sep}$ the inclusion $A \to C([0,1], A)$ as constant functions induces an equivalence $F(C([0,1], A)) \to F(A)$.
2. countable additivity: For a countable family $(A_i)_{i \in I}$ in $\mathbf{Alg}^{sep}$ the canonical map

$$F(\bigoplus_{i \in I} A_i) \to \prod_{i \in I} F(A_i)$$

   is an equivalence.
3. exactness: We have $F(0) \simeq 0$, and for every short exact sequence with a completely positive contractive split (we abbreviate this as cpcs)

$$0 \to I \to A \to Q \to 0$$

of C^*-algebras in $\mathbf{Alg}^{\mathrm{sep}}$ (where I is a closed ideal) we have a pull-back

$$
\begin{array}{ccc}
F(Q) & \longrightarrow & F(A) \\
\downarrow & & \downarrow \\
0 & \longrightarrow & F(I)
\end{array}
$$

If F is homotopy invariant, additive and exact, then we call F a cohomology theory.

In the following we use a spectrum-valued KK-theory functor in order to construct an example of a cohomology theory as in Definition 7.58.

Our starting point is a stable ∞-category $\mathbf{KK}$ whose objects are the separable C^*-algebras as objects. As usual, for A, B in $C^*\mathbf{Alg}^{\mathrm{sep}}$ we let $\mathrm{map}_{\mathbf{KK}}(A, B)$ in $\mathbf{Sp}$ denote the mapping spectrum in $\mathbf{KK}$. This category has the following properties.

1. There is a functor $C^*\mathbf{Alg}^{\mathrm{sep}} \to \mathbf{KK}$ which is the identity on objects.
2. For A, B in $C^*\mathbf{Alg}^{\mathrm{sep}}$ we have an isomorphism of $\mathbb{Z}$-graded groups

$$
\pi_*(\mathrm{map}_{\mathbf{KK}}(A, B)) \cong KK_*(A, B) \,,
$$

where KK_* denotes Kasparovs bivariant K-theory groups (see Blackadar [Bla98]). This isomorphism identifies the composition of morphisms in $\mathbf{KK}$ with the Kasparov product for KK_*.
3. The zero algebra 0 is a zero object in $\mathbf{KK}$.
4. We have an equivalence of spectra $\mathrm{map}_{\mathbf{KK}}(\mathbb{C}, \mathbb{C}) \simeq KU$.
    More generally, for B in $C^*\mathbf{Alg}^{\mathrm{sep}}$ we have $\mathrm{map}_{\mathbf{KK}}(\mathbb{C}, B) \simeq K^{C^*}(B)$, where $K^{C^*}(B)$ is the C^*-algebra K-theory spectrum of B. In fact, one could take this as a definition of $K^{C^*}(B)$. We refer to Sect. 8.4 for further discussion.
5. An cpcs exact sequence of C^*-algebras in $C^*\mathbf{Alg}^{\mathrm{sep}}$ (where I is a closed ideal)

$$
0 \to I \to A \to Q \to 0
$$

gives rise to a push-out diagram

$$
\begin{array}{ccc}
I & \longrightarrow & A \\
\downarrow & & \downarrow \\
0 & \longrightarrow & Q
\end{array}
$$

in $\mathbf{KK}$.
6. If B is a separable C^*-algebra and $(A_i)_{i \in I}$ is a countable family of separable C^*-algebras, then

$$
\mathrm{map}_{\mathbf{KK}}\left(\bigoplus_{i \in I} A_i, B \right) \simeq \prod_{i \in I} \mathrm{map}_{\mathbf{KK}}(A_i, B) \,. \tag{7.13}
$$

Equivalently, $\bigoplus_{i \in I} A_i$ represents the coproduct of the family $(A_i)_{i \in I}$ in **KK**.

7. The upper left corner inclusion $A \to A \otimes \mathbb{K}$ induces an equivalence in **KK**, where $\mathbb{K}$ denotes the compact operators on the Hilbert space $\ell^2(\mathbb{N})$.

8. The inclusion $A \to C([0, 1], A)$ as constant functions induces an equivalence in **KK**.

There are various constructions of such a stable ∞-category [Mah15, BJM17, JJ06] and [LN18, Def. 3.2].

The equivalence (7.13) follows from [Bla98, Thm. 19.7.1] (attributed to J. Rosenberg). In this reference the isomorphism

$$KK_0\left(\bigoplus_{i \in I} A_i, B\right) \cong \prod_{i \in I} KK_0(A_i, B)$$

is shown. The corresponding isomorphism in degree 1 can be obtained by replacing B by its suspension $C_0(\mathbb{R}, B)$. By Bott periodicity these two cases imply Equivalence (7.13).

Let B be a separable C^*-algebra.

Definition 7.59 We define the K-cohomology for C^*-algebras with coefficients in B by

$$K_B := \mathrm{map}_{\mathbf{KK}}(-, B) \colon C^*\mathbf{Alg}^{\mathrm{sep,op}} \to \mathbf{Sp}.$$

Corollary 7.60 *The functor* $K_B \colon \mathbf{Alg}^{\mathrm{sep,op}} \to \mathbf{Sp}$ *is a cohomology theory.*

Proof This functor is homotopy invariant by Property 8 of **KK**, sends countable sums to products by Property 6, and is exact by Property 5. $\square$

7.3.3 Locally Finite Homology Theories from Cohomology Theories for C^*-Algebras

If X is a separable locally compact topological space, then $C_0(X)$ denotes the C^*-algebra defined as the completion of the $*$-algebra of compactly supported functions in the supremum norm. Since X is a separable, this C^*-algebra is separable, too. Recall that morphisms in $\mathbf{Top}^{\mathrm{lc}}$ are proper continuous maps. If $X \to X'$ is a morphism in $\mathbf{Top}^{\mathrm{lc}}$, then the pull-back of functions provides a morphism $C_0(X') \to C_0(X)$ in $C^*\mathbf{Alg}^{\mathrm{sep}}$. We therefore get a functor

$$C_0 \colon \mathbf{Top}^{\mathrm{lc,op}} \to C^*\mathbf{Alg}^{\mathrm{sep}}.$$

This functor has the following properties:

1. If X is in $\mathbf{Top}^{\mathrm{lc}}$ and Y is a closed subset of X, then the inclusion $Y \to X$ is a morphism in $\mathbf{Top}^{\mathrm{lc}}$, and $X \setminus Y$ belongs to $\mathbf{Top}^{\mathrm{lc}}$ (though the inclusion into X is

in general not a morphism). The inclusion induces the second map in the exact sequence

$$0 \to C_0(X \setminus Y) \to C_0(X) \to C_0(Y) \to 0 \,.$$

The first map is given by extension by zero. Since $C_0(Y)$ is separable and nuclear the sequence is cpcs by the Choi-Effros lifting theorem.

2. If $(X_i)_{i \in I}$ is a countable family in $\mathbf{Top}^{lc}$, then $\bigsqcup_{i \in I}^{\text{free}} X_i$ (interpreted in $\mathbf{TopBorn}$) is given by the coproduct of the family in $\mathbf{Top}$ which is again separable and locally compact and therefore an object of $\mathbf{Top}^{lc}$. We have a canonical isomorphism

$$C_0 \Big(\bigsqcup_{i \in I}^{\text{free}} X_i \Big) \cong \bigoplus_{i \in I} C_0(X_i) \,.$$

Let $\mathbf{C}$ be a complete stable ∞-category, and let $F \colon C^*\mathbf{Alg}^{\text{sep,op}} \to \mathbf{C}$ be a functor.

Definition 7.61 We define $F^{\text{an}} := F \circ C_0 \colon \mathbf{Top}^{lc} \to \mathbf{C}$.

Proposition 7.62 *If F is a cohomology theory, then the functor F^{an} has the following properties:*

1. *homotopy invariant,*
2. *excisive (for closed decompositions),*
3. *countably additive,*
4. *locally finite.*

Proof Assertion 1 follows from homotopy invariance of F and the fact that

$$C([0, 1], C_0(X)) \cong C_0([0, 1] \times X) \,.$$

For Assertion 2 we argue as follows. Let X be in $\mathbf{Top}^{lc}$, and let (A, B) be a closed decomposition of X. Then we have an isomorphism $B \setminus (B \setminus A) \cong X \setminus A$ in $\mathbf{Top}^{lc}$ and therefore an isomorphism of C^*-algebras

$$C_0(X \setminus A) \cong C_0(B \setminus (A \cap B)) \,.$$

We consider the following commuting diagram in $C^*\mathbf{Alg}^{\text{sep}}$:

$$
\begin{array}{ccccc}
C_0(B \setminus (A \cap B)) & \longrightarrow & C_0(B) & \longrightarrow & C_0(A \cap B) \\
\cong \uparrow & & \uparrow & & \uparrow \\
C_0(X \setminus A) & \longrightarrow & C_0(X) & \longrightarrow & C_0(A)
\end{array}
$$

The right square comes from a commuting square in $\mathbf{Top}^{\mathrm{lc}}$, and the left square commutes by the explicit description of the left horizontal maps given in Property 1 of C_0 which also asserts that horizontal sequences are exact. We now apply F and get the commuting diagram

$$
\begin{array}{ccccc}
F^{\mathrm{an}}(B \cap A) & \longrightarrow & F^{\mathrm{an}}(B) & \longrightarrow & F^{\mathrm{an}}(B \setminus (A \cap B)) \\
\downarrow & & \downarrow & & \downarrow{\scriptstyle\simeq} \\
F^{\mathrm{an}}(A) & \longrightarrow & F^{\mathrm{an}}(X) & \longrightarrow & F^{\mathrm{an}}(X \setminus A)
\end{array}
$$

Since F is excisive, the horizontal sequences are pieces of fibre sequences. Since the right vertical morphism is an equivalence, the left square is a push-out. This implies closed excision.

We now show Assertion 3. Let $(X_i)_{i \in I}$ be a countable family in $\mathbf{Top}^{\mathrm{lc}}$. Using Property 2 of C_0 and that F is additive we get the chain of equivalences

$$
F^{\mathrm{an}}\Big(\overset{\mathrm{free}}{\coprod_{i \in I}} X_i\Big) \simeq F\Big(\bigoplus_{i \in I} C_0(X_i)\Big) \simeq \prod_{i \in I} F(C_0(X_i)) \simeq \prod_{i \in I} F^{\mathrm{an}}(X_i).
$$

Finally we show Assertion 4. Let X be in $\mathbf{Top}^{\mathrm{lc}}$. Then there exists an open exhaustion $(B_n)_{n \in \mathbb{N}}$ of X such that $\bar{B}_n$ is compact for every n in $\mathbb{N}$. Then every relatively compact subset of X is contained in B_n for sufficiently large n. The same argument as in the proof Lemma 7.33 (using that F^{an} is additive and excisive) shows that $\lim_{n \in \mathbb{N}} F^{\mathrm{an}}(X \setminus B_n) \simeq 0$. The analog of Remark 7.13 for $\mathbf{Top}^{\mathrm{lc}}$ then implies that F^{an} is locally finite.   □

Let $F \colon C^*\mathbf{Alg}^{\mathrm{sep,op}} \to \mathbf{C}$ be a functor and assume that $\mathbf{C}$ is a complete and cocomplete stable ∞-category. The following Definition combines 7.51 and 7.15.

Definition 7.63 We define the functor

$$
F^{\mathrm{an,lf}} := L(F^{\mathrm{an}})^{\mathrm{lf}} \colon \mathbf{TopBorn} \to \mathbf{C}.
$$

Corollary 7.64 *Assume that F is a cohomology theory.*

1. *$F^{\mathrm{an,lf}}$ is a locally finite homology theory.*
2. *$F^{\mathrm{an,lf}}_{|\mathbf{Top}^{\mathrm{lc}}} \simeq F^{\mathrm{an}}$.*

Proof Assertion 1 follows from Propositions 7.62 and 7.56. For 2 we use Lemma 7.57.   □

We can apply Proposition 7.43 to $F^{\mathrm{an,lf}}$. Recall the subset of objects $\mathbf{F}$ of $\mathbf{TopBorn}$ introduced at the beginning of Sect. 7.1.5. By Example 7.42 the set $\mathbf{F}$ contains, e.g. locally finite simplicial complexes with the structures induced from the spherical path metrics. Let X be in $\mathbf{TopBorn}$.

Corollary 7.65 *If X is homotopy equivalent to an object of the subcategory* **F***, then we have a natural equivalence*

$$(F(*) \wedge \Sigma_+^{\infty,\mathrm{top}})^{\mathrm{lf}}(X) \simeq F^{\mathrm{an,lf}}(X).$$

Let B be in $\mathbf{C}^*\mathbf{Alg}^{\mathrm{sep}}$.

Definition 7.66 The analytic locally finite K-homology with coefficients in B is the functor

$$K_B^{\mathrm{an,lf}} : \mathbf{TopBorn} \to \mathbf{Sp}$$

obtained by applying Definition 7.63 to the K-cohomology K_B of C^*-algebras with coefficients in B (Definition 7.59).

In view of the Property 4 of **KK** we have an equivalence $K_B^{\mathrm{an,lf}}(*) \simeq K^{C^*}(B)$. Specializing Corollary 7.65 we get for X in **TopBorn**:

Corollary 7.67 *If X in* **F***, then we have an equivalence*

$$(K^{C^*}(B) \wedge \Sigma_+^{\infty,\mathrm{top}})^{\mathrm{lf}}(X) \simeq K_B^{\mathrm{an,lf}}(X).$$

In particular, if X belongs to **F** and is separable and locally compact, then using Lemma 7.57 and Property 2 of **KK** we get an isomorphism of groups

$$\pi_*(K^{C^*}(B) \wedge \Sigma_+^{\infty,\mathrm{top}})^{\mathrm{lf}}(X) \simeq KK_*(C_0(X), B). \tag{7.14}$$

Remark 7.68 We got the impression that the isomorphism (7.14) is known as folklore, but we were not able to trace down a reference for this fact.

In the global analysis literature it is standard to consider the functor $X \mapsto KK_*(C_0(X), B)$ as the definition of the group-valued locally finite K-homology functor with coefficients in a C^*-algebra B. Its homotopy theoretic version represented by the left-hand side of (7.14) is usually not used.   □

7.4 Coarsification Spaces

In this section we introduce the notion of a coarsification of a bornological coarse space X. A coarsification of X is a simplicial complex with a reference map from X such that the value of a coarsification of a locally finite homology on X can be calculated by applying the locally finite homology theory itself to the coarsification, see Proposition 7.81.

This is actually one of the origins of coarse algebraic topology. Assume that the classifying space BG of a group G has a model which is a finite simplicial complex. Gersten [Ger93, Thm. 8] noticed that for such a group the group cohomology

$H^*(G, \mathbb{Z}G)$ is a quasi-isometry invariant. Using coarse algebraic topology, we may explain this phenomenon as follows. We have the isomorphism $H^*(G, \mathbb{Z}G) \cong H^*_c(EG)$, see [Bro82, Prop. VIII.7.5]. Furthermore, we have an isomorphism $H^*_c(EG) \cong H\mathcal{X}^*(EG)$, where the latter is Roe's coarse cohomology [Roe93b, Prop. 3.33].

Let now G and H are two groups with the property that BG and BH are homotopy equivalent to finite simplicial complexes. We equip G and H with word metrics associated to choices of finite generating sets. Then a quasi-isometry $G \to H$ of the underlying metric spaces translates to a coarse equivalence $EG \to EH$. Since coarse cohomology is invariant under coarse equivalences we get a chain of isomorphisms relating $H^*(G, \mathbb{Z}G)$ and $H^*(H, \mathbb{Z}H)$. Therefore the cohomology group $H^*(G, \mathbb{Z}G)$ is a quasi-isometry invariant of G.

The above observation, translated to homology theories, can be generalized to the fact that a locally finite homology theory evaluated on a finite-dimensional, uniformly contractible space is a coarse invariant of such spaces, see Lemma 7.73. So on such spaces locally finite homology theories behave like coarse homology theories. And actually, this observation originally coined the idea of what a coarse homology theory should be.

Let $f_0, f_1 : X \to X'$ be maps between topological spaces, and let A' be a be a subset of X' containing the images of f_0 and f_1. We say that f_0 and f_1 are homotopic in A' if there exists a homotopy $h : I \times X \to A'$ from f_0 to f_1 whose image is also contained in A'.

Let X be a metric space. For a point x of X and a positive real number R we denote by $B(R, x)$ the ball in X of radius R with center x.

Definition 7.69 ([Gro93, Sec. 1.D]) X is uniformly contractible if for all R in $(0, \infty)$ exists an S in $[R, \infty)$ such that for all x in X the inclusion $B(R, x) \to X$ is homotopic in $B(S, x)$ to a constant map.

Note that the property of being uniformly contractible only depends on the quasi-isometry class of the metric.

In the following a metric space X will be considered as a bornological coarse space X_d with the structures induced by the metric, see Example 2.18. We equip a simplicial complex K with a good metric (see Definition 5.36) and denote the associated bornological coarse space by K_d. If K is finite-dimensional, then the metric on K is independent of the choices up to quasi-isometry. In particular, the bornological coarse space K_d is well-defined up to equivalence.

Definition 7.70 ([Roe96, Def. 2.4]) A coarsification of a bornological coarse space X is a pair (K, f) consisting of a finite-dimensional, uniformly contractible simplicial complex K and an equivalence $f : X \to K_d$ in **BornCoarse**.

Example 7.71 If X is a non-empty bounded bornological coarse space with the maximal coarse structure (e.g., a bounded metric space), then the pair $(*, f : X \to *)$ is a coarsification of X.

A coarsification of $\mathbb{Z}$ is given by $(\mathbb{R}, \iota : \mathbb{Z} \to \mathbb{R})$. Another possible choice is $(\mathbb{R}, -\iota)$.

Let K be a finite-dimensional simplicial complex with a good metric. We assume that K is uniformly contractible. Its the zero skeleton $K^{(0)}$ has an induced metric and gives rise to a bornological coarse space $K_d^{(0)}$. Then the inclusion $\iota : K_d^{(0)} \to K_d$ turns (K_d, ι) into a coarsification of the bornological coarse space $K_d^{(0)}$.

The following example is due to Gromov [Gro93, Ex. 1.D$_1$]: We consider a finitely generated group G and assume that it admits a model for its classifying space BG which is a finite simplicial complex. The action of G on the universal covering EG of BG provides a choice of a map $f : G \to EG$ which depends on the choice of a base point. We equip EG with a good metric. Then (EG_d, f) is a coarsification of G equipped with the bornological coarse structure introduced in Example 2.21.   □

Let K be a simplicial complex, and let A be a subcomplex of K. Furthermore, let X be a metric space, and let $f : K_d \to X_d$ be a morphism of bornological coarse spaces such that $f_{|A}$ is continuous.

Lemma 7.72 *If K is finite-dimensional and X is uniformly contractible, then f is close to a morphism of bornological coarse spaces which extends $f_{|A}$ and is in addition continuous.*

Proof We will define a continuous map $g : K \to X$ by induction over the relative skeleta of K such that it is close to f and satisfies $f_{|A} = g_{|A}$. Then $g : K_d \to X_d$ is also a morphism of bornological coarse spaces.

On the zero skeleton $K^{(0)} \cup A$ (relative to A) of K we define $g_0 := f_{|K^{(0)} \cup A}$.

Let now $n \in \mathbb{N}$ and assume that $g_n : K^{(n)} \cup A \to X$ is already defined such that g_n is continuous and close to the restriction of f to $K^{(n)} \cup A$. Since the metric on K is good there exists a uniform bound of the diameters of $(n+1)$-simplices of K. Since g_n is controlled there exists an R in $(0, \infty)$ such that $g(\partial \sigma)$ is contained in some R-ball for every $(n+1)$-simplex σ of K. Let S in $[R, \infty)$ be as in Definition 7.69.

We now extend g_n to the interiors of the $(n+1)$-simplices separately. Let σ be an $(n+1)$-simplex in $K^{(n+1)} \setminus A$. Using a contraction of the simplex σ to its center and the contractibility of R-balls of X inside S-balls, we can continuously extend $(g_n)_{|\partial \sigma}$ to σ with image in an $(S+R)$-neighbourhood of $g_n(\partial \sigma)$.

In this way we get a continuous extension $g_{n+1} : K^{(n+1)} \cup A \to X$ which is still close to $f_{|K^{(n+1)} \cup A}$.   □

A finite-dimensional simplicial complex K can naturally be considered as a topological bornological space with the bornology of metrically bounded subsets. We will denote this topological bornological space by K_t. A morphism of bornological coarse spaces $K_d \to K'_d$ which is in addition continuous is a morphism $K_t \to K'_t$ of topological bornological spaces.

We consider finite-dimensional simplicial complexes K and K' and a morphism $f : K_d \to K'_d$ of bornological coarse spaces.

Lemma 7.73 *If K and K' are uniformly contractible, and f is an equivalence, then f is close to a homotopy equivalence $K_t \to K'_t$, and any two choices of such a homotopy equivalence are homotopic to each other.*

Proof By Lemma 7.72 we can replace f by a close continuous map which we will also denote by f. We claim that this is the desired homotopy equivalence $f : K_t \to K_t$. Denote by $g : K'_d \to K_d$ an inverse equivalence. By Lemma 7.72 we can assume that g is continuous. The compositions $f \circ g$ and $g \circ f$ are close to the respective identities.

We consider the complex $I \times K$ with the subcomplex $\{0, 1\} \times K$. We define a morphism of bornological coarse spaces $I \times K_d \to K_d$ by id_K on $\{0\} \times K$ and by $g \circ f \circ \mathrm{pr}_K$ on $(0, 1] \times K$. By Lemma 7.72 we can replace this map by a close continuous map which is a homotopy between id_{K_t} and $g \circ f : K_t \to K_t$. This shows that $g \circ f : K_t \to K_t$ and id_{K_t} are homotopic. We argue similarly for $f \circ g$.

Furthermore, we obtain homotopies between different choices of continuous replacements of f by a similar argument.   □

Let X be a bornological coarse space and consider two coarsifications (K, f) and (K', f') of X.

Corollary 7.74 *There is a homotopy equivalence $K_t \to K'_t$ uniquely determined up to homotopy by the compatibility with the structure maps f and f'.*

Proof Let $g : K_d \to X$ be an inverse equivalence of f. Then we apply Lemma 7.73 to the composition $f' \circ g : K_d \to K'_d$.   □

Let X be a bornological coarse space.

Definition 7.75 X has strongly bounded geometry if it has the minimal compatible bornology and for every entourage U of X there exists a uniform finite upper bound on the cardinality of U-bounded subsets of X.

Remark 7.76 In the literature, the above notion of strongly bounded geometry is usually only given for discrete metric spaces and either just called "bounded geometry" (Nowak and Yu [NY12, Def. 1.2.6]) or "locally uniformly finite" (Yu [Yu95, p. 224]).   □

Note that the above notion is not invariant under equivalences. The invariant notion is as follows (cf. Roe [Roe96, Def. 2.3(ii)] or Block and Weinberger [BW92, Sec. 3]):

Definition 7.77 A bornological coarse space has bounded geometry if it is equivalent to a bornological coarse space of strongly bounded geometry.

Let K be a simplicial complex.

Definition 7.78 K has bounded geometry if there exist a uniform finite upper bound on the cardinality of the set of vertices of the stars at all vertices of K.

Note that this condition implies that K is finite-dimensional and that the number of simplices that meet at any vertex is uniformly bounded.

Let K be a simplicial complex. If K has bounded geometry (as a simplicial complex), then the associated bornological coarse space K_d has bounded geometry (as a coarse bornological space). In this case the set of vertices of K with induced structures is an equivalent bornological coarse space of strongly bounded geometry.

Remark 7.79 In general, bounded geometry of K_d does not imply bounded geometry of K as a simplicial complex. Consider, e.g., the disjoint union $K := \bigsqcup_{n \in \mathbb{N}} \Delta^n$, where we equip Δ^n with the metric induced from the standard embedding into S^n. The main point is that the diameters of Δ^n for n in $\mathbb{N}$ are uniformly bounded.   □

Let K be a simplicial complex.

Proposition 7.80 *If K has bounded geometry and is uniformly contractible, then we have a natural equivalence $Q(K_d) \simeq \Sigma^{\infty,\mathrm{lf}}_+ K_t$.*

Proof We consider the inclusion $K^{(0)} \to K$ of the zero skeleton and let $K_d^{(0)}$ be the coarse bornological space with the induced structures. If U is an entourage of $K_d^{(0)}$, then we consider the map

$$K^{(0)} \to P_U(K_d^{(0)}) \tag{7.15}$$

given by Dirac measures. Since the metric on K is good and K is finite-dimensional we can choose U so large that all simplices of K are U-bounded. Using convex interpolation we can extend the map (7.15) to a continuous map

$$f : K \to P_U(K_d^{(0)}) \, .$$

By construction,

$$f : K_d \to P_U(K_d^{(0)})_d$$

is an equivalence of bornological coarse spaces. Since K has bounded geometry, $K_d^{(0)}$ is of strongly bounded geometry, and hence $P_U(K_d^{(0)})$ is a finite-dimensional simplicial complex. Since K_d is uniformly contractible, by Lemma 7.72 we can choose a continuous inverse equivalence $g : P_U(K_d^{(0)})_d \to K_d$.

The following argument is taken from Nowak and Yu [NY12, Proof of Thm. 7.6.2]. Since the coarse structure of $K_d^{(0)}$ is induced by a metric we can choose a cofinal sequence $(U_n)_{n \in \mathbb{N}}$ of entourages of $K_d^{(0)}$. We let

$$f_n : P_{U_n}(K_d^{(0)})_d \hookrightarrow P_{U_{n+1}}(K_d^{(0)})_d$$

be the inclusion maps. Each f_n is an equivalence of bornological coarse spaces. For every n in $\mathbb{N}$ we can choose an inverse equivalence

$$g_n : P_{U_{n+1}}(K_d^{(0)})_d \to K_d$$

to the composition $f_n \circ \cdots \circ f_0 \circ f$ which is in addition continuous. Then the composition

$$g_n \circ (f_n \circ \cdots \circ f_0 \circ f) : K_t \to K_t$$

is homotopic to id_{K_t}. The following picture illustrates the situation.

Note that we are free to replace the sequence $(U_n)_{n \in \mathbb{N}}$ by a cofinal subsequence $(U_{n_k})_{k \in \mathbb{N}}$. We use this freedom in order to ensure that for every $n \in \mathbb{N}$ the map f_n is homotopic to the composition $(f_n \circ \cdots \circ f_0 \circ f) \circ g_{n-1}$.

Since the functor $\Sigma_+^{\infty,\mathrm{lf}}$ is homotopy invariant (see Example 7.40), it follows that the induced map

$$(f_n \circ \cdots \circ f_0 \circ f)_* : \pi_*(\Sigma_+^{\infty,\mathrm{lf}} K_t) \to \pi_*(\Sigma_+^{\infty,\mathrm{lf}} P_{U_{n+1}}(K_d^{(0)})_t)$$

is an isomorphisms onto the image of $(f_n)_*$. So the induced map

$$\Sigma_+^{\infty,\mathrm{lf}} K_t \to \underset{n \in \mathbb{N}}{\mathrm{colim}} \, \Sigma_+^{\infty,\mathrm{lf}} P_{U_n}(K_d^{(0)})_t$$

$$\simeq \underset{U}{\mathrm{colim}} \, \Sigma_+^{\infty,\mathrm{lf}} P_U(K_d^{(0)})_t \overset{\text{Definition 6.29}}{\simeq} Q(K_d^{(0)})$$

induces an isomorphism in homotopy groups. The desired equivalence is now given by

$$\Sigma_+^{\infty,\mathrm{lf}} K_t \simeq Q(K_d^{(0)}) \simeq Q(K_d),$$

where for the last equivalence we use that Q preserves equivalences between bornological coarse spaces and that $K_d^{(0)} \to K_d$ is an equivalence. □

Let $\mathbf{C}$ be a complete and cocomplete stable ∞-category. Let $E : \mathbf{TopBorn} \to \mathbf{C}$ be a locally finite homology theory, and let X be a bornological coarse space.

Proposition 7.81 *If X admits a coarsification (K, f) of bounded geometry, then we have a canonical equivalence $(QE)(X) \simeq E(K_t)$.*

Proof Since $f : X \to K_d$ is an equivalence of bornological coarse spaces, it induces an equivalence $(QE)(X) \simeq (QE)(K_d)$. An argument similar to the proof of Proposition 7.80 (replace $\Sigma_+^{\infty,\mathrm{lf}}$ by E) gives the equivalence $(QE)(K_d) \simeq E(K_t)$. □

Chapter 8
Coarse K-Homology

This final section of the book is devoted to the construction and investigation of coarse K-homology. This coarse homology theory and its applications to index theory, geometric group theory and topology are one of the driving forces of development in coarse geometry. A central result is the coarse Baum–Connes conjecture which asserts that under certain conditions on a bornological coarse space the assembly map from the coarsification of the locally finite version of K-theory to the K-theory of a certain Roe algebra associated to the bornological coarse space is an equivalence.

The classical construction of coarse K-homology works for proper metric spaces and produces K-theory groups. The construction involves the following steps:

1. For a space X one chooses an ample X-controlled Hilbert space.
2. One then defines a Roe subalgebra of the bounded operators on that Hilbert space.
3. Finally one defines the coarse K-homology groups of X as the topological K-theory groups of this Roe algebra.

All these notion will be discussed in detail in the subsequent subsections.

In order to fit coarse K-homology into the general framework developed in the present book we require a spectrum valued coarse K-theory functor which is defined on the category **BornCoarse**. The necessity to choose an ample Hilbert space causes two problems. First of all, as we shall see, ample Hilbert spaces do not always exist. Furthermore, even for proper metric spaces (where ample Hilbert spaces exist) it is not clear how to ensure functoriality on the spectrum level.

Our solution is to define for every X in **BornCoarse** a Roe C^*-category in a functorial way which comprises all possible choices of X-controlled Hilbert spaces. We then define the coarse K-homology spectrum as the topological K-theory spectrum of this C^*-category. This solves the functoriality problem. Furthermore, because the Roe category "absorps" all X-controlled Hilbert spaces we do not need the existence of one ample X-controlled Hilbert space which absorps all other.

U. Bunke, A. Engel, *Homotopy Theory with Bornological Coarse Spaces*,
Lecture Notes in Mathematics 2269, https://doi.org/10.1007/978-3-030-51335-1_8

We will define actually two slightly different versions of coarse K-homology functors. The construction and the verification that these functors satisfy the axioms of a coarse homology theory will be completed in Sect. 8.6. In a subsequent paper [BEe] we vastly generalize the constructions of the present section. We consider the equivariant case and allow coefficients in a C^*-category with group action. These equivariant coarse K-homology theories are then used to obtain injectivity results for assembly maps.

In Sects. 8.1 till 8.5 we discuss the technical preliminaries needed for the definition of coarse K-homology.

In Sect. 8.7 we show that in essentially all cases of interest our new definition of coarse K-homology coincides with the classical one roughly described above.

Since we define different versions of coarse K-homology, the question under which circumstances they coincide immediately arises. A partial answer to this question is given by an application of the comparison result stated as Corollary 6.43.

In Sect. 8.9 we will discuss some simple applications of our setup of coarse K-homology to index theory, and in Sect. 8.10 we construct the assembly map as a natural transformation between the coarsification of locally finite K-homology and coarse K-homology. The comparison theorem then immediately proves the coarse Baum–Connes conjecture for spaces of finite asymptotic dimension—a result first achieved by Yu [Yu98].

8.1 X-Controlled Hilbert Spaces

Let X be a bornological coarse space. We can consider X as a discrete topological space. We let $C(X)$ be the C^*-algebra of all bounded, continuous (a void condition) $\mathbb{C}$-valued functions on X with the supremum-norm and involution given by complex conjugation. For a subset Y of X we denote by χ_Y in $C(X)$ the characteristic function of Y.

Definition 8.1 An X-controlled Hilbert space is a pair (H, ϕ), where H is a Hilbert space and $\phi : C(X) \rightarrow B(H)$ is a unital $*$-representation such that for every bounded subset B of X the subspace $\phi(\chi_B)H$ is separable.

For a subset Y of X we introduce the notation

$$H(Y) := \phi(\chi_Y)H \tag{8.1}$$

for the subspace of vectors of H "supported on Y". The operator $\phi(\chi_Y)$ is the orthogonal projection from H onto $H(Y)$. Since we require ϕ to be unital, we have $H(X) = H$. We will also often use the notation $\phi(Y) := \phi(\chi_Y)$. Note that we do not restrict the size of the Hilbert space H globally. But we require that $H(B)$ is separable for every bounded subset B of X, i.e., H is locally separable on X. We will also need the following more restrictive local finiteness condition.

Let X be a bornological coarse space, and let (H, ϕ) be an X-controlled Hilbert space.

Definition 8.2 (H, ϕ) is called locally finite if for every bounded subset B of X the space $H(B)$ is finite-dimensional.

Let X be a bornological coarse space, and let (H, ϕ) be an X-controlled Hilbert space.

Definition 8.3 (H, ϕ) is determined on points if for every vector h of H the condition $\phi(\{x\})h = 0$ for all points x of X implies $h = 0$.

Let (H, ϕ) be an X-controlled Hilbert space. Note that the direct sums below are taken in the sense of Hilbert spaces and in general involve a completion.

Lemma 8.4 *The following are equivalent:*

1. *(H, ϕ) is determined on points,*
2. *The family $(H(\{x\}) \to H)_{x \in X}$ of canonical inclusions induces an isomorphism $\bigoplus_{x \in X} H(\{x\}) \cong H$*
3. *For every partition $(Y_i)_{i \in I}$ of X the family of natural inclusions $H(Y_i) \to H$ induces an isomorphism $\bigoplus_{i \in I} H(Y_i) \cong H$.*

Proof It is clear that Assertion 3 implies Assertion 2, and that Assertion 2 implies Assertion 1.

Assume now Assertion 1 and let $(Y_i)_{i \in I}$ be a pairwise disjoint partition of X. It is clear that the canonical map $\bigoplus_{i \in I} H(Y_i) \to H$ is injective. We must show that its orthogonal complement is trivial. Assume that h in H is orthogonal to the image. If x is in X, then we choose i in I such that $x \in Y_i$. Since $H(\{x\}) \subseteq H(Y_i)$ we conclude that $\phi(\{x\})h = 0$. Since this holds for every x in X we conclude that $h = 0$. $\qquad\square$

Example 8.5 We consider the bornological coarse space $\mathbb{N}_{min,max}$ given by the set $\mathbb{N}$ with the minimal coarse and the maximal bornological structures, see Example 2.13. We choose a point in the boundary of the Stone–Čech compactification of the discrete space $\mathbb{N}$. The evaluation at this point is a character $\phi : C(\mathbb{N}) \to \mathbb{C} = B(\mathbb{C})$ which annihilates all finitely supported functions. The X-controlled Hilbert space $(\mathbb{C}, \phi)$ is not determined on points. $\qquad\square$

In the following we discuss a construction of X-controlled Hilbert spaces.

Example 8.6 Let X be a bornological coarse space. Let D be a locally countable subset of X (see Definition 2.3). We define the counting measure μ_D on the σ-algebra $\mathcal{P}(X)$ of all subsets of X by

$$\mu_D(Y) := |(Y \cap D)|.$$

We set $H_D := L^2(X, \mu_D)$ and define the representation ϕ_D of $C(X)$ on H_D such that f in $C(X)$ acts as a multiplication operator. Then (H_D, ϕ_D) is an X-controlled Hilbert space which is determined on points.

If D is locally finite (see Definition 2.3), then the Hilbert space (H_D, ϕ_D) is locally finite.

As usual we set $\ell^2 := L^2(\mathbb{N}, \mu_{\mathbb{N}})$. Then $(H_D \otimes \ell^2, \phi_D \otimes \mathrm{id}_{\ell^2})$ is also an X-controlled Hilbert space. If $D \neq \emptyset$, then it is not locally finite. $\qquad\square$

Remark 8.7 Observe that the notion of an X-controlled Hilbert space and the properties of being determined on points or locally finite only depend on the bornology of X. $\qquad\square$

Let X be a bornological coarse space. We consider two X-controlled Hilbert spaces (H, ϕ), (H', ϕ') and a bounded linear operator $A \colon H \to H'$. Let $\mathcal{P}_X$ denote the power set of X.

Definition 8.8

1. The support of A is the subset of $X \times X$ given by

$$\mathrm{supp}(A) := \bigcap_{\{U \subseteq X \times X : (\forall Y \in \mathcal{P}_X : A(H(Y)) \subseteq H(U[Y]))\}} U \,.$$

2. A has controlled propagation if $\mathrm{supp}(A)$ is a coarse entourage of X.

The idea of the support of A is that if a vector h in H is supported on Y, then Ah is supported on $\mathrm{supp}(A)[Y]$. So if A has controlled propagation, it moves supports in a controlled manner.

Let $f \colon X \to X'$ be a morphism between bornological coarse spaces, and let (H, ϕ) be an X-controlled Hilbert space.

Definition 8.9 We define the X'-controlled Hilbert space $f_*(H, \phi)$ to be the Hilbert space H with the control $\phi \circ f^*$.

Indeed, $(H, \phi \circ f^*)$ is an X'-controlled Hilbert space: if B' is a bounded subset of X', then $f^{-1}(B')$ is bounded in X since f is proper. Hence $(\phi \circ f^*)(\chi_{B'})H = \phi(\chi_{f^{-1}(B')})H$ is separable. If (H, ϕ) is locally finite, then so is $f_*(H, \phi)$. Similarly, if (H, ϕ) is determined on points, then so is $f_*(H, \phi)$.

In the following we list some obvious properties of the support of operators between controlled Hilbert spaces The support of operators has the following properties which are easy to check:

1. $\mathrm{supp}(A^*) = \mathrm{supp}(A)^{-1}$.
2. If A, A' are two operators between the same X-controlled Hilbert spaces, and λ is in $\mathbb{C}$, then

$$\mathrm{supp}(A + \lambda A') \subseteq \mathrm{supp}(A) \cup \mathrm{supp}(A) \,.$$

3. If (H, ϕ), (H', ϕ') and (H'', ϕ'') are X-controlled Hilbert spaces and $A \colon H \to H'$ and $B \colon H' \to H''$ are bounded operators, then

$$\mathrm{supp}(B \circ A) \subseteq \mathrm{supp}(B) \circ \mathrm{supp}(A) \,.$$

4. Let $f : X \to X'$ be a morphism in **BornCoarse**. If (H, ϕ) and (H', ϕ') are two X-controlled Hilbert spaces and $A : H \to H'$ is a bounded operator, then we let $f_* A := A$ be considered as an operator between X'-controlled Hilbert spaces. Then we have

$$\mathrm{supp}(f_*) \subseteq f_*(\mathrm{supp}(A)).$$

Let (H, ϕ) be an X-controlled Hilbert space, and let H' a closed subspace of H.

Definition 8.10 H' is locally finite, if it admits an X-control ϕ' such that:

1. (H', ϕ') is locally finite (Definition 8.2).
2. (H', ϕ') is determined on points.
3. The inclusion $H' \to H$ has controlled propagation (Definition 8.8).

If we want to highlight ϕ', we say that it recognizes H' as a locally finite subspace of H.

Example 8.11 We consider $\mathbb{Z}$ as a bornological coarse space with the metric structures. Furthermore, we consider the X-controlled Hilbert space $H := L^2(\mathbb{Z})$ with respect to the counting measure. The control ϕ is given by multiplication operators. The $\mathbb{Z}$-controlled Hilbert space (H, ϕ) is locally finite. For n in $\mathbb{Z}$ we let δ_n in $L^2(\mathbb{Z})$ denote the corresponding basis vector of H.

The one-dimensional subspace H' generated by the vector $\sum_{n \in \mathbb{N}} e^{-|n|} \delta_n$ is not locally finite.

On the other hand, the ∞-dimensional subspace H'' generated by the family $(\delta_{2n})_{n \in \mathbb{Z}}$ is locally finite. $\qquad\square$

Example 8.12 We consider a bornological coarse space X and a locally countable subset D of X. Then we can form the X-controlled Hilbert space $(H_D \otimes \ell^2, \phi_D \otimes \mathrm{id}_{\ell^2})$ as in Example 8.6. Let F be some finite-dimensional subspace of ℓ^2, and let D' be a locally finite subset of D. Then $H' := H_{D'} \otimes F$ considered as a subspace of $H_D \otimes \ell^2$ in the natural way is a locally finite subspace. As control we can take the restriction of $\phi_D \otimes \mathrm{id}_{\ell^2}$. $\qquad\square$

8.2 Ample X-Controlled Hilbert Spaces

Let (H, ϕ) be an X-controlled Hilbert space.

Definition 8.13 (H, ϕ) is ample if it is determined on points and there exists an entourage U of X such that $H(U[x])$ is ∞-dimensional for every x in X.

Note that an ample X-controlled Hilbert space is not locally finite.

Remark 8.14 Our notion of ampleness differs from the usual notion where one also demands that no non-zero function in $C(X)$ acts by compact operators.

Consider for example the bornological coarse space $\mathbb{Q}$ with the structures induced from the embedding into the metric space $\mathbb{R}$. The $\mathbb{Q}$-controlled Hilbert space $(H_{\mathbb{Q}}, \phi_{\mathbb{Q}})$ (see Example 8.6) is ample. But the characteristic functions of points (they are continuous since we consider $\mathbb{Q}$ with the discrete topology) act by rank-one operators and are in particular compact.   □

In the following we collect some facts about ample X-controlled Hilbert spaces. Let X be a bornological coarse space, D be a subset of X, and U be an entourage of X.

Definition 8.15

1. D is dense in X, if there exists an entourage V of X such that $V[D] = X$. Sometimes we say that D is V-dense.
2. D is U-separated if for every two distinct points d, d' in D we have $d' \notin U[d]$.
3. X is locally countable (or locally finite, respectively) if it admits an entourage V such that every V-separated subset of X is locally countable (or locally finite, respectively).

Remark 8.16 One should not confuse the above notion of local finiteness of X as a bornological coarse space with the notion of local finiteness of X (as a subset of itself) as given in Definition 2.3. But if X is locally finite in the latter sense, then it is locally finite in the sense of Definition 8.15.3.   □

Example 8.17 Let D be a dense and locally countable subset of X (see Definition 2.3). Then the X-controlled Hilbert space $(H_D \otimes \ell^2, \phi_D \otimes \mathrm{id}_{\ell^2})$ constructed in Example 8.6 is ample.   □

Example 8.18 Let $f : X \to X'$ be a morphism between bornological coarse spaces, and (H, ϕ) be an ample X-controlled Hilbert space. Then $f_*(H, \phi)$ is ample if and only if $f(X)$ is dense in X'.   □

Let X be a bornological coarse space.

Lemma 8.19 *For every symmetric entourage U of X there exists a U-separated and U-dense subset of X.*

Proof The U-separated subsets of X are partially ordered by inclusion. We can use the Lemma of Zorn to get a maximal U-separated subset D of X. Due to maximality of D and since U is symmetric we can conclude that $U[D] = X$.   □

Let X be a bornological coarse space.

Proposition 8.20 *X admits an ample X-controlled Hilbert space if and only if X is locally countable.*

Proof Assume that X is locally countable. Then we can find an entourage U of X such that every U-separated subset of X is locally countable. After enlarging U if necessary we can in addition assume that U is symmetric.

Using Lemma 8.19 we get a U-separated subset D with $U[D] = X$. By our assumption on U the subset D is locally countable. We then construct the ample X-controlled Hilbert space as in Example 8.17.

We now show the converse. Let (H, ϕ) be an ample X-controlled Hilbert space and let U be an entourage such that $H(U[x])$ is ∞-dimensional for all x in X. Consider a $U \circ U^{-1}$-separated subset Y. We must show that $Y \cap B$ is at most countable for every bounded subset B of X.

We have an inclusion

$$H\left(\bigcup_{y \in Y \cap B} U[y] \right) \subseteq H(U[B]) .$$

Since $U[B]$ is bounded we know that $H(U[B])$ is separable. Hence $H(\bigcup_{y \in Y \cap B} U[y])$ is separable, too. On the other hand $U[y] \cap U[y'] = \emptyset$ for all distinct pairs y, y' in Y. Hence

$$H\left(\bigcup_{y \in Y \cap B} U[y] \right) \cong \bigoplus_{y \in Y \cap B} H(U[y]) .$$

Since this is a separable Hilbert space and all $H(U[y])$ are non-zero, the index set $Y \cap B$ of the sum is at most countable.   □

Let X be a bornological coarse space and (H, ϕ) and (H', ϕ') be two X-controlled Hilbert spaces.

Lemma 8.21 *If (H, ϕ) and (H', ϕ') are ample, then there exists a unitary operator $H \to H'$ with controlled propagation.*

Proof The bornological coarse space X is locally countable by Lemma 8.20. Let U be a symmetric entourage such that every U-separated subset is locally countable. After enlarging U if necessary we can also assume that the Hilbert spaces $H(U[x])$ and $H'(U[x])$ are ∞-dimensional for every x in X. Choose a maximal U^2-separated subset D of X as in Lemma 8.19. Consider the ample X-controlled Hilbert space $(H_D \otimes \ell^2, \phi_D \otimes \mathrm{id}_{\ell^2})$ from Example 8.17.

We can construct unitary operators $u_H : H \to H_D \otimes \ell^2$ and $u_{H'} : H' \to H_D \otimes \ell^2$ of controlled propagation. Then $u_{H'}^* u_H : H \to H'$ is the desired unitary operator.

In order to construct, e.g., u_H we proceed as follows. Since D is U^2-separated, $(U[d])_{d \in D}$ is a pairwise disjoint family of subsets of X. Because D is maximal the family of subsets, $(U^2[d])_{d \in D}$ covers X. We choose a well-ordering on D and using transfinite induction we define a partition $(B_d)_{d \in D}$ of X such that

$$U[d] \subseteq B(d) \subseteq U^2[d]$$

for all d in D. To this end we set

$$B_{d_0} := U^2[d_0] \setminus \bigcup_{d \in D, d_0 < d} U[d]$$

for the smallest element d_0 of D. Let now d in D and assume that $B_{d'}$ has been constructed for all d' in D with $d' < d$. Then we set

$$B_d := U^2[d] \setminus \left(\bigcup_{d' \in D, d' < d} B_{d'} \cup \bigcup_{d'' \in D, d < d''} U[d''] \right).$$

Because $U[d] \subseteq B_d$ and B_d is bounded, the Hilbert space $H(B_d)$ is both separable and ∞-dimensional. We now construct a unitary $u_H : H \to H_D \otimes \ell^2$ by choosing a unitary $H(B_d) \to (H_D \otimes \ell^2)(\{d\})$ for every d in D. The propagation of u_H is controlled by U^2.   □

Next we show that an ample X-controlled Hilbert space absorbs every other X-controlled Hilbert space. In view of later applications we will show a slightly stronger result.

Let X be a bornological coarse space and (H, ϕ), (H'', ϕ'') be X-controlled Hilbert spaces. Let $H' \subseteq H$ be a closed subspace.

Lemma 8.22 *Assume:*

1. *(H, ϕ) is ample.*
2. *H' is locally finite.*
3. *(H'', ϕ'') is determined on points.*

Then there exists an isometric inclusion $H'' \to H \ominus H'$ of controlled propagation.

(The notation $H \ominus H'$ stands for the orthogonal complement of H' in H.)

Proof Below we will construct an ample X-controlled Hilbert space (H_1, ϕ_1) and an isometric inclusion of controlled propagation $H_1 \to H \ominus H'$. By the arguments in the proof of Lemma 8.21 we can find an isometric inclusion $H'' \to H_1$ of controlled propagation. Then the composition $H'' \to H_1 \to H \ominus H'$ is an isometric inclusion of controlled propagation.

If view of the proof of Lemma 8.21 we can assume that (H, ϕ) is of the form $(H_D \otimes \ell^2, \phi_D \otimes \ell^2)$ for some locally countable U-dense subset D, where U is some entourage of X. Since H' is locally finite, for every d in D the subspace

$$H_1(\{d\}) := H_D\{d\} \otimes \ell^2 \cap H'^{,\perp}$$

is ∞-dimensional. We define the closed subspace

$$H_1 := \bigoplus_{d \in D} H_1(\{d\}) \subseteq H \ominus H'$$

with the control $\phi_1 := (\phi_D \otimes \ell^2)_{|H_1}$. The X-controlled Hilbert space (H_1, ϕ_1) is ample and the propagation of the inclusion $H_1 \to H$ is controlled by diag_X. □

Example 8.23 Let M be a Riemannian manifold, and let E be a hermitian vector bundle on M. Then we can consider the Hilbert space $H := L^2(M, E)$ of square integrable measurable sections of E, where M is equipped with the Riemannian volume measure. This measure is defined on the Borel σ-algebra of M or some completion of it, but not on $\mathcal{P}(M)$ except if M is zero-dimensional.

We consider M as a bornological coarse space with the bornological coarse structure induced by the Riemannian distance. By our convention $C(M)$ is the C^*-algebra of all bounded $\mathbb{C}$-valued functions on M and therefore contains non-measurable functions. Therefore it can not act simply as multiplication operators on H. To get around this problem we can apply the following general procedure.

We call a measure space $(Y, \mathcal{R}, \mu)$ separable if the metric space $(\mathcal{R}, d)$ with the metric $d(A, B) := \mu(A \triangle B)$ is separable. Note that $L^2(Y, \mu)$ is a separable Hilbert space if $(Y, \mathcal{R}, \mu)$ is a separable measure space.

Let $(X, \mathcal{C}, \mathcal{B})$ be a bornological coarse space such that the underlying set X is also equipped with the structure of a measure space $(X, \mathcal{R}, \mu)$. We want that the Hilbert space $H := L^2(X, \mu)$ becomes X-controlled by equipping it with a representation of $C(X)$ which is derived from the representation by multiplication operators.

Assume the following:

1. $\mathcal{B} \cap \mathcal{R}$ is cofinal in $\mathcal{B}$.
2. A subset Y of X is measurable if the intersection $Y \cap B$ is measurable for every B in $\mathcal{B} \cap \mathcal{R}$.
3. For every measurable, bounded subset B of X the measure space $(B, \mathcal{R} \cap B, \mu_{|\mathcal{R} \cap B})$ is separable.
4. There exists an entourage U in $\mathcal{C}$ and a partition $(D_\alpha)_{\alpha \in A}$ of X into non-empty, pairwise disjoint, measurable, U-bounded subsets of X with the property that the set $\{\alpha \in A : D_\alpha \cap B \neq \emptyset\}$ is countable for every bounded subset B of X.

For every α in A we choose a point d_α in D_α.

We consider the Hilbert space

$$H := L^2(X, \mu).$$

We furthermore define a map

$$\phi : C(X) \to B(H), \qquad f \mapsto \phi(f) := \sum_{\alpha \in A} f(d_\alpha) \chi_{D_\alpha}$$

where we understand the right-hand side as a multiplication operator by a function which we will also denote by $\phi(f)$ in the following. First note that $\phi(f)$ is a measurable function. Indeed, for every B in $\mathcal{B} \cap \mathcal{R}$ the restriction $\phi(f)|_B$ is a countable sum of measurable functions (due to Assumption 4) and hence measurable. By Assumption 2 we then conclude that $\phi(f)$ is measurable.

Since $(D_\alpha)_{\alpha \in A}$ is a partition we see that ϕ is a homomorphism of C^*-algebras. If B is a bounded subset of X, then

$$H(B) = \phi(\chi_B)H \subseteq \chi_{U[B]}H \subseteq \chi_V H.$$

where V is some measurable subset of X which contains $U[B]$ and is bounded. Such a V exists due to Assumption 1 and the compatibility of $\mathcal{B}$ and $\mathcal{C}$. Because of Assumption 3 we know that $\chi_V H$ is separable, and therefore $H(B)$ is separable. So $L^2(X, \mu)$ is an X-controlled Hilbert space. By construction it is determined on points.

We continue the example of our Riemannian manifold. We show that we can apply the construction above in order to turn $H = L^2(M, E)$ into an M-controlled Hilbert space. The measure space $(M, \mathcal{R}^{Borel}, \mu)$ with the Borel σ-algebra $\mathcal{R}^{Borel}$ and the Riemannian volume measure μ clearly satisfies the Assumptions 1, 2 and 3.

We discuss Assumption 4. We consider the entourage U_1 in $\mathcal{C}$ (see (2.1)). By Lemma 8.19 we can find a countable U_1-separated (Definition 8.15) subset A of M such that $U_1[A] = M$. We fix a bijection of A with $\mathbb{N}$ which induces an ordering of A. Using induction we can construct a partition $(D_\alpha)_{\alpha \in A}$ of M such that $D_\alpha \subseteq U_1[\alpha]$ for every α in A. If α is the smallest element of A, then we set $D_\alpha := U_1[\alpha]$. Assume now that α is in A and D_β has been defined already for every β in A with $\beta < \alpha$. Then we set

$$D_\alpha := U_1[\alpha] \setminus \bigcup_{\{\beta \in A : \beta < \alpha\}} D_\beta.$$

This partition satisfies Assumption 4 for the entourage U_1. It consists of Borel measurable subsets since $U_1[\alpha] \subseteq M$ is open for every α in A and hence measurable, and the set D_α is obtained from the collection of $U_1[\beta]$ for β in A using countably many set-theoretic operations. Since A was U_1-separated we have $\alpha \in D_\alpha$ for every α in A and therefore D_α is non-empty. $\square$

8.3 Roe Algebras

Let X be a bornological coarse space, and let (H, ϕ) be an X-controlled Hilbert space. In the present section we associate to this data various C^*-algebras, all of which are versions of the Roe algebra. They differ by the way the conditions of controlled propagation and local finiteness are implemented.

We start with the local finiteness conditions. Recall Definition 8.10 for the notion of a locally finite subspace.

We consider two X-controlled Hilbert spaces $(H, \phi), (H', \phi')$ and a bounded linear operator $A: (H, \phi) \to (H', \phi')$.

Definition 8.24 A is locally finite if there exist orthogonal decompositions $H = H_0 \oplus H_0^{\perp}$ and $H' = H_0' \oplus H_0'^{\perp}$ and a bounded operator $A_0 : H_0 \to H_0'$ such that:

1. H_0 and H_0' are locally finite.
2.

$$A = \begin{pmatrix} A_0 & 0 \\ 0 & 0 \end{pmatrix}.$$

Definition 8.25 A is locally compact if for every bounded subset B of X the products $\phi'(B)A$ and $A\phi(B)$ are compact.

It is clear that a locally finite operator is locally compact.

Remark 8.26 One could think of defining local finiteness of A similarly as local compactness by requiring that $\phi'(B)A$ and $A\phi(B)$ are finite-dimensional for every bounded subset of X. It is not clear that this definition is equivalent to Definition 8.24. Our motivation for using Definition 8.24 is that it works in the Comparison Theorem 8.88.   □

We now introduce propagation conditions.

Let X be a bornological coarse space. We consider two subsets B and B' and an entourage U of X.

Definition 8.27 B' is U-separated from B

$$U[B] \cap B' = \emptyset.$$

Let (H, ϕ) and (H', ϕ') be two X-controlled Hilbert spaces and $A : H \to H'$ be a bounded linear operator, and let U be an entourage of X.

Definition 8.28

1. A is U-quasi-local if for every ε in $(0, \infty)$ there exists an integer n in $\mathbb{N}$ such that for any two mutually U^n-separated subsets B and B' of X we have $\|\phi'(B')A\phi(B)\| \leq \varepsilon$.
2. A is quasi-local if it is U-quasi-local for some entourage U of X.

It is clear that an operator with controlled propagation quasi-local. Indeed, if A is U-controlled, the $\phi'(B')A\phi(B) = 0$ provided B' is U-separated from B.

Another possible definition of the notion of quasi-locality is the following. We add the word "weakly" in order to distinguish it from Definition 8.28

Definition 8.29 A is weakly quasi-local if for every ε in $(0, \infty)$ there exists an entourage U of X such that for any two mutually U-separated subsets B and B' of X we have $\|\phi'(B')A\phi(B)\| \leq \varepsilon$.

It is obvious that quasi-locality implies weak quasi-locality. The following is also clear.

Corollary 8.30 *If the coarse structure of X is generated by a single entourage, then the notions of weak quasi-locality and quasi-locality over X coincide.*

Remark 8.31 In general, being weakly quasi-local is strictly weaker than being quasi-local as the following example shows.

We equip the set $X := \mathbb{Z} \times \{0, 1\}$ with the minimal bornology and the coarse structure $\mathcal{C}\langle(U_n)_{n\in\mathbb{N}}\rangle$, where for n in $\mathbb{Z}$ the entourage U_n is given by

$$U_n := \mathrm{diag}_X \cup \{((n, 0), (n, 1)), ((n, 1), (n, 0))\}.$$

Let (H, ϕ) be the X-controlled Hilbert space as in Example 8.6 with $D = X$. We further let A in $B(H)$ be the operator which sends $\delta_{(n,0)}$ to $e^{-|n|}\delta_{(n,1)}$ and is zero otherwise.

We claim that A is weakly quasi-local, but not quasi-local. Every entourage U of X is contained in a finite union of generators. Hence there exists n in $\mathbb{Z}$ such that

$$U \cap (\{n\} \times \{0, 1\})^2 = \{(n, n)\} \times \mathrm{diag}_{\{0,1\}}.$$

The points $(n, 0)$ and $(n, 1)$ are U^k-separated for all k in $\mathbb{N}$. But

$$\|\phi(\{(n, 1)\})A\phi(\{(n, 0)\})\| = e^{-|n|}$$

independently of k. So the operator A is not quasi-local in the sense of Definition 8.28.

On the other hand, given ε in $(0, \infty)$, we can choose the entourage

$$U := \bigcup_{n\in\mathbb{Z}, e^{|-n|}\geq\varepsilon} U_n$$

of X. If B and B' are mutually U-separated subsets, then one easily checks that

$$\|\phi'(B')A\phi(B)\| \leq \varepsilon.$$

This shows that A is weakly quasi-local.   □

Let (H, ϕ) and (H', ϕ') be two X-controlled Hilbert spaces, and U be an entourage of X.

Lemma 8.32

1. *The space of locally compact operators is a closed subspace of $B(H, H')$.*
2. *The space of U-quasi-local operators is a closed subspace of $B(H, H')$.*
3. *The space of weakly quasi-local operators is a closed subspace of $B(H, H')$.*

Proof Assertion 1 is clear since the compact operators on H or H' are closed in the corresponding spaces of bounded operators.

We now show Assertion 2. Assume that $A : H \to H'$ is in the closure of the U-quasi-local operators. We must show that A itself is U-quasi-local. We fix ε in $(0, \infty)$. Then there exists a U-quasi-local operator $A' : H \to H'$ such that $\|A - A'\| \leq \varepsilon/2$. We can now choose n in $\mathbb{N}$ sufficiently large X such that for all pairs B, B' of mutually U^n-separated subsets of X we have $\|\phi(B')A'\phi(B)\| \leq \varepsilon/2$. Then we have

$$\|\phi'(B')A\phi(B)\| \leq \|\phi'(B')(A - A')\phi(B)\| + \|\phi(B')A'\phi(B)\| \leq \varepsilon.$$

For Assertion 3 we argue similarly. Assume that $A : H \to H'$ is in the closure of the weakly quasi-local operators. We must show that A itself is weakly quasi-local. We fix ε in $(0, \infty)$. Then there exists a quasi-local operator $A' : H \to H'$ such that $\|A - A'\| \leq \varepsilon/2$. We can now choose an entourage U of X such that for all pairs B, B' of mutually U-separated subsets of X we have $\|\phi(B')A'\phi(B)\| \leq \varepsilon/2$. Then we have

$$\|\phi'(B')A\phi(B)\| \leq \|\phi'(B')(A - A')\phi(B)\| + \|\phi(B')A'\phi(B)\| \leq \varepsilon. \qquad \square$$

Our motivation to use the notion of quasi-locality in the sense of Definition 8.28 in contrast to the weak quasi-locality as in Definition 8.29 is that it ensures the u-continuity of the coarse K-theory functor $K\mathcal{X}_{ql}$, see Proposition 8.86.

Remark 8.33 Note that by Lemma 8.32 the weakly quasi-local operators form a closed subspace of all operators. In contrast, the space of quasi-local operators defined as in Definition 8.28 is not closed in general. But by Corollary 8.30 it is closed if the coarse structure is generated by a single entourage. $\qquad \square$

Let X be a bornological coarse space and (H, ϕ) an X-controlled Hilbert space. In the following we define four versions of Roe algebras. They are all closed C^*-subalgebras of $B(H)$ which are generated by operators with appropriate local finiteness and propagation conditions.

Definition 8.34

1. $C^*(X, H, \phi)$ is generated by the operators which have controlled propagation and are locally finite.
2. $C^*_{lc}(X, H, \phi)$ is generated by the operators which have controlled propagation and are locally compact.
3. $C^*_{ql}(X, H, \phi)$ is generated by the operators which are quasi-local and locally finite.
4. $C^*_{lc,ql}(X, H, \phi)$ is generated by the operators which are quasi-local and locally compact.

Remark 8.35 If (H, ϕ) is determined on points and if the coarse structure of X is generated by a single entourage, then in view of Lemmas 8.32 and 8.75 we can replace the phrase "generated by the" by "is the algebra of" in the case of $C^*_{lc,ql}(X, H, \phi)$. $\qquad \square$

We have the following obvious inclusions of C^*-algebras:

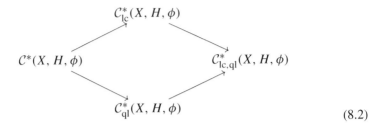

$$(8.2)$$

Let (H, ϕ) and (H', ϕ') be two X-controlled Hilbert spaces, and let $u \colon H \to H'$ be an isometry of controlled propagation. Then we get induced injective homomorphisms

$$\mathcal{C}_?^*(X, H, \phi) \to \mathcal{C}_?^*(X, H', \phi') \qquad (8.3)$$

given in all four cases by the formula $A \mapsto uAu^*$. For justification in the case $? \in \{\emptyset, \mathrm{ql}\}$ note that if H'' is a locally finite subspace of H, then $u(H'')$ is a locally finite subspace of H'.

Remark 8.36 Classically the Roe algebra is defined as the closure of the locally compact, controlled propagation operators. The reason why we use the locally finite version as our standard version of Roe algebra (it does not come with any subscript in its notation) is because the functorial definition of coarse K-homology in Sect. 8.6 is modeled on the locally finite version and we want the Comparison Theorem 8.88.

Proposition 8.40 shows that in almost all cases of interest (Riemannian manifolds and countable discrete groups endowed with a left-invariant, proper metric) the locally finite version of the Roe algebra coincides with its locally compact version.

But note that we do not know whether the analogous statement of Proposition 8.40 for the quasi-local case is true, i.e., under which non-trivial conditions on X the equality

$$\mathcal{C}_{\mathrm{ql}}^*(X, H, \phi) \overset{?}{=} \mathcal{C}_{\mathrm{lc,ql}}^*(X, H, \overset{*}{\phi})$$

holds true.   □

Remark 8.37 It is an interesting question under which non-trivial conditions on X we have equalities

$$C^*(X, H, \phi) = \mathcal{C}_{\mathrm{ql}}^*(X, H, \phi), \quad \mathcal{C}_{\mathrm{lc}}^*(X, H, \phi) = \mathcal{C}_{\mathrm{lc,ql}}^*(X, H, \phi).$$

As far as the authors know the only non-trivial space (i.e., not a bounded one) for which this was known is $\mathbb{R}^n$ [LR85, Prop. 1.1]. The conjecture was that these equalities hold true for any space X of finite asymptotic dimension; see Roe [Roe96,

Rem. on Page 20] where this is claimed (but no proof is given). Meanwhile,[1] Špakula and Tikuisis [ŠT19] have shown the second equality under the even weaker condition that X has finite straight decomposition complexity. □

Let X be a bornological coarse space, and let (H, ϕ) be an X-controlled Hilbert space. If (H, ϕ) is locally finite (Definition 8.2), then it is obvious that

$$C^*(X, H, \phi) = C^*_{\mathrm{lc}}(X, H, \phi).$$

The following proposition gives a generalization of this equality to cases where (H, ϕ) is not necessarily locally finite.

Let X be a bornological coarse space.

Definition 8.38 X is separable if it admits an entourage U for which there exists a countable U-dense (Definition 8.15) subset.

Example 8.39 If (X, d) is a separable metric space, then the bornological coarse space $(X, \mathcal{C}_d, \mathcal{B})$ is separable, where $\mathcal{C}_d$ is the coarse structure (2.2) induced by the metric and $\mathcal{B}$ is any compatible bornology. □

Let X be a bornological coarse space, and let (H, ϕ) be an X-controlled Hilbert space.

Proposition 8.40 *Assume:*

1. *X is separable and locally finite (Definition 8.15).*
2. *(H, ϕ) is determined on points (Definition 8.3).*

Then $C^*(X, H, \phi) = C^*_{\mathrm{lc}}(X, H, \phi)$.

Proof Since X is separable we can choose an entourage U' and a countable subset D' of X such that D' is U'-dense. Since X is locally finite we can enlarge U' and in addition assume that every U'-separated subset of X is locally finite. Let D be a maximal U'-separated subset of D' and set $U := U'^{,2}$. Then D is U-dense and locally finite. For simplicity of notation we only consider the case that D is infinite. Let $\mathbb{N} \ni n \mapsto d_n \in D$ be a bijection. We define inductively $B_0 := U[d_0]$ and

$$B_n := U[d_n] \setminus \bigcup_{m < n} B_m.$$

Then $(B_n)_{n \in \mathbb{N}}$ is a partition of X into U-bounded subsets. After deleting empty members and renumbering we can assume that B_n is not empty for every n in $\mathbb{N}$.

---

[1] After the appearance of the first version of the present paper.

We consider an operator A in $C^*_{\mathrm{lc}}(X, H, \phi)$. We will show that for every given ε in $(0, \infty)$ there exists a locally finite subspace H' of H and an operator $A' : H' \to H'$ such that

1. $\|A - A'\| \leq \varepsilon$ (we implicitly extend A' by zero on $H'^{,\perp}$).
2. A' has controlled propagation.

Note that we verify a condition that is stronger than Definition 8.24 since we take the same subspace as domain and target of A'.

As a first step we choose a locally compact operator A_1 of controlled propagation such that $\|A - A_1\| \leq \varepsilon/2$. For every n in $\mathbb{N}$ we define the subset

$$B'_n := \mathrm{supp}(A_1)[B_n] \cup B_n .$$

We then have $A_1 B_n = B'_n A_1 B_n$ for every n in $\mathbb{N}$, where for a subset B of X we abbreviate the operator $\phi(B)$ by B. Since A_1 is locally compact we can now find for every n in $\mathbb{N}$ a finite-dimensional projection P_n on $H(B'_n)$ such that

$$\|P_n B'_n A_1 B_n P_n - A_1 B_n\| \leq \varepsilon \cdot 2^{-n-2} .$$

Then we define H' to be the subspace of H generated by the images of P_n. Let H_n be the subspace of H' generated by the images of $P_1, \ldots, P_n$. We define Q_n to be the orthogonal projection onto the subspace $H_n \ominus H_{n-1}$. For every n in $\mathbb{N}$ we choose a point b_n in B_n. Then we define the control on H' by $\phi'(f) := \sum_{n \in \mathbb{N}} f(b_n) Q_n$. With these choices the inclusion $(H', \phi') \to (H, \phi)$ has controlled propagation. Moreover, (H', ϕ') is a locally finite X-controlled Hilbert space. Indeed, if B is a bounded subset of X, then $U^{-1}[B] \cap D$ is finite since D is locally finite. Hence the set $\{n \in \mathbb{N} \mid B \cap B_n \neq \emptyset\}$ is finite. Similarly, $\{n \in \mathbb{N} \mid B \cap B'_n \neq \emptyset\}$ is finite. This shows that H' is a locally finite subspace of H.

Furthermore,

$$\left\| \sum_{n \in \mathbb{N}} P_n B'_n A_1 B_n P_n - A_1 \right\| \leq \varepsilon/2 ,$$

where the sum converges in the strong operator topology. The operator

$$A' := \sum_{n \in \mathbb{N}} P_n B'_n A_1 B_n P_n$$

is an operator on H' which has controlled propagation (when considered as an operator on (H, ϕ) after extension by zero).

Finally, we observe that

$$\|A' - A\| \leq \varepsilon ,$$

which completes the proof. $\square$

Let X be a bornological coarse space, and let $(H, \phi), (H', \phi')$ be two X-controlled Hilbert spaces.

Lemma 8.41 *If both (H, ϕ) and (H', ϕ') are ample (see Definition 8.13), then we have isomorphisms of C^*-algebras (for all four possible choices for ? from Definition 8.34)*

$$C_?^*(X, H, \phi) \cong C_?^*(X, H', \phi')$$

given by conjugation with a unitary operator $H \to H'$ of controlled propagation.

Proof This follows from Lemma 8.21 together with (8.3).   □

Finally, let $f : X \to X'$ be a morphism between bornological coarse spaces, and (H, ϕ) be an X-controlled Hilbert space. Then we can form the X'-controlled Hilbert space $f_*(H, \phi) := (H, \phi \circ f^*)$ (see Definition 8.9). The Roe algebras for (H, ϕ) and $f_*(H, \phi)$ are closed subalgebras of $B(H)$.

Lemma 8.42 *We have inclusions*

$$C_?^*(X, H, \phi) \subseteq C_?^*(X, H, \phi \circ f^*)$$

for all four possible choices for ? from Definition 8.34.

Proof We just show that the local finiteness and propagation conditions imposed on A in $B(H)$ using the X-control ϕ are stronger than the ones imposed with the X'-control $\phi \circ f^*$. In greater detail one argues as follows:

If A is locally compact with respect to the X-control ϕ, then it is locally compact with respect to the X'-control $\phi \circ f^*$. This follows from the properness of f.

If H' is a locally finite subspace of H with respect to the X-control ϕ, then it is so with respect to the X'-control $\phi \circ f^*$. Indeed, if ϕ' is an X-control of H' such that (H', ϕ') is locally finite and $H' \to H$ has finite propagation, then $\phi' \circ f^*$ is an X'-control for H' which can be used to recognize H' as a locally finite subspace of H with respect to the X'-control $\phi \circ f^*$. This reasoning implies that if A is locally finite on (H, ϕ), then it is so on $f_*(H, \phi)$.

For the propagation conditions we argue similarly. If A is of U-controlled propagation (or U-quasi-local) with respect to the X-control ϕ, then it is of $(f \times f)(U)$-controlled propagation (or $(f \times f)(U)$-quasi-local) with respect to the X'-control $\phi \circ f^*$.   □

Remark 8.43 Let (H, ϕ) be an X-controlled Hilbert space. By definition this means that ϕ is a unital *-representation of all bounded functions on X. But usually, e.g., if X is a Riemannian manifold with a Hermitian vector bundle E, we want to use the Hilbert space $L^2(E, X)$ and the representation should be the one given by multiplication operators. But this representation is not defined on all bounded functions on X, but only on the measurable ones. To get around this we can construct a new representation as in Example 8.23. But now we have to argue why the Roe algebra we get by using this new representation coincides with the Roe algebra that we get from using the usual representation by multiplication operators. We will explain this now in a slightly more general context than Riemannian manifolds but focus on ample Hilbert spaces.

Let (Y, d) be a separable proper metric space, and let Y_d be the associated bornological coarse space. We let Y_t denote the underlying locally compact topological space of Y which we consider as a topological bornological space, see Sect. 7.3.1. Let furthermore (H, ϕ) be an ample Y_d-controlled Hilbert space (in the sense of Definition 8.13) and $\rho : C_0(Y_t) \to B(H)$ be a $*$-representation (see Sect. 7.3.3 for C_0) such that the following conditions are satisfied:

1. If $\rho(f)$ is compact, then $f = 0$.
2. There exists an entourage U of Y_d such that

$$\phi(U[\mathrm{supp}(f)])\rho(f) = \rho(f)\phi(U[\mathrm{supp}(f)]) = \rho(f) .$$

3. There exists an entourage U of Y_d such that for every bounded subset B of Y_d there exists f in $C_0(Y_t)$ with $\mathrm{supp}(f) \subseteq U[B]$ such that $\rho(f)\phi(B) = \phi(B)\rho(f) = \phi(B)$.

Since Y_d has a countable exhaustion by bounded subsets the Hilbert space H is separable. Note that the third condition implies that ρ is non-degenerate. Together with the first condition it ensures that H is an ample or standard Hilbert space in the sense of Roe [Roe93b, Sec. 4.1], [Roe03, Sec. 4.4] and Higson–Roe [HR95]. The second and third conditions together imply that the local finiteness, local compactness, finite propagation and quasi-locality conditions imposed using ρ or ϕ are equivalent.

Returning to our example of a Riemannian manifold X with a Hermitian vector bundle E, we denote by H the Hilbert space $L^2(E, X)$ and by ρ the representation of $C_0(X)$ by multiplication operators. Using the construction from Example 8.23 we get a representation ϕ of all bounded functions on X. If X has no zero-dimensional component, then (H, ϕ) is an ample X-controlled Hilbert space. In this case the three conditions above are satisfied and we have the equalities $C_?^*(X, H, \phi) = C_?^*(X, H, \rho)$ as desired.   □

8.4 K-Theory of C^*-Algebras

Let $\mathbf{C^*Alg}$ denote the category of not necessarily unital C^*-algebras and not necessarily unit preserving homomorphisms. The construction of coarse K-homology theory uses a K-theory functor

$$K^{C^*} : \mathbf{C^*Alg} \to \mathbf{Sp} \tag{8.4}$$

for *C**-algebras (not to be confused with the *K*-cohomology functor introduced in Definition 7.59). In the following we list the properties of this functor which we will use in the constructions later. The functor K^{C^*} has the following properties.

1. K^{C^*} represents *C**-algebra *K*-theory groups: For every *C**-algebra *A* the spectrum $K^{C^*}(A)$ represents the topological *K*-theory of *A*, i.e., the homotopy groups $\pi_* K^{C^*}(A)$ are naturally isomorphic to the *K*-theory groups $K_*(A)$. References for the group-valued *K*-theory of *C**-algebras are Blackadar [Bla98] and Higson–Roe [HR00b].
2. We have $K^{C^*}(\mathbb{C}) \simeq KU$.
3. K^{C^*} preserves filtered colimits: Given a filtered family $(A_i)_{i \in I}$ in **C*Alg** we can form the colimit $A := \operatorname{colim}_{i \in I} A_i$ in **C*Alg**. The natural morphism

$$\operatorname*{colim}_{i \in I} K^{C^*}(A_i) \to K^{C^*}(A)$$

   is an equivalence.

   For example, in the case of a family subalgebras of $B(H)$ the colimit *A* is the closure of the union $\bigcup_{i \in I} A_i$.
4. K^{C^*} preserves sums: If $(A_i)_{i \in I}$ is a family of *C**-algebras, then the canonical morphism

$$\bigoplus_{i \in I} K^{C^*}(A_i) \to K^{C^*}\left(\bigoplus_{i \in I} A_i\right)$$

   is an equivalence.
5. K^{C^*} is exact: We have $K^{C^*}(0) \simeq 0$ and K^{C^*} sends exact sequences of *C**-algebras $0 \to I \to A \to Q \to 0$ to cocartesian squares

$$\begin{array}{ccc} K^{C^*}(I) & \longrightarrow & K^{C^*}(A) \\ \downarrow & & \downarrow \\ 0 & \longrightarrow & K^{C^*}(Q) \end{array} \qquad (8.5)$$

   in **Sp**. Equivalently, we have a fibre sequence

$$\cdots \to K^{C^*}(I) \to K^{C^*}(A) \to K^{C^*}(Q) \to \Sigma K^{C^*}(I) \to \cdots .$$

6. K^{C^*} is stable: For every *A* in **C*Alg** the left upper corner inclusion $A \to A \otimes \mathbb{K}(\ell^2)$ induces an equivalence $K^{C^*}(A) \to K^{C^*}(A \otimes \mathbb{K}(\ell^2))$. Here $\mathbb{K}(\ell^2)$ denotes the compact operators on the Hilbert space ℓ^2.
7. K^{C^*} is homotopy invariant: For every *A* in **C*Alg** the inclusion $A \to C([0, 1], A)$ as constant functions induces an equivalence $K^{C^*}(A) \to K^{C^*}(C([0, 1], A))$.

8. K^{C^*} is Bott periodic: For every A in **C*Alg** there exists a natural equivalence $K^{C^*}(A) \to \Sigma^2 K^{C^*}(A)$. In order to check that a morphism $K^{C^*}(A) \to K^{C^*}(B)$ is an equivalence it suffices to show that $\pi_i K^{C^*}(A) \to \pi_i K^{C^*}(B)$ is an isomorphism for $i = 0, 1$.

9. Assume that $h \colon A \to B$ is a homomorphism between C^*-algebras, and that u is a partial isometry in the multiplier algebra of B such that $hu^*u = h$. Then $h' := uhu^* \colon A \to B$ is a homomorphism of C^*-algebras, and we have an equality of induced maps $\pi_* K^{C^*}(h) = \pi_* K^{C^*}(h') \colon \pi_* K^{C^*}(A) \to \pi_* K^{C^*}(B)$.

A construction of a K-theory functor (8.4) was provided by Joachim [Joa03, Def. 4.9]. He verifies [Joa03, Thm. 4.10] that his spectrum indeed represents C^*-algebra K-theory. In the following we argue that this functor has the properties listed above, which are not stated explicitly in this reference.

Concerning homotopy invariance note that every spectrum-valued functor representing C^*-algebra K-theory has this property. This follows from the fact that the group-valued functor sends the homomorphism $A \to C([0, 1], A)$ to an isomorphism.

Similarly, continuity with respect to filtered colimits is implied by the corresponding known property of the group-valued K-theory functor.

Finally, the group-valued K-theory functor sends a sequence (8.5) to a long exact sequence of K-theory groups (which will be a six-term sequence due to Bott periodicity). So once we know that the composition $K(I) \to K(A) \to K(A/I)$ is the zero map in spectra we can produce a morphism to $K(I) \to \mathrm{Fib}(K(A) \to K(A/I))$ and conclude, using the Five Lemma, that it is an equivalence. An inspection of [Joa03, Def. 4.9] shows that the composition $K(I) \to K(A) \to K(A/I)$ is level-wise the constant map to the base point.

The Bott periodicity is actually a formal consequence of the other properties. The periodicity morphism of spectra can be constructed from the long exact sequence of a so-called Toeplitz extension and an equivalence $\Sigma K^{C^*}(C_0(\mathbb{R}, A)) \simeq K^{C^*}(A)$.

An alternative option is to construct K^{C^*} as a left Kan-extension

$$
\begin{array}{ccc}
\mathbf{C^*Alg}^{\mathrm{sep}} & \xrightarrow{\ \mathrm{map}_{\mathbf{KK}}(\mathbb{C}, -)\ } & \mathbf{Sp} \\
{\scriptstyle \mathrm{incl}}\big\downarrow & \nearrow {\scriptstyle K^{C^*}} & \\
\mathbf{C^*Alg} & &
\end{array}
\qquad (8.6)
$$

where **KK** is the stable ∞-category discussed in Sect. 7.3.2. The point wise formula for the evaluation of this left Kan extension yields

$$
K^{C^*}(A) := \operatorname*{colim}_{B \subseteq A} \mathrm{map}_{\mathbf{KK}}(\mathbb{C}, B),
$$

where the colimit runs over all separable subalgebras B of A. We will use this description later to define the assembly map. All properties except 3 of K^{C^*} are then consequences of the properties of **KK** and the functor $C^*\mathbf{Alg}^{\mathrm{sep}} \to \mathbf{KK}$ listed in Sect. 7.3.2.

Remark 8.44 The Property 9 of K^{C^*} is not so easy to find in the literature. Hence it is reasonable to provide the argument.

Since u is a partial isometry we have two projections $p := u^*u$ and $q := uu^*$ in the multiplier algebra of B. We consider the C^*-algebra $C := pBp \oplus qBq$ and the homomorphism $d : C \to M_2(B)$ given by $(b, b') \mapsto \mathtt{diag}(b, b')$. The assumption on h is equivalent to the fact that h has a factorization

$$h : A \xrightarrow{\tilde{h}} pBp \to B ,$$

where the second homomorphism is the embedding. We then have an equality

$$i_{11} \circ h = d \circ \tilde{h} : A \to M_2(B) ,$$

where i_{11} is the left-upper corner inclusion. The matrix

$$v := \begin{pmatrix} 0 & u^* \\ u & 0 \end{pmatrix}$$

can be considered as a unitary in the multiplier algebra of C. We now calculate that

$$d \circ (v\tilde{h}v^*) = i_{22} \circ h' ,$$

where i_{22} is the lower right corner inclusion and $h' := uhu^*$. By [HR95, Lem. 4.6.1] we have the equality

$$\pi_* K^{C^*}(v\tilde{h}v^*) = \pi_* K^{C^*}(\tilde{h}) .$$

This implies by functoriality of $\pi_* K^{C^*}$ that $\pi_* K^{C^*}(i_{11} \circ h) = \pi_* K^{C^*}(i_{22} \circ h')$. Since we also have the equality of isomorphisms $\pi_* K^{C^*}(i_{11}) = \pi_* K^{C^*}(i_{22})$ we conclude that $\pi_* K^{C^*}(h) = \pi_* K^{C^*}(h')$. $\qquad\qquad\square$

8.5 C^*-Categories and Their K-Theory

We have seen in Sect. 8.3 that we can associate a Roe algebra (with various decorations) to a bornological coarse space X equipped with an X-controlled Hilbert space. Morally, coarse K-homology of X should be coarse K-theory of this Roe algebra. This definition obviously depends on the choice of the X-controlled Hilbert space. In good cases X admits an ample X-controlled Hilbert space. Since any two choices of an ample Hilbert space are isomorphic (Lemma 8.21), the Roe algebra is independent of the choice of the ample X-controlled Hilbert space up to isomorphism.

It is not good enough to associate to X the K-theory spectrum of the Roe algebra for some choice of an X-controlled ample Hilbert space. First of all this would not give a functor. Moreover, we want the coarse K-theory to be defined for all bornological coarse spaces, also for those which do not admit an X-controlled ample Hilbert space (Lemma 8.20).

Remark 8.45 The naive way out would be to define the coarse K-theory spectrum of X as the colimit over all choices of X-controlled Hilbert spaces of the K-theory spectra of the associated Roe algebras. The problem is that the index category of this colimit is not a filtered poset. It would give a wrong homotopy type of the coarse K-theory spectrum of X even in the presence of ample X-controlled Hilbert spaces.

One could try the following solution: one could consider the category of X-controlled Hilbert spaces as a topological category, interpret the functor from X-controlled Hilbert spaces to spectra as a topologically enriched functor and take the colimit in this realm.

One problem with this approach is the following: in order to show functoriality, given a morphism $X \to Y$ and X-controlled, resp. Y-controlled Hilbert spaces (H, ϕ) and (H', ϕ'), we must choose an isometry $V : H \to H'$ with a certain propagation condition on its support (cf. Higson–Roe–Yu [HRY93, Sec. 4]). Then one uses conjugation by V in order to construct a map from the Roe algebra $C^*(X, H, \phi)$ to $C^*(Y, H', \phi')$. Now two different choices of isometries V and V' are related by an inner conjugation. Therefore, on the level on K-theory groups of the Roe algebras conjugation by V and V' induce the same map. But inner conjugations do not necessarily act trivially on the K-theory spectra. When we pursue this route further we would need the statement that the space of such isometries V is contractible. But up to now we were only able to show that this space is connected (see Theorem 8.99).

In the present paper we pursue therefore the following alternative idea. Morally we perform the colimit over the Hilbert spaces on the level of C^*-algebras before going to spectra. Technically we consider the X-controlled Hilbert spaces as objects of a C^*-category whose morphism spaces are the analogues of the Roe algebras. The idea of using C^*-categories in order to produce a functorial K-theory spectrum is not new and has been previously employed in different situations , e.g., by Davis–Lück [DL98], Hambleton–Pedersen [HP04], and Joachim [Joa03].   □

In the present section we review C^*-categories a their K-theory. The details of the application to coarse K-homology will be given in Sect. 8.6.

Remark 8.46 Motivated by the usage of C^*-categories in the present section several follow-up papers dealing with aspects of the theory of C^*-categories appeared after the first version of the book was finished.

1. [Bun19]: homotopy theory for unital C^*-categories
2. [Bunb]: categorial properties of the category of non-unital C^*-categories and crossed products

3. [BEa]: infinite orthogonal sums in C^*-categories and properties of the K-theory
   for C^*-categories, in particular that it preserves infinite products
4. [BEe]: coarse K-homology with coefficients in a C^*-category, a generalization
   of the theory of the present section to the equivariant case and with the category
   of Hilbert spaces replaced by a general C^*-category   □

In the sections below we will consider C^*-algebras and spectra in the small
universe (as opposed to the very small universe). In this case we add a superscript
$(-)^{\text{la}}$ to the notation for the categories, but keep the notation for the functors.

8.5.1 Definition of C*-Categories

In the present paper we will consider small C^*-categories. The category $\mathbf{C^*Cat}$ is
a large category. Before we give the details of the definition note that our basic
example is the small category of very small Hilbert spaces and bounded operators.

Definition 8.47 A C^*-category $\mathbf{C}$ is given by the following structures:

1. a (possibly non-unital) category $\mathbf{C}$,
2. for every two objects c, c' in $\mathbf{C}$ a complex Banach space structure on the set
   $\mathrm{Hom}_{\mathbf{C}}(c, c')$,
3. an involution $* : \mathbf{C}^{\mathrm{op}} \to \mathbf{C}$.

These structures must satisfy the following conditions:

1. The involution fixes objects and acts isometrically and anti-linearly on the
   morphism spaces.
2. The C^*-condition is satisfied: For every morphism $A : c \to c'$ in $\mathbf{C}$ we have

$$\|A^*A\|_{\mathrm{Hom}_{\mathbf{C}}(c,c)} = \|A\|^2_{\mathrm{Hom}_{\mathbf{C}}(c,c')} .$$

3. For every morphism $A : c \to c'$ in $\mathbf{C}$ the operator A^*A is a non-negative operator
   in the C^*-algebra $\mathrm{End}_{\mathbf{C}}(c)$.

   Note that we will usually just write $\| - \|$ instead of $\| - \|_{\mathrm{Hom}_{\mathbf{C}}(c,c')}$.

Definition 8.48 A functor between C^*-categories is a (not necessarily unit-
preserving) functor between $\mathbb{C}$-linear categories which is compatible with the
$*$-operation.

Remark 8.49 In view of Definition 8.47 it would be very natural to require a
condition of compatibility of a functor with the norms on the morphism spaces.
But it turns out that that a functor in the sense of Definition 8.48 is automatically
norm decreasing [Mit02, Prop. 2.14]. Hence such a condition is redundant.

   We refer to [Bunb] for an alternative, but equivalent definition of a C^*-category
as a $\mathbb{C}$-linear $*$-category with additional properties (in contrast to additional Banach
space structures).   □

Definition 8.50

1. A C^*-category is unital if every object has an identity morphism.
2. A functor between unital C^*-categories is unital if it preserves identity morphisms.

Remark 8.51 In the papers listed in Remark 8.46 the notation $\mathbf{C^*Cat}$ is reserved for the category of unital C^*-categories and unital functors. The notation for the non-unital case is $\mathbf{C^*Cat}^{nu}$. In the present book we use the notation $\mathbf{C^*Cat}$ for the non-unital case and say explicitly when we require a C^*-category or a functor to be unital. Since we only consider the large category of small C^*-categories we do not add the superscript "la".   □

A good reference for C^*-categories is [Mit02]. We furthermore refer to Davis–Lück [DL98, Sec. 2] and Joachim [Joa03, Sec. 2].

8.5.2 From C*-Categories to C*-Algebras and K-Theory

We let $\mathbf{C^*Alg}^{la}$ denote the large category of (possibly non-unital) small C^*-algebras and (not necessarily unit preserving) homomorphisms. Note that in contrast $\mathbf{C^*Alg}$ is the small full subcategory of $\mathbf{C^*Alg}^{la}$ of very small C^*-algebras.

A C^*-algebra can be considered as a C^*-category with a single object. This provides an inclusion $\mathbf{C^*Alg}^{la} \to \mathbf{C^*Cat}$. It fits into an adjunction

$$A^f : \mathbf{C^*Cat} \leftrightarrows \mathbf{C^*Alg}^{la} : \mathrm{incl} \,. \tag{8.7}$$

The functor A^f has first been introduced by Joachim [Joa03]. Even if $\mathbf{C}$ is a unital C^*-category the C^*-algebra $A^f(\mathbf{C})$ is non-unital in general. This happens, e.g. in the case when $\mathbf{C}$ has infinitely many objects. Because of this fact we need the context of non-unital C^*-categories in order to present A^f as a left-adjoint.

Remark 8.52 In the following we describe A^f explicitly. The functor A^f associates to a C^*-category $\mathbf{C}$ a C^*-algebra $A^f(\mathbf{C})$. The unit of the adjunction is a functor $\mathbf{C} \to A^f(\mathbf{C})$ between C^*-categories. The superscript f stands for free which distinguishes the algebra from another C^*-algebra $A(\mathbf{C})$ associated to $\mathbf{C}$ which is simpler, but not functorial in $\mathbf{C}$. We will discuss $A(\mathbf{C})$ later.

We first construct the free $*$-algebra $\tilde{A}^f(\mathbf{C})$ generated by the morphisms of $\mathbf{C}$. If A is a morphism in $\mathbf{C}$, then we let (A) in $\tilde{A}^f(\mathbf{C})$ denote the corresponding generator. We then form the quotient by an equivalence relation which reflects the sum, the composition, and the $*$-operator already present in $\mathbf{C}$.

1. If c, c' are in $\mathrm{Ob}(\mathbf{C})$, A, B are in $\mathrm{Hom}_{\mathbf{C}}(c, c')$, and λ, μ are in $\mathbb{C}$, then we have the relation $\lambda(A) + \mu(B) \sim (\lambda A + \mu B)$.
2. If c, c', c'' are in $\mathrm{Ob}(\mathbf{C})$, A is in $\mathrm{Hom}_{\mathbf{C}}(c, c')$, and B is in $\mathrm{Hom}_{\mathbf{C}}(c', c'')$, then we have the relation $(B)(A) \sim (B \circ A)$.
3. For every morphism A in $\mathbf{C}$ we have the relation $(A^*) \sim (A)^*$.

We denote the quotient $*$-algebra of $\tilde{A}^f(\mathbf{C})$ by this set of relations by $A^f(\mathbf{C})_0$. We equip $A^f(\mathbf{C})_0$ with the maximal C^*-norm and let $A^f(\mathbf{C})$ be the closure. In general, this C^*-algebra is only small (as opposed to very small) since $\mathbf{C}$ is small. This is the reason for considering large C^*-algebras. The natural functor

$$\mathbf{C} \to A^f(\mathbf{C}) \tag{8.8}$$

sends every object of $\mathbf{C}$ to the single object of $A^f(\mathbf{C})$ and a morphism A of $\mathbf{C}$ to the image of the generator (A) in $A^f(\mathbf{C})$ under the quotient map and the natural map into the completion.

It has been observed by Joachim [Joa03] that this construction has the following universal property: For every C^*-algebra $\mathbf{B}$ (considered as a C^*-category with a single object) a functor $\mathbf{C} \to \mathbf{B}$ in $\mathbf{C}^*\mathbf{Cat}$ has a unique extension to a homomorphism $A^f(\mathbf{C}) \to \mathbf{B}$ of C^*-algebras.

Let now $\mathbf{C} \to \mathbf{C}'$ be a functor between C^*-categories. Then we get a functor $\mathbf{C} \to A^f(\mathbf{C}')$ by composition with (8.8). It now follows from the universal property of $A^f(\mathbf{C})$ that this functor extends uniquely to a homomorphism of C^*-algebras

$$A^f(\mathbf{C}) \to A^f(\mathbf{C}')\,.$$

This finishes the construction of the functor A^f on functors between C^*-categories. The universal property of A^f is equivalent to the adjunction (8.7).   □

To a C^*-category $\mathbf{C}$ we can also associate a simpler C^*-algebra $A(\mathbf{C})$. In order to construct this C^*-algebra we start with the description of a $*$-algebra $A(\mathbf{C})_0$. The underlying $\mathbb{C}$-vector space of $A(\mathbf{C})_0$ is the algebraic direct sum

$$A(\mathbf{C})_0 := \bigoplus_{c,c' \in \mathrm{Ob}(\mathbf{C})} \mathrm{Hom}_{\mathbf{C}}(c, c')\,. \tag{8.9}$$

The product $(g, f) \mapsto gf$ on $A_0(\mathbf{C})$ is defined by

$$gf := \begin{cases} g \circ f & \text{if } g \text{ and } f \text{ are composable}, \\ 0 & \text{otherwise}. \end{cases}$$

The $*$-operation on $\mathbf{C}$ induces an involution on $A(\mathbf{C})_0$ in the obvious way. We again equip $A(\mathbf{C})_0$ with the maximal C^*-norm and define $A(\mathbf{C})$ as the closure of $A(\mathbf{C})_0$ with respect to this norm.

Example 8.53 If $\mathbf{C}$ is the category of very small Hilbert spaces, then the vector space $A(\mathbf{C})_0$ is the sum over all spaces of bounded operators between two very small Hilbert spaces. This $*$-algebra belongs to the small universe and $A(\mathbf{C})$ is not very small.   □

Note that the assignment $\mathbf{C} \mapsto A(\mathbf{C})$ is not a functor since non-composable morphisms in a C^*-category may become composable after applying a C^*-functor. However, $\mathbf{C} \mapsto A(\mathbf{C})$ is functorial for C^*-functors that are injective on objects.

There is a canonical C^*-functor $\mathbf{C} \to A(\mathbf{C})$. By the universal property of A^f it extends uniquely to a homomorphism $\alpha_{\mathbf{C}} \colon A^f(\mathbf{C}) \to A(\mathbf{C})$ of C^*-algebras.

Lemma 8.54 *If $f \colon \mathbf{C} \to \mathbf{D}$ is a functor between C^*-categories which is injective on objects, then the following square commutes:*

$$
\begin{array}{ccc}
A^f(\mathbf{C}) & \xrightarrow{A^f(f)} & A^f(\mathbf{D}) \\
\alpha_{\mathbf{C}} \downarrow & & \downarrow \alpha_{\mathbf{D}} \\
A(\mathbf{C}) & \xrightarrow{A(f)} & A(\mathbf{D})
\end{array}
$$

Proof We consider the diagram

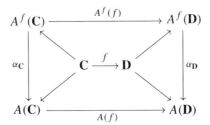

The left and the right triangles commute by construction of the transformation $\alpha_{...}$. The upper and the lower middle square commute by construction of the functors $A^f(-)$ and $A(-)$. Therefore the outer square commutes. □

Joachim [Joa03, Prop. 3.8] showed that if $\mathbf{C}$ is unital and has only countably many objects, then $\alpha_{\mathbf{C}} \colon A^f(\mathbf{C}) \to A(\mathbf{C})$ is a stable homotopy equivalence and therefore induces an equivalence in K-theory. We will generalize this now to our more general setting. Let $K^{C^*} \colon \mathbf{C^*Alg}^{\mathrm{la}} \to \mathbf{Sp}^{\mathrm{la}}$ be the K-theory functor for small C^*-algebras discussed in Sect. 8.4.

Proposition 8.55 *The homomorphism $\alpha_{\mathbf{C}} \colon A^f(\mathbf{C}) \to A(\mathbf{C})$ induces an equivalence of spectra*

$$
K^{C^*}(\alpha_{\mathbf{C}}) \colon K^{C^*}(A^f(\mathbf{C})) \to K^{C^*}(A(\mathbf{C})) .
$$

Proof If $\mathbf{C}$ is unital and has countable may objects, then the assertion of the proposition has been shown by Joachim [Joa03, Prop. 3.8].

First assume that $\mathbf{C}$ has countably many objects, but is possibly non-unital. In the following we argue that the arguments from the proof of [Joa03, Prop. 3.8] are applicable and show that the canonical map $\alpha_{\mathbf{C}} \colon A^f(\mathbf{C}) \to A(\mathbf{C})$ is a stable homotopy equivalence. In the following we recall the construction of the stable inverse which is a homomorphism

$$\beta : A(\mathbf{C}) \to A^f(\mathbf{C}) \otimes \mathbb{K}$$

of C^*-algebras, where $\mathbb{K} := \mathbb{K}(H)$ are the compact operators on the Hilbert space

$$H := \ell^2(\mathrm{Ob}(\mathbf{C}) \cup \{e\}) \,,$$

where e is an artificially added point. The assumption on the cardinality of $\mathrm{Ob}(\mathbf{C})$ is made since we want that $\mathbb{K}$ is the algebra of compact operators on a separable Hilbert space. Two points x, y in $\mathrm{Ob}(\mathbf{C}) \cup \{e\}$ provide a rank-one operator $\Theta_{y,x}$ in $\mathbb{K}(H)$ which sends the basis vector corresponding to x to the vector corresponding to y, and is zero otherwise. By definition, the homomorphism β sends A in $\mathrm{Hom}_\mathbf{C}(x, y)$ to

$$\beta(A) := A \otimes \Theta_{y,x} \,.$$

If A and B are composable morphisms, then the relation $\Theta_{z,y}\Theta_{y,x} = \Theta_{z,x}$ implies that $\beta(B \circ A) = \beta(B)\beta(A)$. Moreover, if A,B are not composable, then $\beta(B)\beta(A) = 0$. Finally, $\beta(A)^* = \beta(A^*)$ since $\Theta_{y,x}^* = \Theta_{x,y}$. It follows that β is a well-defined *-homomorphism.

The argument now proceeds by showing that the composition $(\alpha_\mathbf{C} \otimes \mathrm{id}_{\mathbb{K}(H)}) \circ \beta$ is homotopic to $\mathrm{id}_{A(\mathbf{C})} \otimes \Theta_{e,e}$, and that the composition $\beta \circ \alpha_\mathbf{C}$ is homotopic to $\mathrm{id}_{A^f(\mathbf{C})} \otimes \Theta_{e,e}$. Note that in our setting $\mathbf{C}$ is not necessarily unital. In the following we directly refer to the proof of [Joa03, Prop. 3.8]. The only step in the proof of that proposition where the identity morphisms are used is the definition of the maps denoted by $u_x(t)$ in the reference. But they in turn are only used to define the map denoted by Ξ later in that proof. The crucial observation is that we can define this map Ξ directly without using any identity morphisms in $\mathbf{C}$.

We conclude that the morphism

$$K^{C^*}(\alpha_\mathbf{C}) : K^{C^*}(A^f(\mathbf{C})) \to K^{C^*}(A(\mathbf{C}))$$

is an equivalence for C^*-categories $\mathbf{C}$ with countably many objects.

In order to extend this to all C^*-categories, we use that

$$\mathbf{C} \cong \operatorname*{colim}_{\mathbf{C}'} A^f(\mathbf{C}')$$

in $\mathbf{C}^*\mathbf{Cat}$, where the colimit runs over the small filtered poset of full subcategories with countably many objects. This can be shown easily by verifying the universal property of a colimit, see also [Bunb, Lem. 4.5].

Since A^f is a left-adjoint it preserves colimits and we get an isomorphism

$$A^f(\mathbf{C}) \cong \operatorname*{colim}_{\mathbf{C}'} A^f(\mathbf{C}').$$

The connecting maps of the indexing family are functors which are injections on objects. By an inspection of the definition we see that the canonical map is an isomorphism

$$\operatorname*{colim}_{\mathbf{C}'} A(\mathbf{C}')_0 \cong A(\mathbf{C})_0,$$

where the colimit is interpreted in the category of pre-C^*-algebras [Bun19, Def. 2.14]. The completion with respect to the maximal norm is a left-adjoint [Bun19, (3.17)] and preserves colimits. Hence the canonical morphism is an isomorphism[2]

$$\operatorname*{colim}_{\mathbf{C}'} A(\mathbf{C}') \cong A(\mathbf{C}). \tag{8.10}$$

Recall that the index category of these colimits is a filtered poset. By Property 3 of K^{C^*} saying that this functor preserves filtered colimits the morphism

$$K^{C^*}(\alpha_{\mathbf{C}}) : K^{C^*}(A^f(\mathbf{C})) \to K^{C^*}(A(\mathbf{C}))$$

is equivalent to the morphism

$$\operatorname*{colim}_{\mathbf{C}'} K^{C^*}(\alpha_{\mathbf{C}'}) : \operatorname*{colim}_{\mathbf{C}'} K^{C^*}(A^f(\mathbf{C}')) \to \operatorname*{colim}_{\mathbf{C}'} K^{C^*}(A(\mathbf{C}')).$$

Since the categories $\mathbf{C}'$ appearing in the colimit now have at most countable many objects we have identified $K^{C^*}(\alpha_{\mathbf{C}})$ with a colimit of equivalences. Hence this morphism itself is an equivalence.   □

Definition 8.56 We define the topological K-theory functor for C^*-categories as the composition

$$\mathrm{K}^{C^*Cat} : \mathbf{C^*Cat} \xrightarrow{A^f} \mathbf{C^*Alg}^{\mathrm{la}} \xrightarrow{K^{C^*}} \mathbf{Sp}^{\mathrm{la}}.$$

In the subsequent subsections we show the basic properties of this K-theory functor for C^*-categories.

---

[2]One could also check this isomorphism directly as it was the approach in the first version of this book. But the theory developed in the reference formalizes some of the arguments nicely.

8.5.3 K-Theory Preserves Filtered Colimits

Note that the functor $A^f : \mathbf{C^*Cat} \to \mathbf{C^*Alg}^{la}$ is a left-adjoint, see (8.7). Hence it preserves all small colimits. Furthermore, the functor $K^{C^*} : \mathbf{C^*Alg}^{la} \to \mathbf{Sp}^{la}$ preserves small filtered colimits by Property 3.

Corollary 8.57 *The functor* K^{C^*Cat} *preserves small filtered colimits.*

As shown in [Bunb, Thm. 4.1] the category $\mathbf{C^*Cat}$ admits all small colimits. But in general it might be complicated to calculate a filtered colimit of C^*-categories explicitly. In the present book we only need the special case formulated in Proposition 8.58.

We consider a C^*-category $\mathbf{C}$ and a filtered family $(\mathbf{D}_i)_{i \in I}$ of subcategories of $\mathbf{C}$ with the following special properties:

1. These subcategories all have the same set of objects as $\mathbf{C}$.
2. On the level of morphisms the inclusion $\mathbf{D}_i \hookrightarrow \mathbf{C}$ is an isometric embedding of a closed subspace.

In this situation we can define a full subcategory

$$\mathbf{D} := \operatorname*{colim}_{i \in I} \mathbf{D}_i \tag{8.11}$$

of $\mathbf{C}$. It again has the same set of objects as $\mathbf{C}$. For two objects c, c' of $\mathrm{Ob}(\mathbf{C})$ we define $\mathrm{Hom}_{\mathbf{D}}(c, c')$ to be the closure in $\mathrm{Hom}_{\mathbf{C}}(c, c')$ of the linear subspace $\bigcup_{i \in I} \mathrm{Hom}_{\mathbf{D}_i}(c, c')$. One checks easily that $\mathbf{D}$ is again a C^*-category.

Proposition 8.58 *The canonical morphism*

$$\operatorname*{colim}_{i \in I} K^{C^*Cat}(\mathbf{D}_i) \to K^{C^*Cat}(\mathbf{D})$$

is an equivalence.

Proof One easily checks by verifying the universal property that $\mathbf{D}$ represents the colimit of the family $(\mathbf{D}_i)_{i \in I}$. For a detailed argument see also [Bunb, Lem. 4.5]. We can now apply Corollary 8.57. □

8.5.4 K-Theory Preserves Unitary Equivalences

Let $\mathbf{D}$ be a C^*-category, and let $u : d \to d'$ be a morphism in $\mathbf{D}$.

Definition 8.59

1. u is a partial isometry if uu^* and u^*u are projections.
2. u is an isometry if it is a partial isometry and $u^*u = \mathrm{id}_d$.

3. u is unitary, if it is an isometry and u^* is an isometry, too.

If $u : d \to d'$ is a unitary, then d and d' admit unit endomorphisms.

Let $f, g : \mathbf{C} \to \mathbf{D}$ be two functors in $\mathbf{C}^*\mathbf{Cat}$.

Definition 8.60 A unitary isomorphism $u : f \to g$ is a natural transformation $u = (u_c)_{c \in \mathbf{C}}$ such that $u_c : f(c) \to g(c)$ is unitary for every object c in $\mathbf{C}$.

Let $f, g : \mathbf{C} \to \mathbf{D}$ be two functors between unital C^*-categories.

Lemma 8.61 *Assume:*

1. $\mathbf{D}$ *is unital.*
2. f *and* g *are unitarily isomorphic.*
3. $K^{C^*Cat}(f) : K^{C^*Cat}\mathbf{C}) \to K^{C^*Cat}(\mathbf{D})$ *is an equivalence.*

Then $K^{C^*Cat}(g) : K^{C^*Cat}(\mathbf{C})) \to K^{C^*Cat}(\mathbf{D})$ *is an equivalence.*

Proof We first assume that $\mathbf{C}$ and $\mathbf{D}$ have countably many objects. The difficulty of the proof is that a unitary isomorphism between the functors f and g is not well-reflected on the level of algebras $A^f(\mathbf{C})$ and $A^f(\mathbf{D})$. But it induces an isomorphism between the induced functors f_* and g_* on the module categories of the C^*-categories $\mathbf{C}$ and $\mathbf{D}$.

The following argument provides this transition. We use that the K-theory of $A^f(\mathbf{C})$ considered as a C^*-category with a single object is the same as the K-theory of $A^f(\mathbf{C})$ considered as a C^*-algebra. Since the unit of the adjunction (8.7) is a natural transformation we have a commutative diagram

$$
\begin{array}{ccc}
\mathbf{C} & \xrightarrow{\ h\ } & \mathbf{D} \\
\downarrow & & \downarrow \\
A^f(\mathbf{C}) & \xrightarrow{\ A^f(h)\ } & A^f(\mathbf{D})
\end{array}
$$

of C^*-categories, where h is in $\{f, g\}$. We apply K-theory $K_* = \pi_* K^{C^*}$ and get the commutative square:

$$
\begin{array}{ccc}
K_*(\mathbf{C}) & \xrightarrow{\ K_*(h)\ } & K_*(\mathbf{D}) \\
\downarrow & & \downarrow \\
K_*(A^f(\mathbf{C})) & \xrightarrow{\ K_*(A^f(h))\ } & K_*(A^f(\mathbf{D}))
\end{array}
$$

By [Joa03, Cor. 3.10] the vertical homomorphism are isomorphisms. By assumption $K_*(A^f(f))$ is an isomorphism. We conclude that $K_*(f)$ is an isomorphism. The proof of [Joa03, Lem. 2.7] can be modified to show that the functors f_* and g_* between the module categories of $\mathbf{C}$ and $\mathbf{D}$ are equivalent. So if $K_*(f)$ is an isomorphism, then so is $K_*(g)$. Finally, we conclude that $K_*(A^f(g))$ is an isomorphism. This gives the result in the case of countable many objects.

The general case then follows by considering filtered colimits over subcategories with countably many objects as in the proof of Proposition 8.55.   □

Let $f : \mathbf{C} \to \mathbf{D}$ be a functor between unital C^*-categories.

Definition 8.62 f is a unitary equivalence if there exists a functor $g : \mathbf{D} \to \mathbf{C}$ such that $f \circ g$ and $g \circ f$ are unitary isomorphic to the respective identity functors.

The functor g is called an inverse of f.

Let $f : \mathbf{C} \to \mathbf{D}$ be a functor between unital C^*-categories.

Corollary 8.63 *If f is a unitary equivalence, then* $K^{C^*Cat}(f) : K^{C^*Cat}(\mathbf{C}) \to K^{C^*Cat}(\mathbf{D})$ *is an equivalence.*

Proof Let $g : \mathbf{D} \to \mathbf{C}$ be an inverse of f. Then $f \circ g$ (or $g \circ f$, respectively) is unitarily isomorphic to $\mathrm{id}_{\mathbf{D}}$ (or $\mathrm{id}_{\mathbf{C}}$, respectively). By Lemma 8.61 the morphisms $K^{C^*Cat}(g \circ f)$ and $K^{C^*Cat}(f \circ g)$ are equivalences of spectra. Since K^{C^*Cat} is a functor this implies that $K^{C^*}(f)$ is an equivalence of spectra.   □

8.5.5 Exactness of K-Theory

Let $\mathbf{D}$ be a C^*-category. We furthermore consider a wide C^*-subcategory $\mathbf{C}$ of $\mathbf{D}$, i.e., $\mathbf{C}$ has the same objecs as $\mathbf{D}$, and the morphism spaces of $\mathbf{C}$ are closed subspaces of the morphism spaces of $\mathbf{D}$ which are preserved by the $*$-operation.

Definition 8.64 $\mathbf{C}$ is a closed ideal in $\mathbf{D}$ if for all triples x, y, z of objects of $\mathbf{D}$, f in $\mathrm{Hom}_{\mathbf{D}}(x, y)$ and g in $\mathrm{Hom}_{\mathbf{C}}(y, z)$ we have $gf \in \mathrm{Hom}_{\mathbf{C}}(x, z)$.

The condition in Definition 8.64 is the analogue of being a right-ideal. Since $\mathbf{C}$ is preserved by the involution this condition also implies an analogous left-ideal condition.

If $\mathbf{C} \to \mathbf{D}$ is a closed ideal we can form the quotient category $\mathbf{D}/\mathbf{C}$ by declaring

$$\mathrm{Ob}(\mathbf{D}/\mathbf{C}) := \mathrm{Ob}(\mathbf{D}) \quad \text{and} \quad \mathrm{Hom}_{\mathbf{D}/\mathbf{C}}(x, y) := \mathrm{Hom}_{\mathbf{D}}(x, y)/\mathrm{Hom}_{\mathbf{C}}(x, y),$$

where the latter quotient is taken in the sense of Banach spaces.

The following has been verify, e.g. in [Mit02, Cor. 4.8].

Lemma 8.65 *If $\mathbf{C} \to \mathbf{D}$ is a closed ideal, then the quotient category $\mathbf{D}/\mathbf{C}$ is a C^*-category.*

We will depict this situation by

$$0 \to \mathbf{C} \to \mathbf{D} \to \mathbf{D}/\mathbf{C} \to 0$$

and call this an exact sequence of C^*-categories.

Remark 8.66 In [Bunb, Thm. 4.1] it is shown that **C*Cat** is cocomplete, hence in particular admits push-outs. The quotient **D/C** fits into a square

$$
\begin{array}{ccc}
\mathbf{C} & \longrightarrow & \mathbf{D} \\
\downarrow & & \downarrow \\
0[\mathrm{Ob}(\mathbf{D})] & \longrightarrow & \mathbf{D/C}
\end{array}
$$

where $0[X]$ denotes the C^*-category with the set of objects X and only zero morphisms. This square is a push-out and a pull-back at the same time.   □

Proposition 8.67 *The functor* $\mathrm{K}^{C^*\mathrm{Cat}}$ *sends exact sequences of C^*-categories to fibre sequences.*

The next lemma prepares the proof of this proposition.

Lemma 8.68 *If* $\mathbf{C} \to \mathbf{D}$ *is a closed ideal, then we have a short exact sequence of C^*-algebras*

$$
0 \to A(\mathbf{C}) \to A(\mathbf{D}) \to A(\mathbf{D/C}) \to 0 \,.
$$

Proof Recall that we define the C^*-algebra $A(\mathbf{C})$ as the closure of the pre-C^*-algebra $A(\mathbf{C})_0$ given by (8.9) using the maximal C^*-norm. Since taking the closure with respect to the maximal C^*-norm is a left-adjoint it preserves surjections. Consequently the obvious surjection $A(\mathbf{D})_0 \to A(\mathbf{D/C})_0$ induces a surjection $A(\mathbf{D}) \to A(\mathbf{D/C})$.

The homomorphism $A(\mathbf{C})_0 \to A(\mathbf{D})_0$ of pre-*-algebras induces a morphism $A(\mathbf{C}) \to A(\mathbf{D})$ of C^*-algebras. We must show that this is the injection of a closed ideal which is precisely the kernel of the surjection $A(\mathbf{D}) \to A(\mathbf{D/C})$.

In the following we argue that $A(\mathbf{C}) \to A(\mathbf{D})$ is isometric. We consider a finite subset of the set of objects of $\mathbf{D}$ and the full subcategories $\mathbf{C}'$ and $\mathbf{D}'$ of $\mathbf{C}$ and $\mathbf{D}$, respectively, on these objects. It has been observed in the proof of [Joa03, Lem. 3.6] that $A(\mathbf{D}')_0$ with its topology as a finite sum (8.9) of Banach spaces is closed in $A(\mathbf{D})$. Consequently, the inclusion $A(\mathbf{D}')_0 \hookrightarrow A(\mathbf{D})$ induces the maximal norm on $A(\mathbf{D}')_0$. We now use that $A(\mathbf{C}')_0$ is closed in $A(\mathbf{D}')_0$. Consequently, the maximal C^*-norm on $A(\mathbf{C}')_0$ is the norm induced from the inclusion $A(\mathbf{C}')_0 \hookrightarrow A(\mathbf{D}')_0 \hookrightarrow A(\mathbf{D})$.

The norm induced on $A(\mathbf{C})_0$ from the inclusion $A(\mathbf{C})_0 \hookrightarrow A(\mathbf{D})$ is bounded above by the maximal norm but induces the maximal norm on $A(\mathbf{C}')_0$. Since every element of $A(\mathbf{C})_0$ is contained in $A(\mathbf{C}')_0$ for some full subcategory $\mathbf{C}'$ on a finite subset of objects, we conclude that the inclusion $A(\mathbf{C})_0 \to A(\mathbf{D})$ induces the maximal norm.

It follows that the *-homomorphism $A(\mathbf{C}) \to A(\mathbf{D})$ is isometric. As an isometric inclusion the homomorphism of C^*-algebras $A(\mathbf{C}) \to A(\mathbf{D})$ is injective.

The composition

$$A(\mathbf{C}) \to A(\mathbf{D}) \to A(\mathbf{D}/\mathbf{C})$$

vanishes by the functoriality of the process of taking completions with respect to maximal C^*-norms. So it remains to show that this sequence is exact. We start with the obvious exact sequence

$$A(\mathbf{C})_0 \to A(\mathbf{D})_0 \xrightarrow{\pi} A(\mathbf{D}/\mathbf{C})_0.$$

If we restrict to the subcategories with finite objects we get an isomorphism

$$A(\mathbf{D}')_0 / A(\mathbf{C}')_0 \to A(\mathbf{D}'/\mathbf{C}')_0.$$

This isomorphism is an isometry when we equip all algebras with their maximal C^*-norms. Furthermore, by the same arguments as above the inclusion $A(\mathbf{D}')_0 / A(\mathbf{C}')_0 \to A(\mathbf{D}/\mathbf{C})$ is an isometry. Using the arguments from above we conclude that $A(\mathbf{D})_0 / A(\mathbf{C})_0 \to A(\mathbf{D}/\mathbf{C})$ is an isometry. This assertion implies that the closure of $A(\mathbf{C})_0$ in the topology induced from $A(\mathbf{D})$ is the kernel of $A(\mathbf{D}) \to A(\mathbf{D}/\mathbf{C})$. But above we have seen that this closure is exactly $A(\mathbf{C})$. □

Note that Lemma 8.68 implies an isomorphism of C^*-algebras

$$A(\mathbf{D}/\mathbf{C}) \cong A(\mathbf{D})/A(\mathbf{C}). \tag{8.12}$$

Proof of Proposition 8.67 Let

$$0 \to \mathbf{C} \to \mathbf{D} \to \mathbf{D}/\mathbf{C} \to 0$$

be an exact sequence of C^*-categories. Then we get the following commuting diagram

$$
\begin{array}{ccc}
K^{C^*Cat}(\mathbf{C}) \longrightarrow K^{C^*Cat}(\mathbf{D}) \longrightarrow K^{C^*Cat}(\mathbf{D}/\mathbf{C}) \\
\simeq \downarrow K^{C^*}(\alpha_{\mathbf{C}}) \qquad \simeq \downarrow K^{C^*}(\alpha_{\mathbf{D}}) \qquad \simeq \downarrow K^{C^*}(\alpha_{\mathbf{D}/\mathbf{C}}) \\
K^{C^*}(A(\mathbf{C})) \longrightarrow K^{C^*}(A(\mathbf{D})) \longrightarrow K^{C^*}(A(\mathbf{D}/\mathbf{C}))
\end{array}
$$

The vertical morphisms are equivalences by Proposition 8.55 and the definition $K^{C^*Cat} := K^{C^*} \circ A^f$. Using (8.12) and Property 5 of K^{C^*} we see that the lower part of the diagram is a segment of a fibre sequence. Hence the upper part is a segment of a fibre sequence, too. □

8.5.6   Additivity of K-Theory

Let **D** be a C^*-category, and let d, d' be objects in **D**.

Definition 8.69 An orthogonal sum of d and d' is triple $(d \oplus d', u, u')$ of an object $d \oplus d'$ of **D** and isometries $u : d \to d \oplus d'$ and $u' : d' \to d \oplus d'$ such that $u'u^* = 0$ and $uu^* + u'u'^{,*} = \mathrm{id}_{d \oplus d'}$.

The isometries u and u' are called the canonical inclusions. If an orthogonal sum exists, then it is unique up to unique unitary isomorphism.

Remark 8.70 In [BEa] we discuss the concept of infinite orthogonal sums in C^*-categories. In the present book we only need the obvious abstract Definition 8.69 of finite orthogonal sums. We will encounter infinite orthogonal sums as well but only in the context of C^*-categories built from Hilbert spaces where these sums can be understood directly.   □

An object d in a C^*-category **D** is called a zero object if it $\mathrm{End}_{\mathbf{D}}(d) \cong 0$.

Definition 8.71 **D** is called additive if it admits a zero object and orthogonal sums for all pairs of objects.

Note that an additive C^*-category is automatically unital.

Let **C** be a second C^*-category, and let $\phi, \phi' : \mathbf{C} \to \mathbf{D}$ be two functors. Then we can define a new functor $\phi \oplus \phi' : \mathbf{C} \to \mathbf{D}$. It sends an object c in **C** to a choice of a sum $\phi(c) \oplus \phi(c)$. Furthermore, it sends a morphism $A : c_0 \to c_1$ in **C** to the morphism $u_1 \phi(A) u_0^* + u'_1 \phi'(A) u_0'^{,*}$, where u_i and u'_i for $i = 0, 1$ are the canonical inclusions for the sums $\phi(c_i) \oplus \phi'(c_i)$.

The functor is uniquely determined up to unique unitary isomorphism by the choices adopted on the level of objects.

The following proposition says that the K-theory functor K^{C^*Cat} preserves sums of functors.

Let **C** and **D** be in **C*****Cat**, and let $\phi, \phi' : \mathbf{C} \to \mathbf{D}$ be two functors.

Proposition 8.72 *If* **D** *is additive, then we have an equivalence*

$$\mathrm{K}^{C^*Cat}(\phi \oplus \phi') \simeq \mathrm{K}^{C^*Cat}(\phi) + \mathrm{K}^{C^*Cat}(\phi') : \mathrm{K}^{C^*Cat}(\mathbf{C}) \to \mathrm{K}^{C^*Cat}(\mathbf{D}) .$$

Before we start with the actual proof of the proposition we show that the K-theory functor K^{C^*Cat} preserves some products of C^*-categories.

Let **C** and **D** be in **C*****Cat**.

Lemma 8.73 *Assume:*

1. **C** *and* **D** *are unital.*
2. **C** *and* **D** *admit zero objects.*

Then the morphism

$$\mathrm{K}^{C^*Cat}(\mathrm{pr}_{\mathbf{C}}) \oplus \mathrm{K}^{C^*Cat}(\mathrm{pr}_{\mathbf{D}}) : \mathrm{K}^{C^*Cat}(\mathbf{C} \times \mathbf{D}) \to \mathrm{K}^{C^*Cat}(\mathbf{C}) \oplus \mathrm{K}^{C^*Cat}(\mathbf{D})$$
$$(8.13)$$

is an equivalence.

Proof For a set X let $0[X]$ be the category with the set of objects X and with only zero morphisms. We consider the commuting diagram

$$\begin{array}{ccccc}
0[\mathrm{Ob}(\mathbf{C})] & \xrightarrow{\;\omega_{\mathbf{C}}\;} & \mathbf{C} & \xrightarrow{\;=\;} & \mathbf{C} \\
\downarrow & & \downarrow{\scriptstyle z_{\mathbf{C}}} & & \downarrow{\scriptstyle z'_{\mathbf{C}}} \\
0[\mathrm{Ob}(\mathbf{C})] \times \mathbf{D} & \xrightarrow{\omega_{\mathbf{C}} \times \mathrm{id}_{\mathbf{D}}} & \mathbf{C} \times \mathbf{D} & \longrightarrow & \mathbf{C} \times 0[\mathrm{Ob}(\mathbf{D})]
\end{array} \qquad (8.14)$$

in $\mathbf{C}^*\mathbf{Cat}$. The functor $\omega_{\mathbf{C}}$ is the canonical inclusion. The vertical functors send an object c of $\mathbf{C}$ to the object $(c, 0_{\mathbf{D}})$ of $\mathbf{C} \times \mathbf{D}$, where $0_{\mathbf{D}}$ is a chosen zero object in $\mathbf{D}$. The action of the left vertical functor on morphisms is clear. Finally, the middle and right vertical functor send a morphism $A : c \to c'$ in $\mathbf{C}$ to the morphism $(A, 0) : (c, 0_{\mathbf{D}}) \to (c', 0_{\mathbf{D}})$ in $\mathbf{C} \times \mathbf{D}$. The horizontal lines are exact sequences of C^*-categories. It is furthermore easy to check that the right vertical functor $z'_{\mathbf{C}}$ is a unitary equivalence. Indeed, it is obviously fully faithful and essentially surjective. We now apply the functor K^{C^*Cat} and get the commuting diagram

$$\begin{array}{ccccc}
\mathrm{K}^{C^*Cat}(0[\mathrm{Ob}(\mathbf{C})]) & \xrightarrow{\mathrm{K}^{C^*Cat}(\omega_{\mathbf{C}})} & \mathrm{K}^{C^*Cat}(\mathbf{C}) & \longrightarrow & \mathrm{K}^{C^*Cat}(\mathbf{C}) \\
\downarrow & & \downarrow{\scriptstyle \mathrm{K}^{C^*Cat}(z_{\mathbf{C}})} & & \simeq\downarrow{\scriptstyle \mathrm{K}^{C^*Cat}(z'_{\mathbf{C}})} \\
\mathrm{K}^{C^*Cat}(0[\mathrm{Ob}(\mathbf{C})] \times \mathbf{D}) & \xrightarrow{\mathrm{K}^{C^*Cat}(\omega_{\mathbf{C}} \times \mathrm{id}_{\mathbf{D}})} & \mathrm{K}^{C^*Cat}(\mathbf{C} \times \mathbf{D}) & \longrightarrow & \mathrm{K}^{C^*Cat}(\mathbf{C} \times 0[\mathrm{Ob}(\mathbf{D})])
\end{array}$$
$$(8.15)$$

in $\mathbf{Sp}^{\mathrm{la}}$. Using the observations made above we conclude with Proposition 8.67 that the horizontal sequences are segments of fibre sequence, and with Corollary 8.63 that left vertical morphism is an equivalence. Hence the left square is a push-out.

We now use that there is a unitary equivalence $z'_{\mathbf{D}} : \mathbf{D} \to 0[\mathrm{Ob}(\mathbf{C})] \times \mathbf{D}$. Consequently we have an identification $\mathrm{K}^{C^*Cat}(z'_{\mathbf{D}}) : \mathrm{K}^{C^*Cat}(\mathbf{D}) \xrightarrow{\cong} \mathrm{K}^{C^*Cat}(0[\mathrm{Ob}(\mathbf{C})] \times \mathbf{D})$ for the lower left corner. Finally note that the left upper corner is the K-theory of a category consisting only of zero objects and therefore vanishes by Property 5 of K^{C^*} and the fact that $A^f(0[X]) \cong 0$. Using stability in order to identify coproducts with products we conclude that

$$\mathrm{K}^{C^*Cat}(z_{\mathbf{C}}) + \mathrm{K}^{C^*Cat}(z_{\mathbf{D}}) : \mathrm{K}^{C^*Cat}(\mathbf{D}) \oplus \mathrm{K}^{C^*Cat}(\mathbf{D}) \to \mathrm{K}^{C^*Cat}(\mathbf{C} \times \mathbf{D}) \qquad (8.16)$$

is an equivalence. We now observe that $\mathrm{pr}_\mathbf{C} \circ z_\mathbf{C} = \mathrm{id}_\mathbf{C}$ and $\mathrm{pr}_\mathbf{D} \circ z_\mathbf{D} = \mathrm{id}_\mathbf{D}$, and that $\mathrm{pr}_\mathbf{C} \circ z_\mathbf{D}$ and $\mathrm{pr}_\mathbf{C} \circ z_\mathbf{D}$ are zero morphisms. This immediately implies that (8.13) is an equivalence inverse to (8.16).   □

Proof of Proposition 8.72 Since $\mathbf{D}$ is additive we can choose a zero object $0_\mathbf{D}$ in $\mathbf{D}$. We consider the diagram in $\mathbf{Sp}^{\mathrm{la}}$

$$\begin{array}{ccc}
 & \mathrm{K}^{C^*Cat}(\mathbf{D}\times\mathbf{D}) & \\
\mathrm{K}^{C^*Cat}(\mathrm{pr}_0)\oplus\mathrm{K}^{C^*Cat}(\mathrm{pr}_1)\nearrow & & \searrow \mathrm{K}^{C^*Cat}(\oplus) \\
\mathrm{K}^{C^*Cat}(\mathbf{D})\oplus\mathrm{K}^{C^*Cat}(\mathbf{D}) & \xrightarrow{\quad+\quad} & \mathrm{K}^{C^*Cat}(\mathbf{D})
\end{array}$$
$\underset{\simeq}{} \qquad\qquad\qquad\qquad\qquad\qquad\qquad\qquad\qquad\qquad\qquad\qquad (8.17)$

where the left vertical morphism is an equivalence by Lemma 8.73. We claim that (8.17) naturally commutes. Using the explicit inverse (8.16) of $\mathrm{K}^{C^*Cat}(\mathrm{pr}_0) \oplus \mathrm{K}^{C^*Cat}(\mathrm{pr}_1)$ and the universal property of $+$ it suffices to show that the compositions

$$\mathrm{K}^{C^*Cat}(\mathbf{D}) \xrightarrow{\iota_i} \mathrm{K}^{C^*Cat}(\mathbf{D}) \oplus \mathrm{K}^{C^*Cat}(\mathbf{D}) \xrightarrow{\mathrm{K}^{C^*Cat}(z_0)+\mathrm{K}^{C^*Cat}(z_1)}$$

$$\mathrm{K}^{C^*Cat}(\mathbf{D}\times\mathbf{D}) \xrightarrow{\mathrm{K}^{C^*Cat}(\mathrm{id}_\mathbf{D}\oplus\mathrm{id}_\mathbf{D})} \mathrm{K}^{C^*Cat}(\mathbf{D})$$

are equivalent to the identity, where $\iota_i : \mathrm{K}^{C^*Cat}(\mathbf{D}) \xrightarrow{\iota_i} \mathrm{K}^{C^*Cat}(\mathbf{D}) \oplus \mathrm{K}^{C^*Cat}(\mathbf{D})$ denote the canonical inclusions for $i = 0, 1$.

In the case $i = 0$ this composition is induced by applying K^{C^*Cat} to the endofunctor $s : \mathbf{D} \to \mathbf{D}$ which is given as follows:

1. objects: s sends an object D to the representative $D \oplus 0_\mathbf{D}$.
2. morphisms: s sends a morphism $f : D \to D'$ to the morphism $f \oplus 0 : D \oplus 0_\mathbf{D} \to D' \oplus 0_\mathbf{D}$.

We have a unitary equivalence $u : \mathrm{id}_\mathbf{D} \to s$ given by the family $(u_D)_{D\in\mathrm{Ob}(\mathbf{D})}$, where $u_D : D \to D \oplus 0_\mathbf{D}$ is the canonical inclusion. Hence $\mathrm{K}^{C^*Cat}(s) \simeq \mathrm{K}^{C^*Cat}(\mathrm{id}_\mathbf{D})$. The case $i = 1$ is analogous.

We have the following diagram in $\mathbf{Sp}^{\mathrm{la}}$

$$\begin{array}{ccc}
 & \mathrm{K}^{C^*Cat}(\mathbf{C}) & \\
\mathrm{K}^{C^*Cat}(\mathrm{diag}_\mathbf{C})\nearrow & & \searrow \mathrm{diag}_{\mathrm{K}^{C^*Cat}(\mathbf{C})} \\
\mathrm{K}^{C^*Cat}(\mathbf{C}\times\mathbf{C}) \xrightarrow[\simeq]{\mathrm{K}^{C^*Cat}(\mathrm{pr}_0)\oplus\mathrm{K}^{C^*Cat}(\mathrm{pr}_1)} & & \mathrm{K}^{C^*Cat}(\mathbf{C}) \oplus \mathrm{K}^{C^*Cat}(\mathbf{C}) \\
\downarrow \mathrm{K}^{C^*Cat}(\phi\times\phi') & & \downarrow \mathrm{K}^{C^*Cat}(\phi)\oplus\mathrm{K}^{C^*Cat}(\phi') \\
\mathrm{K}^{C^*Cat}(\mathbf{D}\times\mathbf{D}) \xrightarrow[\simeq]{\mathrm{K}^{C^*Cat}(\mathrm{pr}_0)\oplus\mathrm{K}^{C^*Cat}(\mathrm{pr}_1)} & & \mathrm{K}^{C^*Cat}(\mathbf{D}) \oplus \mathrm{K}^{C^*Cat}(\mathbf{D}) \\
\mathrm{K}^{C^*Cat}(\oplus)\searrow & & \swarrow + \\
 & \mathrm{K}^{C^*Cat}(\mathbf{D}) &
\end{array}$$

$$\qquad\qquad\qquad\qquad\qquad\qquad\qquad\qquad\qquad\qquad\qquad\qquad\qquad (8.18)$$

The lower triangle commutes by (8.17). The upper triangle and the middle square obviously commute. The left top-down path is the map $K^{C^*Cat}(\phi \oplus \phi')$, while the right top-down path is $K^{C^*Cat}(\phi) + K^{C^*Cat}(\phi')$. The filler of (8.18) provides the desired equivalence.

$\square$

8.6 Coarse K-Homology

In this section we define the coarse K-homology functors $K\mathcal{X}$ and $K\mathcal{X}_{ql}$. Our main result is Theorem 8.79 stating that these functors are coarse homology theories.

Let X be a bornological coarse space. Note that Hilbert spaces below are assumed to be very small. The Roe categories also defined below will be small C^*-categories and therefore objects of $\mathbf{C}^*\mathbf{Cat}$. Recall the Definitions 8.1, 8.2, 8.3 and 8.8.

We consider the $\mathbb{C}$-linear $*$-category $\mathbf{C}^*(X)_0$ whose objects are the locally finite X-controlled Hilbert spaces which are determined on points, and whose morphisms are the bounded operators of controlled propagation. Using the properties of the support listed in Sect. 8.1 and the closure property of the coarse structure of X (Definition 2.4) we check that $\mathbf{C}^*(X)_0$ is indeed a $\mathbb{C}$-linear $*$-category. If (H, ϕ) and (H', ϕ') are objects of $\mathbf{C}^*(X)_0$, then the norm of the Banach space of bounded operators $B(H, H')$ induces a norm on $\mathrm{Hom}_{\mathbf{C}^*(X)_0}((H, \phi), (H', \phi'))$. We can form the completion of the morphism spaces of $\mathbf{C}^*(X)_0$ with respect to these norms.

Definition 8.74 We define the Roe category $\mathbf{C}^*(X)$ as follows:

1. The objects of $\mathbf{C}^*(X)$ are the locally finite X-controlled Hilbert spaces which are determined on points.
2. The morphism spaces of $\mathbf{C}^*(X)$ are obtained from the morphism spaces of $\mathbf{C}^*(X)_0$ by completing with respect to the norms described above.

It is clear that $\mathbf{C}^*(X)$ is a well-defined C^*-category in the sense of Definition 8.47.

For the following recall also Definition 8.28. Let (H, ϕ), (H', ϕ') and (H'', ϕ'') be X-controlled Hilbert spaces, and let $A : H \to H'$ and $B : H' \to H''$ be quasi-local operators.

Lemma 8.75 *If (H', ϕ') is determined on points, then $B \circ A$ is quasi-local.*

Proof If an operator is U-quasi-local, then it is also U'-quasi-local for every bigger entourage U'. We can therefore assume that A and B are both U-quasi-local for some entourage U of X. We will show, that then BA is also U-quasi-local.

We consider ε in $(0, \infty)$. Then we can find an integer n in $\mathbb{N}$ such that

$$\|\phi''(Y')B\phi'(Y)\| \le \frac{\varepsilon}{2(1 + \|A\|)} \quad \text{and} \quad \|\phi'(Y')A\phi(Y)\| \le \frac{\varepsilon}{2(1 + \|B\|)} \qquad (8.19)$$

whenever Y' and Y are mutually U^n-separated subsets of X. Assume now that Z and Z' are two mutually U^{2n}-separated subsets of X. Since (H', ϕ') is determined on points, by Lemma 8.4 we have

$$\phi'(U^n[Z]) + \phi'(X \setminus U^n[Z]) = \mathrm{id}_{H'}.$$

This gives the equality

$$\phi''(Z')BA\phi(Z) = \phi''(Z')B\phi'(U^n[Z])A\phi(Z) + \phi''(Z')B\phi'(X \setminus U^n[Z])A\phi(Z).$$

We now note that Z' and $U^n[Z]$ are mutually U^n-separated, and that also $X \setminus U^n[Z]$ and Z are mutually $U1n$-separated. Using the estimates in (8.19) we conclude that

$$\|\phi''(Z')BA\phi(Z)\| \le \|\phi''(Z')B\phi'(U[Z])A\phi(Z)\| + \|\phi''(Z')B\phi'(X \setminus U^n[Z])A\phi(Z)\| \le \varepsilon.$$

$\square$

We consider the category $\mathbf{C}^*_{\mathrm{ql}}(X)_0$ whose objects are all locally finite X-controlled Hilbert spaces on X which are determined on points, and whose morphisms are the quasi-local operators. It is clear that the adjoint of a quasi-local operator is quasi-local. A linear combination of a U-quasi-local and a U'-quasi-local operator is $U \cup U'$-quasi-local. By Lemma 8.32 the composition of two quasi-local operators is again quasi-local. This shows that $\mathbf{C}^*_{\mathrm{ql}}(X)_0$ is a $\mathbb{C}$-linear $*$-category. If (H, ϕ) and (H', ϕ') are objects of $\mathbf{C}^*_{\mathrm{ql}}(X)_0$, then the norm of the Banach space of bounded operators $B(H, H')$ induces a norm on $\mathrm{Hom}_{\mathbf{C}^*_{\mathrm{ql}}(X)_0}((H, \phi), (H', \phi'))$. We can form the completion of the morphism spaces of $\mathbf{C}^*_{\mathrm{ql}}(X)_0$ with respect these norms.

Definition 8.76 We define the quasi-local Roe category $\mathbf{C}^*_{\mathrm{ql}}(X)$ as follows:

1. The objects of $\mathbf{C}^*_{\mathrm{ql}}(X)$ are the locally finite X-controlled Hilbert spaces which are determined on points.
2. The morphism spaces of $\mathbf{C}^*_{\mathrm{ql}}(X)$ are obtained from the morphism spaces of $\mathbf{C}_{\mathrm{ql}}(X)_0$ by completing with respect to the norms described above.

It is again clear that $\mathbf{C}^*_{\mathrm{ql}}(X)$ is a well-defined C^*-category in the sense of Definition 8.47.

The categories $\mathbf{C}^*(X)$ and $\mathbf{C}^*_{\mathrm{ql}}(X)$ are C^*-categories which have the same set of objects. Since operators of controlled propagation are quasi-local we conclude that morphisms of $\mathbf{C}^*(X)$ form a subset of the morphisms of $\mathbf{C}^*_{\mathrm{ql}}(X)$.

The definition of the Roe categories is closely related to the definitions of the corresponding Roe algebras $C^*(X, H, \phi)$ and $C^*_{\mathrm{ql}}(X, H, \phi)$ from Definition 8.34. For every object (H, ϕ) in $\mathbf{C}^*(X)$ by definition we have isomorphisms

$$\mathrm{End}_{\mathbf{C}^*(X)}((H, \phi)) \cong C^*(X, H, \phi), \quad \mathrm{End}_{\mathbf{C}^*_{\mathrm{ql}}(X)}((H, \phi), (H, \phi)) \cong C^*_{\mathrm{ql}}(X, H, \phi).$$

Remark 8.77 Our reason for using the locally finite versions of the Roe algebras to model coarse K-homology is that the Roe categories as defined above are unital.

The obvious alternative to define the Roe categories to consist of all X-controlled Hilbert spaces, and to take locally compact versions of the operators as morphisms, failed, as we will explain now.

Recall that Corollary 8.63 says that equivalent C^*-categories have equivalent K-theories. For non-unital C^*-algebras it would be necessary to redefine the notion of equivalence using a generalization to C^*-categories of the notion of a multiplier algebra of a C^*-algebra. But we were not able to fix all details of an argument which extends Corollary 8.63 from the unital to the non-unital case.

This corollary is used in an essential way to show that the coarse K-homology functor is coarsely invariant and satisfies excision. ☐

As a next step we extend the Definitions 8.74 and 8.76 to functors

$$\mathbf{C}^*, \mathbf{C}^*_{\mathrm{ql}} : \mathbf{BornCoarse} \to \mathbf{C}^*\mathbf{Cat} .$$

Let $f : X \to X'$ be a morphism between bornological coarse spaces. If (H, ϕ) is a locally finite X-controlled Hilbert space, then the X'-controlled Hilbert space $f_*(H, \phi)$ defined in Definition 8.9 is again locally finite (since f is proper). Similarly, if (H, ϕ) is determined on points, then $f_*(H, \phi)$ is determined on points. Here we use the conclusion $1 \Rightarrow 3$ in Lemma 8.4 for the partition $(f^{-1}(\{x'\}))_{x' \in X'}$ of X. On the level of objects the functors $\mathbf{C}^*(f)$ and $\mathbf{C}^*_{\mathrm{ql}}(f)$ send (H, ϕ) to $f_*(H, \phi)$.

Let (H, ϕ) and (H', ϕ') be objects of $\mathbf{C}^*(X)$, and let A be in $\mathrm{Hom}_{\mathbf{C}^*(X)}((H, \phi),$ $(H', \phi'))$ (or $\mathrm{Hom}_{\mathbf{C}^*_{\mathrm{ql}}(X)}((H, \phi), (H', \phi'))$, respectively). Then by definition A can be approximated by operators of controlled propagation (or quasi-local operators, respectively) with respect to the controls ϕ and ϕ'. As in the proof of Lemma 8.42 we see that A can also be approximated by operators of controlled propagation (or quasi-local operators, respectively) with respect to the controls $\phi \circ f^*$ and $\phi' \circ f^*$. We can thus define the functor $\mathbf{C}^*(f)$ (or $\mathbf{C}^*_{\mathrm{ql}}(f)$, respectively) on objects such that it sends A to A.

It is obvious that this description of $\mathbf{C}^*_?(f)$ is compatible with the composition and the $*$-operation.

In the following we use the K-theory functor K^{C^*Cat} for C^*-categories of Definition 8.56.

Definition 8.78 We define the coarse K-homology functors

$$K\mathcal{X}, K\mathcal{X}_{\mathrm{ql}} : \mathbf{BornCoarse} \to \mathbf{Sp}$$

as compositions

$$\mathbf{BornCoarse} \xrightarrow{\mathbf{C}^*, \mathbf{C}^*_{\mathrm{ql}}} \mathbf{C}^*\mathbf{Cat} \xrightarrow{K^{C^*Cat}} \mathbf{Sp}^{\mathrm{la}} .$$

Theorem 8.79 *The functors* $K\mathcal{X}$ *and* $K\mathcal{X}_{\mathrm{ql}}$ *are coarse homology theories.*

Proof In the following we verify the conditions listed in Definition 4.22. The Propositions 8.80, 8.82, 8.85 and 8.86 together will imply the theorem.   □

Proposition 8.80 *The functors* $K\mathcal{X}$ *and* $K\mathcal{X}_{\mathrm{ql}}$ *are coarsely invariant.*

Proof Let $f, g : X \to X'$ be morphisms between bornological coarse spaces.   □

Lemma 8.81 *If* f *and* g *are close (Definition 3.13), then the functors* $\mathbf{C}^*(f)$ *and* $\mathbf{C}^*(g)$ *(or* $\mathbf{C}^*_{\mathrm{ql}}(f)$ *and* $\mathbf{C}^*_{\mathrm{ql}}(g)$*) are unitarily isomorphic.*

Proof We define a unitary isomorphism $u : \mathbf{C}^*(f) \to \mathbf{C}^*(g)$. The isomorphism u is given by the family of unitaries $(u_{(H,\phi)})_{(H,\phi) \in \mathrm{Ob}(\mathbf{C}^*(X))}$, where

$$u_{(H,\phi)} := \mathrm{id}_H : (H, \phi \circ f^*) \to (H, \phi \circ g^*).$$

Since f and g are close, the identity id_H has indeed controlled propagation. The quasi-local case is analogous.   □

We can now finish the proof of the proposition. Let X be a bornological coarse space. We consider the projection $\pi : \{0, 1\} \otimes X \to X$ and the inclusion $i_0 : X \to \{0, 1\} \otimes X$, where $\{0, 1\}$ has the maximal structures. Then $\pi \circ i_0 = \mathrm{id}_X$ and $i_0 \circ \pi$ is close to the identity of $\{0, 1\} \otimes X$. By Lemma 8.81 the functor $\mathbf{C}^*(\pi)$ (or $\mathbf{C}^*_{\mathrm{ql}}(\pi)$, respectively) is an equivalence of C^*-categories. By Corollary 8.63 the morphism $K\mathcal{X}(\pi)$ (or $K\mathcal{X}_{\mathrm{ql}}(\pi)$, respectively) is an equivalence of spectra.   □

Proposition 8.82 *The functors* $K\mathcal{X}$ *and* $K\mathcal{X}_{\mathrm{ql}}$ *satisfy excision.*

Proof We will discuss the quasi-local case in detail. For $K\mathcal{X}$ the argument is similar.[3] Let $(Z, \mathcal{Y})$ be a complementary pair on a bornological coarse space X. We must show that

$$
\begin{array}{ccc}
K\mathcal{X}_{\mathrm{ql}}(Z \cap \mathcal{Y}) & \longrightarrow & K\mathcal{X}_{\mathrm{ql}}(Z) \\
\downarrow & & \downarrow \\
K\mathcal{X}_{\mathrm{ql}}(\mathcal{Y}) & \longrightarrow & K\mathcal{X}_{\mathrm{ql}}(X)
\end{array}
$$

(8.20)

is a push-out square.

If $f : Z \to X$ is an inclusion of a subset with induced structures, then we let $\mathbf{D}(Z)$ be the C^*-category defined as follows:

1. The objects of $\mathbf{D}(Z)$ are the objects of $\mathbf{C}^*_{\mathrm{ql}}(X)$.
2. The morphisms $\mathrm{Hom}_{\mathbf{D}(Z)}((H, \phi), (H', \phi'))$ are the quasi-local bounded operators $(H(Z), \phi_Z) \to (H'(Z), \phi'_Z)$.

---

[3] A detailed argument in this case can also be found in [BEe].

Here the Z-control of $H(Z)$ is defined by $\phi_Z(f) := \phi(\tilde{f})_{|H(Z)}$, where $\tilde{f}$ in $C(X)$ is any extension of f in $C(Z)$ to X.

We have a natural inclusion $\mathbf{D}(Z) \to \mathbf{C}^*_{\mathrm{ql}}(X)$ as a closed subcategory given by the identity on objects and the canonical inclusions

$$\mathrm{Hom}_{\mathbf{D}(Z)}((H, \phi), (H', \phi')) \to \mathrm{Hom}_{\mathbf{C}^*_{\mathrm{ql}}(X)}((H, \phi), (H', \phi'))$$

on the level of morphisms.

We have a fully-faithful functor

$$\iota_Z : \mathbf{D}(Z) \to \mathbf{C}^*_{\mathrm{ql}}(Z). \tag{8.21}$$

It sends the object (H, ϕ) of $\mathbf{D}(Z)$ to the object $(H(Z), \phi_Z)$ of $\mathbf{C}^*_{\mathrm{ql}}(Z)$. On morphisms it is given by the identification

$$\mathrm{Hom}_{\mathbf{D}(Z)}((H, \phi), (H', \phi')) = \mathrm{Hom}_{\mathbf{C}^*_{\mathrm{ql}}(Z)}((H(Z), \phi_Z), (H'(Z), \phi'_Z)).$$

This functor is essentially surjective. Indeed, if (H, ϕ) is in $\mathrm{Ob}(\mathbf{C}^*_{\mathrm{ql}}(Z))$, then

$$f_*(H, \phi) = (H, \phi \circ f^*) \in \mathrm{Ob}(\mathbf{C}^*_{\mathrm{ql}}(X))$$

can be considered as an object of $\mathbf{D}(Z)$, and we have an unitary isomorphism

$$(H(Z), (\phi \circ f^*)_Z) \cong (H, \phi)$$

in $\mathbf{C}^*_{\mathrm{ql}}(Z)$. Consequently, $\iota_Z : \mathbf{D}(Z) \to \mathbf{C}^*_{\mathrm{ql}}(Z)$ is a unitary equivalence of C^*-categories. By Corollary 8.63 the induced morphism

$$K^{C^*Cat}(\mathbf{D}(Z)) \to K\mathcal{X}_{\mathrm{ql}}(Z) \tag{8.22}$$

an equivalence of spectra.

The functor $f_* : \mathbf{C}^*_{\mathrm{ql}}(Z) \to \mathbf{C}^*_{\mathrm{ql}}(X)$ identifies $\mathbf{C}^*_{\mathrm{ql}}(Z)$ isomorphically with the full subcategory of all objects (H, ϕ) of $\mathbf{D}(Z)$ with $H(Z) = H$. The inclusion $\mathbf{C}^*_{\mathrm{ql}}(Z) \to \mathbf{D}(Z)$ is a unitary equivalence of C^*-categories with inverse ι_Z. We therefore have the following factorization

$$
\begin{array}{ccc}
K\mathcal{X}_{\mathrm{ql}}(Z) & \xrightarrow{\quad K\mathcal{X}_{\mathrm{ql}}(f) \quad} & K\mathcal{X}_{\mathrm{ql}}(X) \\
& \underset{K^{C^*Cat}(\iota_Z)}{\searrow} \;\; \overset{\simeq}{\nearrow} & \nearrow \\
& K^{C^*Cat}(\mathbf{D}(Z)) &
\end{array}
\tag{8.23}
$$

We consider a big family $\mathcal{Y} = (Y_i)_{i \in I}$ in X. We get an increasing family $(\mathbf{D}(Y_i))_{i \in I}$ (defined as above with Z replaced by Y_i) of subcategories of $\mathbf{C}^*_{\mathrm{ql}}(X)$.

We let $\mathbf{D}(\mathcal{Y})$ be the closure of the union of this family in $\mathbf{C}^*_{\mathrm{ql}}(X)$ in the sense of C^*-categories, see (8.11). □

Lemma 8.83 $\mathbf{D}(\mathcal{Y})$ *is a closed ideal in* $\mathbf{C}^*_{\mathrm{ql}}(X)$.

Proof We consider a morphism A in $\mathbf{D}(\mathcal{Y})$ between two objects. Let moreover Q be a morphism in $\mathbf{C}^*_{\mathrm{ql}}(X)$ which is left composable with A. We must show that $QA \in \mathbf{D}(\mathcal{Y})$. Since $\mathbf{D}(\mathcal{Y})$ is closed it suffices to show this for quasi-local operators Q. In the following we will assume that Q is quasi-local.

The morphism A can be approximated by a family $(A_i)_{i \in I}$ of morphisms A_i in $\mathbf{D}(Y_i)$ between the same objects.

We consider ε in $(0, \infty)$ and choose i in I such that

$$\|A_i - A\| \leq \frac{\varepsilon}{2(\|Q\| + 1)}.$$

Here the norms are the Banach norms on the respective morphism spaces. For an X-controlled Hilbert space (H, ϕ) and a subset B of X in the following we abbreviate the action of $\phi(\chi_B)$ on H by B. Since Q is weakly quasi-local we can find an entourage U of X such that

$$\|B' Q B\| \leq \frac{\varepsilon}{2(\|A_i\| + 1)}$$

whenever B, B' are mutually U-separated. Using that the family of subsets $\mathcal{Y}$ is big we now choose j in I so that $X \setminus Y_j$ and Y_i are mutually U-separated. Then we get the inequality

$$\|(X \setminus Y_j) Q A_i\| \leq \varepsilon/2.$$

Here we use implicitly the relation $A_i = A_i Y_i$. Now $Y_j Q A_i$ belongs to $\mathbf{D}(\mathcal{Y})$, and

$$\|Y_j Q A_i - QA\| \leq \varepsilon.$$

Since we can take ε arbitrary small and $\mathbf{D}(\mathcal{Y})$ is closed we get $QA \in \mathbf{D}(\mathcal{Y})$. We can analogously conclude that $AQ \in \mathbf{D}(\mathcal{Y})$. □

Since $\mathbf{D}(\mathcal{Y})$ is a closed ideal we can form the quotient C^*-category $\mathbf{C}^*_{\mathrm{ql}}(X)/\mathbf{D}(\mathcal{Y})$, see Lemma 8.65.

We now consider a complementary pair $(Z, \mathcal{Y})$. The family $(\mathbf{D}(Z \cap Y_i))_{i \in I}$ is an increasing family of subcategories of $\mathbf{D}(Z)$. We let $\mathbf{D}(Z \cap \mathcal{Y})$ be the closure of the union of these subcategories. The same argument as for Lemma 8.83 shows that $\mathbf{D}(Z \cap \mathcal{Y})$ is a closed ideal in $\mathbf{D}(Z)$. Alternatively one could use the unitary equivalence ι_Z in (8.21) and its inverse f_* in order to reduce the assertion formally to this lemma.

Lemma 8.84 *The inclusion* $\mathbf{D}(Z) \to \mathbf{C}^*_{\mathrm{ql}}(X)$ *induces an isomorphism of quotients*

$$\psi : \mathbf{D}(Z)/\mathbf{D}(Z \cap \mathcal{Y}) \to \mathbf{C}^*_{\mathrm{ql}}(X)/\mathbf{D}(\mathcal{Y}) . \tag{8.24}$$

Proof Note that the inclusion is the identity on objects. We construct an inverse functor

$$\lambda : \mathbf{C}^*_{\mathrm{ql}}(X)/\mathbf{D}(\mathcal{Y}) \to \mathbf{D}(Z)/\mathbf{D}(Z \cap \mathcal{Y}) .$$

It is again the identity on objects. On morphisms it sends a class $[A]$ in $\mathbf{C}^*_{\mathrm{ql}}(X)/\mathbf{D}(\mathcal{Y})$ to

$$\lambda([A]) := [ZAZ] .$$

We now argue that λ is well-defined and an inverse of ψ.

We first show that λ is well-defined as a map between morphism sets. We fix two objects of $\mathbf{C}^*_{\mathrm{ql}}(X)$ and consider morphisms between these objects. We assume that A in $\mathbf{D}(\mathcal{Y})$ is such a morphism. Then we can obtain A as a limit of a family $(A_i)_{i \in I}$ of morphisms A_i in $\mathbf{D}(Y_i)$. Now ZA_iZ belongs to the image of $\mathbf{D}(Z \cap Y_i)$. Hence for the limit we get $ZAZ \in \mathbf{D}(Z \cap \mathcal{Y})$.

It is clear that λ preserves the involution and is compatible with the $\mathbb{C}$-linear structure on the morphism spaces. It remains to check the compatibility with compositions. We consider two composable morphisms A and B in $\mathbf{C}^*_{\mathrm{ql}}(X)$. We must show that

$$\lambda([AB]) = \lambda([A])\lambda([B]) .$$

To this end it suffices to show that $ZABZ - ZAZBZ \in \mathbf{D}(Z \cap \mathcal{Y})$. We can write this difference in the form ZAZ^cBZ, where $Z^c := X \setminus Z$. Let i in I be such that $Z \cup Y_i = X$. Then $Z^c \subseteq Y_i$. Since the ideal is closed we assume that A and B are quasi-local.

For every U in the coarse structure $\mathcal{C}$ of X there exists j_U in I with $i \le j_U$ such that $U[Y_i] \subseteq Y_{j_U}$. In this case the sets Z^c and $Y^c_{j_U}$ are mutually U-separated. Since A and B are weakly quasi-local we have

$$\lim_{U \in \mathcal{C}} \|ZAZ^cBZ - Y_{j_U}ZAZ^cBY_{j_U}Z\| = 0 .$$

Note that the projection $Y_{j_U}Z$ belongs to $\mathbf{D}(Z \cap \mathcal{Y})$. Since this is an ideal we have $Y_{j_U}ZAZ^c$, $Z^cBY_{j_U}Z \in \mathbf{D}(Z \cap \mathcal{Y})$. Since the ideal is closed we finally conclude that $ZAZ^cBZ \in \mathbf{D}(Z \cap \mathcal{Y})$ as required.

Finally we show that λ is an inverse of ψ. It is easy to see that $\lambda \circ \psi = \mathrm{id}$. We claim that $\psi \circ \lambda = \mathrm{id}$. We consider the difference $[A] - \psi(\lambda([A]))$ which is represented by $A - ZAZ$. Then we study the term

$$(A - ZAZ) - Y_i(A - ZAZ)Y_i . \tag{8.25}$$

Note that the second term belongs to $\mathbf{D}(\mathcal{Y})$. Hence in order to show the claim it suffices to see that (8.25) can be made arbitrary small by choosing i in I appropriately. We have the identity

$$(A - ZAZ) = Z^c A + ZAZ^c .$$

This gives

$$(A - ZAZ) - Y_i(A - ZAZ)Y_i = Y_i^c Z^c A + Y_i^c ZAZ^c + Y_i Z^c AY^c + Y_i ZAZ^c Y_i^c .$$

Using that $X = Y_i \cup Z$ for sufficiently large i in I we get $Y_i^c Z^c = 0$. Therefore if i in I is large it remains to consider

$$Y_i^c ZAZ^c + Y_i Z^c AY_i^c .$$

Since $\mathbf{D}(\mathcal{Y})$ is closed we can assume that A is quasi-local. For every U in $\mathcal{C}$ we can choose i_U in I so large that $Y_{i_U}^c$ and Z^c are U-separated. Then we have

$$\lim_{U \in \mathcal{C}} \| Y_{i_U}^c ZAZ^c \| = 0 , \quad \lim_{U \in \mathcal{C}} \| Y_{i_U} Z^c AY_{i_U}^c \| = 0 .$$

$\square$

After these preparations we can now turn to the actual proof of Proposition 8.82. By Proposition 8.67 stating that K^{C^*Cat} is exact we obtain the following commuting diagram

$$\begin{array}{ccccc}
\mathrm{K}^{C^*Cat}(\mathbf{D}(Z \cap \mathcal{Y})) & \longrightarrow & \mathrm{K}^{C^*Cat}(\mathbf{D}(Z)) & \longrightarrow & \mathrm{K}^{C^*Cat}(\mathbf{D}(Z)/\mathbf{D}(Z \cap \mathcal{Y})) \\
\downarrow & & \downarrow & & \mathrm{K}^{C^*Cat}(\psi) \downarrow \simeq \\
\mathrm{K}^{C^*Cat}(\mathbf{D}(\mathcal{Y})) & \longrightarrow & \mathrm{K}^{C^*Cat}(\mathbf{C}_{ql}^*(X)) & \longrightarrow & \mathrm{K}^{C^*Cat}(\mathbf{C}_{ql}^*(X)/\mathbf{D}(\mathcal{Y}))
\end{array}$$

$$\tag{8.26}$$

where the horizontal sequences are segments of fibre sequences. The vertical morphisms are induced by the inclusions and the right vertical morphism is an equivalence by Lemma 8.84. We conclude that the left square is a push-out square in $\mathbf{Sp}^{la}$.

We now show that this square is equivalent to the square (8.20) which is then also a push-out square. The equivalence is induced by the fillers of the following big diagram:

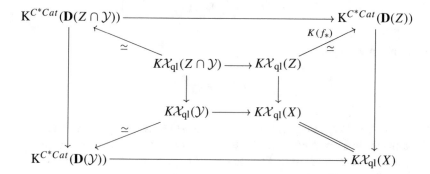

We explain the lower left diagonal equivalence. The inclusions $Y_i \to X$ induce unitary equivalences $\mathbf{C}^*_{\mathrm{ql}}(Y_i) \to \mathbf{D}(Y_i)$ which are natural in i. Consequently we have an equivalence

$$K\mathcal{X}_{\mathrm{ql}}(\mathcal{Y}) \simeq \operatorname*{colim}_{i \in I} K^{C^*Cat}(\mathbf{C}^*_{\mathrm{ql}}(Y_i)) \simeq \operatorname*{colim}_{i \in I} K^{C^*Cat}(\mathbf{D}(Y_i)) \stackrel{\mathrm{Prop.\ 8.58}}{\simeq} K^{C^*Cat}(\mathbf{D}(\mathcal{Y})).$$

The triangle (8.23) yields the filler of the right square. Furthermore, the family of triangles (8.23) for Y_i in place of Z yields the filler of the lower square. One gets the upper left diagonal equivalence and the upper square in a similar manner. Finally the filler left square is induced from the family of commuting squares

$$
\begin{array}{ccc}
\mathbf{D}(Z \cap Y_i) & \longleftarrow & \mathbf{C}^*_{\mathrm{ql}}(Z \cap Y_i) \\
\downarrow & & \downarrow \\
\mathbf{D}(Y_i) & \longleftarrow & \mathbf{C}^*_{\mathrm{ql}}(Y_i)
\end{array}
$$

This completes the proof of Proposition 8.82.   □
Let X be a bornological coarse space.

Proposition 8.85 *If X is flasque, then $K\mathcal{X}(X) \simeq 0$ and $K\mathcal{X}_{\mathrm{ql}}(X) \simeq 0$.*

Proof We discuss the quasi-local case. For $K\mathcal{X}$ the argument is similar.[4]
    Let $f : X \to X$ be a morphism which implements flasqueness (see Definition 3.21). We define a functor

$$S : \mathbf{C}^*_{\mathrm{ql}}(X) \to \mathbf{C}^*_{\mathrm{ql}}(X)$$

---

[4]A detailed argument in this case can also be found in [BEe].

as follows:

1. On objects S is given by

$$S(H, \phi) := \left(\bigoplus_{n \in \mathbb{N}} H, \bigoplus_{n \in \mathbb{N}} \phi \circ f^{n,*} \right).$$

2. On morphisms we set

$$S(A) := \bigoplus_{n \in \mathbb{N}} A.$$

Condition 3 in Definition 3.21 ensures that $S(H, \phi)$ is locally finite. One furthermore checks in a straightforward manner that $S(H, \phi)$ is determined on points.

If A is U-quasi-local, then $S(A)$ is $\bigcup_{n \in \mathbb{N}} (f^n \times f^n)(U)$-quasi-local. By Condition 2 of Definition 3.21 the operator $S(A)$ belongs to $\mathbf{C}^*_{\mathrm{ql}}(X)$.

We now note that the C^*-category $\mathbf{C}^*_{\mathrm{ql}}(X)$ is additive. The sum of two objects (H, ϕ) and (H', ϕ') is represented by $(H \oplus H', \phi \oplus \phi')$. As explained in Sect. 8.5.6 we can form the direct sum of functors with values in $\mathbf{C}^*_{\mathrm{ql}}(X)$.

We note the following relation

$$\mathrm{id}_{\mathbf{C}^*_{\mathrm{ql}}(X)} \oplus f_* \circ S \cong S. \tag{8.27}$$

of endofunctors of $\mathbf{C}^*_{\mathrm{ql}}(X)$. We now apply K^{C^*Cat} and use its additivity shown Proposition 8.72 in order to get

$$\mathrm{id}_{K\mathcal{X}_{\mathrm{ql}}(X)} + K\mathcal{X}_{\mathrm{ql}}(f_*) \circ \mathrm{K}^{C^*Cat} \circ S \simeq \mathrm{K}^{C^*Cat} \circ S.$$

Since f_* is close to id_X by Condition 1 in Definition 3.21 and $K\mathcal{X}_{\mathrm{ql}}$ is coarsely invariant by Proposition 8.80 we conclude that $K\mathcal{X}_{\mathrm{ql}}(f_*) \simeq \mathrm{id}_{\mathrm{id}_{K\mathcal{X}_{\mathrm{ql}}(X)}}$. The resulting relation

$$\mathrm{id}_{K\mathcal{X}_{\mathrm{ql}}(X)} + \mathrm{K}^{C^*Cat} \circ S \simeq \mathrm{K}^{C^*Cat} \circ S \tag{8.28}$$

in $\mathrm{End}_{\mathbf{Sp}^{\mathrm{la}}}(K\mathcal{X}_{\mathrm{ql}}(X))$ implies that $K\mathcal{X}_{\mathrm{ql}}(X) \simeq 0$.   □

Proposition 8.86 *The functors $K\mathcal{X}$ and $K\mathcal{X}_{\mathrm{ql}}$ are u-continuous.*

Proof We consider the quasi-local case. For $K\mathcal{X}$ the argument is similar.[5]

Let X be a bornological coarse space with coarse structure $\mathcal{C}$. We observe that the C^*-categories $\mathbf{C}^*_{\mathrm{ql}}(X)$ and $\mathbf{C}^*_{\mathrm{ql}}(X_U)$ have the same set of objects, see Remark 8.7. Let (H, ϕ) and (H', ϕ') be X-controlled Hilbert spaces.

---

[5] A detailed argument in this case can also be found in [BEe].

Then $\mathrm{Hom}_{\mathbf{C}^*_{\mathrm{ql}}(X_U)}((H, \phi), (H', \phi'))$ can be identified with the subspace of $\mathrm{Hom}_{\mathbf{C}^*_{\mathrm{ql}}(X)}((H, \phi), (H', \phi'))$ of all U-quasi-local operators. The C^*-category $\mathbf{C}^*_{\mathrm{ql}}(X)$ is then equal to the completion of the union of these subcategories for all U in $\mathcal{C}$. By Proposition 8.58 we conclude that

$$\operatorname*{colim}_{U \in \mathcal{C}} K\mathcal{X}_{\mathrm{ql}}(X_U) \simeq K\mathcal{X}_{\mathrm{ql}}(C)$$

as required.   □

This finishes the proof of Theorem 8.79.

A priori the condition of controlled propagation is stronger than quasi-locality. Hence for every bornological coarse space X we have a natural inclusion

$$\mathbf{C}^*(X) \to \mathbf{C}^*_{\mathrm{ql}}(X)$$

of C^*-categories. It induces a natural transformation between the coarse K-theory functors

$$K\mathcal{X} \to K\mathcal{X}_{\mathrm{ql}} . \tag{8.29}$$

Lemma 8.87 *If X is discrete, then $K\mathcal{X}(X) \to K\mathcal{X}_{\mathrm{ql}}(X)$ is an equivalence.*

Proof For a discrete bornological coarse space X the natural inclusion $\mathbf{C}^*(X) \to \mathbf{C}^*_{\mathrm{ql}}(X)$ is an equality. Indeed, the diagonal is the maximal entourage of such a space. Consequently, the condition of quasi-locality reduces to the condition of controlled (in this case, actually zero) propagation.   □

8.7 Comparison with the Classical Definition

We consider a bornological coarse space X and an X-controlled Hilbert space (H, ϕ). By $C^*(X, H, \phi)$ and $C^*_{\mathrm{ql}}(X, H, \phi)$ we denote the Roe algebras (Definition 8.34) associated to this data. Recall that these are the versions generated by locally finite operators of finite propagation (or quasi-local operators, respectively).

We let $K_* := \pi_* K^{C^*} : \mathbf{C}^*\mathbf{Alg}^{\mathrm{la}} \to \mathbf{Ab}^{\mathrm{la}, \mathbb{Z}\mathrm{gr}}$ denote the $\mathbb{Z}$-graded abelian group-valued K-theory functor for C^*-algebras (see Sect. 8.4), and we let $K\mathcal{X}_{\mathrm{ql},*} := \pi_* K\mathcal{X}_{\mathrm{ql}}$ and $K\mathcal{X}_* := \pi_* K\mathcal{X}$ denote the $\mathbb{Z}$-graded abelian group-valued functors derived from the coarse homology functors.

Theorem 8.88 (Comparison Theorem) *If (H, ϕ) is ample, then we have canonical isomorphisms*

$$K_*(C^*(X, H, \phi)) \cong K\mathcal{X}_*(X) \text{ and } K_*(C^*_{\mathrm{ql}}(X, H, \phi)) \cong K\mathcal{X}_{\mathrm{ql},*}(X) .$$

Remark 8.89 In Theorem 8.99 below (and also [BEd, Thm. 6.1] in the equivariant case) we refine the left isomorphism to a natural equivalence of spectra.   □

Proof The actual proof of Theorem 8.88 requires a few preparations. We discuss the case of the controlled-propagation Roe algebra. The quasi-local case is analogous.

We consider a bornological coarse space and an X-controlled Hilbert space (H, ϕ). Recall Definition 8.10 of the notion of a locally finite subspace. Let H' and H'' be locally finite subspaces of H. Then we can form the closure of their algebraic sum in H for which we use the notation

$$H' \bar{+} H'' := \overline{H' + H''} \subseteq H .$$ □

Lemma 8.90 *If* (H, ϕ) *is determined on points, then* $H' \bar{+} H''$ *is a locally finite subspace of* H.

Proof By assumption we can find an entourage V of X and X-controls ϕ' and ϕ'' on H' and H'' such that (H', ϕ') and (H'', ϕ'') are locally finite X-controlled Hilbert spaces and the inclusions $H' \to H$ and $H'' \to H$ have propagation controlled by V^{-1}.

Fix a symmetric entourage U of X. By Lemma 8.19 we can choose a U-separated subset D of X with $U[D] = X$.

By the axiom of choice we choose a well-ordering of D. We will construct the X-control ψ on $H' \bar{+} H''$ by a transfinite induction. In the following argument f denotes a generic element of $C(X)$.

We start with the smallest element d_0 of D. We set

$$B_{d_0} := U[d_0] .$$

Because of the inclusion

$$\phi(B_{d_0})H' \subseteq \phi'(V[B_{d_0}])H'$$

the subspace $\phi(B_{d_0})H'$ is finite-dimensional. Analogously we conclude that $\phi(B_{d_0})H''$ is finite-dimensional. Consequently, $\phi(B_{d_0})(H' + H'') = \phi(B_{d_0})(H' \bar{+} H'')$ is finite-dimensional. We define the subspace

$$H_{d_0} := \phi(B_{d_0})(H' + H'') \subseteq H$$

and let Q_{d_0} be the orthogonal projection onto H_{d_0}. We further define the control ψ_{d_0} of H_{d_0} by $\psi_{d_0}(f) := f(d_0)Q_{d_0}$. Using ψ_{d_0} we recognize $H_{d_0} \subseteq H$ as a locally finite subspace.

Let $\lambda + 1$ in D be a successor ordinal and suppose that the subset B_λ of $U[\lambda]$, the subspace H_λ of H, the projection Q_λ, and $\psi_\lambda : C(X) \to B(H_\lambda)$ have already been constructed such that ψ_λ recognizes H_λ as a locally finite subspace of H. Then we define the subset

$$B_{\lambda+1} := U[\lambda + 1] \setminus \bigcup_{\mu \leq \lambda} B_\mu$$

of X. As above we observe that $\phi(B_{\lambda+1})(H' \bar{+} H'') \subseteq H$ is finite-dimensional. We define the closed subspace

$$H_{\lambda+1} := H_\lambda + \phi(B_{\lambda+1})(H' \bar{+} H'')$$

of H. We let $Q_{\lambda+1}$ be the orthogonal projection onto $H_{\lambda+1} \ominus H_\lambda$ and define the control $\psi_{\lambda+1}$ of $H_{\lambda+1}$ by

$$\psi_{\lambda+1}(f) := \psi_\lambda(f) + f(d_{\lambda+1})Q_{\lambda+1}.$$

We conclude easily that it recognizes $H_{\lambda+1}$ as a locally finite subspace of H.

Let λ be a limit ordinal. Then we set $B_\lambda := \emptyset$ and $Q_\lambda := 0$. We define the closed subspace

$$H_\lambda := \overline{\bigcup_{\mu < \lambda} H_\mu}$$

of H. We furthermore define the control ψ_λ of H_λ by

$$\psi_\lambda(f) := \sum_{\mu < \lambda} f(d_\mu)Q_\mu.$$

We argue that the sum describes a well-defined operator on H_λ. The sum has a well-defined interpretation on H_μ for all $\mu < \lambda$.

Let $(h_k)_{k \in \mathbb{N}}$ be a sequence in $\bigcup_{\mu < \lambda} H_\mu$ converging to h. For every k in $\mathbb{N}$ there is a μ in D with $\mu < \lambda$ such that $h_k \in H_\mu$. We set $\psi_\lambda(f)h_k := \psi_\mu(f)h_k$. This definition does not depend on the choice of μ. Given ε in $(0, \infty)$ we can find k_0 in $\mathbb{N}$ such that $\|h_k - h\| \leq \varepsilon$ for all k in $\mathbb{N}$ with $k \geq k_0$. Then for integers k, k' in $\mathbb{N}$ with $k, k' \geq k_0$ we have $\|\psi_\lambda(f)h_k - \psi_\lambda(f)h_{k'}\| \leq \varepsilon\|f\|_\infty$. This shows that $(\psi_\lambda(f)h_k)_{k \in \mathbb{N}}$ is a Cauchy sequence. We define

$$\psi_\lambda(f)h := \lim_{k \to \infty} \psi_\lambda(f)h_k.$$

The estimates $\|\psi_\lambda(f)h_k\| = \|\psi_\mu(f)h_k\| \leq \|f\|_\infty\|h_k\|$ for all k in $\mathbb{N}$ imply the estimate $\|\psi_\lambda(f)h\| \leq \|f\|_\infty\|h\|$ and hence show that $\psi_\lambda(f)$ is continuous. One furthermore checks, using the mutual orthogonality of the Q_μ for $\mu < \lambda$, that $f \mapsto \psi_\lambda(f)$ is a C^*-algebra homomorphism from $C(X)$ to $B(H_\lambda)$.

We claim that ψ_λ recognizes H_λ as a locally finite subspace of H. If B is a bounded subset of X, then the inclusion $\psi_\lambda(B)H_\lambda \subseteq \phi(U[B])(H' \dotplus H'')$ shows that $\psi_\lambda(B)H_\lambda$ is finite-dimensional. Furthermore, the propagation of the inclusion of H_λ into H is controlled by U.

By construction $(H_\lambda, \phi_\lambda)$ is determined on points. $\square$

Let (H, ϕ) be an X-controlled Hilbert space. The set of locally finite subspaces of (H, ϕ) is partially ordered by inclusion. Lemma 8.90 has the following corollary.

Corollary 8.91 *If (H, ϕ) is determined on points, then the partially ordered set of locally finite subspaces of an X-controlled Hilbert space is filtered.*

If H' is a locally finite subspace of H, then the Roe algebra $\mathcal{C}^*(X, H', \phi')$ (as a subalgebra of $B(H')$) and the inclusion $\mathcal{C}^*(X, H', \phi') \to \mathcal{C}^*(X, H, \phi)$ do not depend on the choice of ϕ' recognizing H' as a locally finite subspace. This provides the connecting maps of the colimit in the statement of the following lemma.

Lemma 8.92 *We have a canonical isomorphism*

$$K_*(\mathcal{C}^*(X, H, \phi)) \cong \operatorname*{colim}_{H'} K_*(\mathcal{C}^*(X, H', \phi'))$$

where the colimit runs over the filtered partially ordered set of locally finite subspaces H' of H.

Proof If A is a bounded linear operator on H', then the conditions of having controlled propagation as an operator on (H, ϕ) or on (H', ϕ') are equivalent. It follows that $\mathcal{C}^*(X, H', \phi')$ is exactly the subalgebra generated by operators coming from H'. So by definition of the Roe algebra

$$\mathcal{C}^*(X, H, \phi) \cong \operatorname*{colim}_{H'} \mathcal{C}^*(X, H', \phi'),$$

where the colimit is taken over the filtered partially ordered set of locally finite subspaces of H and interpreted in the category of C^*-algebras. Since the K-theory functor for C^*-algebras preserves filtered colimits by Property 3 of K^{C^*} we conclude the desired isomorphisms. $\square$

We now continue with the proof of Theorem 8.88. By Proposition 8.55 and Definition 8.78 we have an isomorphism

$$K\mathcal{X}_*(X) \cong K_*(A(\mathbf{C}^*(X))). \qquad (8.30)$$

The same argument as for the isomorphism (8.10) shows that we have a canonical isomorphism

$$\operatorname*{colim}_{\mathbf{D}} A(\mathbf{D}) \cong A(\mathbf{C}^*(X)), \qquad (8.31)$$

where **D** runs over the filtered subset of subcategories of $\mathbf{C}^*(X)$ with finitely many objects. Since the K_*- commutes with filtered colimits we get the canonical isomorphism

$$\operatorname*{colim}_{\mathbf{D}} K_*(A(\mathbf{D})) \cong K_*(A(\mathbf{C}^*(X))).$$

If H' is a locally finite subspace of H, then we can consider $C(X, H', \phi')$ as a subcategory of $\mathbf{C}^*(X)$ with a single object (H', ϕ'). These inclusions induce the two up-pointing homomorphisms in Diagram (8.32) below.

Let H' and H'' be closed locally finite subspaces of H such that $H' \subseteq H''$.

Lemma 8.93 *The following diagram commutes:*

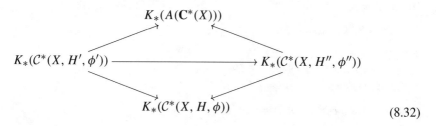

$$(8.32)$$

Proof The commutativity of the lower triangle is clear since it is induced from a commutative triangle of Roe algebras. We now consider the upper triangle. Let $i : C^*(X, H', \phi') \to A(\mathbf{C}^*(X))$, $j : C^*(X, H'', \phi'') \to A(\mathbf{C}^*(X))$, and $k : C^*(X, H', \phi') \to (X, H'', \phi''))$ denote the inclusions. We consider the inclusion $u : H' \to H''$ in $\operatorname{Hom}_{\mathbf{C}^*(X)}((H', \phi'), (H'', \phi''))$ as a morphism in $A(\mathbf{C}^*(X))$ in the natural way, it lives in single summand of (8.9). Then we have equalities of homomorphisms $i = iu^*u$ and $i' := uiu^* = j \circ k$ from $C^*(X, H', \phi')$ to $A(\mathbf{C}^*(X))$. By Property 9 of K^{C^*} the maps i and i' induce the same maps on the level of K-theory groups. Hence the upper triangle commutes as well. □

We continue with the proof of Theorem 8.88. By a combination of Lemmas 8.92 and 8.93 we get a canonical homomorphism

$$\alpha : K_*(C^*(X, H, \phi)) \to K_*(A(\mathbf{C}^*(X))). \qquad (8.33)$$

Lemma 8.94 *The homomorphism* (8.33) *is an isomorphism.*

Proof Let x in $K_*(C^*(X, H, \phi))$ be given such that $\alpha(x) = 0$. Then there exists a locally finite subspace H' of H such that x is realized by some x' in $K_*(C^*(X, H', \phi'))$. Furthermore there exists a subcategory **D** of $\mathbf{C}^*(X)$ with finitely many objects containing (H', ϕ') such that x' maps to zero in $K_*(A(\mathbf{D}))$. Using ampleness of (H, ϕ), by Lemma 8.22 we can extend the embedding $H' \to H$ to an embedding $H'' := \bigoplus_{(H_1, \phi_1) \in \mathrm{Ob}(\mathbf{D})} H_1 \to H$ as a locally finite subspace containing H'. Then x' maps to zero in $K_*(C(X, H'', \phi''))$. Hence $x = 0$.

Let now y in $K_*(A(\mathbf{C}^*(X)))$ be given. Then there exists a subcategory with finitely many objects $\mathbf{D}$ of $\mathbf{C}^*(X)$ and a class y' in $K(A(\mathbf{D}))$ mapping to y. By Lemma 8.22 we can choose an embedding of $H' := \bigoplus_{(H_1,\phi_1)\in\mathrm{Ob}(\mathbf{D})} H_1 \to H$ as a locally finite subspace. Then y' determines an element x in $K_*(\mathcal{C}^*(X, H, \phi))$ such that $\alpha(x) = y$.   □

The combination of Lemma 8.94 with Eq. (8.30) finishes the proof of Theorem 8.88.   □

Remark 8.95 A version of the Comparison Theorem 8.88 has been discussed in [HP04, Sec. 2.2].   □

In combination with Proposition 8.40 the Theorem 8.88 implies the following comparison. Let X be a bornological coarse space and (H, ϕ) an X-controlled Hilbert space.

Corollary 8.96 *Assume that*

1. *X is separable (Definition 8.38),*
2. *X is locally finite (Definition 8.15), and*
3. *(H, ϕ) is ample (Definition 8.13).*

Then we have a canonical isomorphism

$$K_*(\mathcal{C}^*_{\mathrm{lc}}(X, H, \phi)) \simeq K\mathcal{X}_*(X).$$

Example 8.97 Combining the Properties 2 and 6 of K^{C^*} we conclude $K^{C^*}(\mathbb{K}(\ell^2)) \simeq KU$.

We can consider ℓ^2 as an ample $*$-controlled Hilbert space on the space $*$. In this case all four versions of the Roe algebras coincide with $\mathbb{K}(\ell^2)$. We get

$$K\mathcal{X}(*) \simeq K\mathcal{X}_{\mathrm{ql}}(*) \simeq K^{C^*}(\mathbb{K}(\ell^2))$$

and hence $K\mathcal{X}(*)$ and $K\mathcal{X}_{\mathrm{ql}}(*)$ are equivalent to KU.   □

Let X be a bornological coarse space.

Theorem 8.98 *Assume that X has weakly finite asymptotic dimension (Definition 5.58). Then:*

1. *The canonical transformation $K\mathcal{X}(X) \to K\mathcal{X}_{\mathrm{ql}}(X)$ is an equivalence.*
2. *If (H, ϕ) is an ample X-controlled Hilbert space, then the inclusion of Roe algebras $\mathcal{C}^*(X, H, \phi) \hookrightarrow \mathcal{C}^*_{\mathrm{ql}}(X, H, \phi)$ induces an equivalence*

$$K(\mathcal{C}^*(X, H, \phi)) \to K(\mathcal{C}^*_{\mathrm{ql}}(X, H, \phi))$$

of K-theory spectra.

Proof By Theorem 8.79 the functors $K\mathcal{X}$ and $K\mathcal{X}_{ql}$ are coarse homology theories in the sense of Definition 4.22. Assertion 1 follows from Lemma 8.87 together with Corollary 6.43.

Assertion 2 follows from 1 and Theorem 8.88. □

For the study of secondary or higher invariants in coarse index theory it is important to refine the isomorphism given in Theorem 8.88 to the spectrum level. This is the content of Theorem 8.99.1 below.

Let X, X' be bornological coarse spaces. Let furthermore (H, ϕ) be an ample X-controlled Hilbert space and (H', ϕ') be an ample X'-controlled Hilbert space. Assume that we are given a morphism $f : X' \to X$ and an isometry $V : H' \to H$. In the following in order to interpret $\mathrm{supp}(V)$ we view V as an operator from $f_*(H, \phi)$ to (H', ϕ'). We assume that there exists an entourage U of X such that $\mathrm{supp}(V) \subseteq U$. Then we define a morphism of C^*-algebras by

$$v : C^*(X', H', \phi') \to C^*(X, H, \phi), \quad v(A) := VAV^*.$$

If A has W-controlled propagation, then VAV^* has $U \circ (f \times f)(W) \circ U^{-1}$-controlled propagation. One furthermore checks that the conjugation $V - V^*$ preserves local finiteness. The only relation between v and f is the support condition on V. If V_1, V_2 are two isometries $H' \to H$ satisfying $\mathrm{supp}(V_1) \subseteq U$ and $\mathrm{supp}(V_2) \subseteq U$, then the induced maps $K_*(v_1)$ and $K_*(v_2)$ on K-theory groups are the same. To see this, following [HRY93, Lem. 3 in Sec. 4], we note that $V_2 V_1^*$ is a multiplier on $C^*(X, H, \phi)$, and that $v_1(V_2 V_1^*)^*(V_2 V_1^*) = v_1$ and $v_2 = (V_2 V_1^*)v_1(V_2 V_1^*)^*$. We now use Property 9 of K^{C^*}. The refinement of the equality $K_*(v_1) = K_*(v_2)$ to the spectrum level is content of Point 2 in the next theorem.

Theorem 8.99

1. *There is a canonical (up to equivalence) equivalence of spectra*

$$\kappa_{(X,H,\phi)} : K^{C^*}(C^*(X, H, \phi)) \to K\mathcal{X}(X).$$

2. *We have a commuting diagram*

$$\begin{array}{ccc}
K^{C^*}(C^*(X', H', \phi')) & \xrightarrow{\kappa_{(X',H',\phi')}} & K\mathcal{X}(X') \\
\downarrow{\scriptstyle K^{C^*}(v)} & & \downarrow{\scriptstyle f_*} \\
K^{C^*}(C^*(X, H, \phi)) & \xrightarrow{\kappa_{(X,H,\phi)}} & K\mathcal{X}(X)
\end{array} \tag{8.34}$$

Proof We start with the proof of the first assertion. We consider the category $\mathbf{C}^*(X, H, \phi)$ whose objects are triples (H', ϕ', U),[6] where (H', ϕ') is an object of $\mathbf{C}^*(X)$ and U is a controlled isometric embedding $U : H' \to H$ as a locally

---

[6]This interpretation of the symbols H' and ϕ' is local in the proof of the first part of the theorem.

finite subspace. A morphism $A : (H'_0, \phi'_0, U_0) \to (H'_1, \phi'_1, U_1)$ is an operator in $\mathrm{Hom}_{\mathbf{C}^*(X)}((H'_0, \phi'_0), (H'_1, \phi'_1))$.

We consider the Roe algebra $C^*(X, H, \phi)$ as a non-unital C^*-category with a single object. Then we have functors between non-unital C^*-categories

$$\mathbf{C}^*(X) \xleftarrow{F} \mathbf{C}^*(X, H, \phi) \xrightarrow{I} C^*(X, H, \phi).$$

The functor F forgets the inclusions. The action of I on objects is clear, and it maps a morphism $A : (H', \phi', U') \to (H'', \phi'', U'')$ in $\mathbf{C}^*(X, H, \phi)$ to the morphism $I(A) := U'' A U'^{,*}$ of $C^*(X, H, \phi)$.

We get an induced diagram of K-theory spectra

$$\mathrm{K}^{C^*Cat}(\mathbf{C}^*(X)) \xleftarrow{\mathrm{K}^{C^*Cat}(F)} \mathrm{K}^{C^*Cat}(\mathbf{C}^*(X, H, \phi)) \xrightarrow{\mathrm{K}^{C^*Cat}(I)} K^{C^*}(C^*(X, H, \phi)).$$

$\square$

Lemma 8.100 $K(F)$ *is an equivalence.*

Proof We claim that F is a unitary equivalence of C^*-categories. Then $\mathrm{K}^{C^*Cat}(F)$ is an equivalence of spectra by Corollary 8.63.

Note that F is fully faithful. Furthermore, since (H, ϕ) is ample, for every object (H', ϕ') of $\mathbf{C}^*(X)$ there exist a controlled isometric embedding $U : H' \to H$ by Lemma 8.22. Hence the object (H', ϕ') is in the image of F. So F is surjective on objects. $\square$

Lemma 8.101 $K^{C^*}(I)$ *is an equivalence.*

Proof We show that $K_*(I)$ is isomorphism on homotopy groups.

Surjectivity Let x be in $K_*(C^*(X, H, \phi))$. By Lemma 8.92 there exists a locally finite subspace H' of (H, ϕ) and a class x' in $K_*(C^*(X, H', \phi'))$ such that $K_*(u')(x') = x$ for the canonical homomorphism $u' : C^*(X, H', \phi') \to C^*(X, H, \phi)$ which sends the operator A in its domain to $U' A U'^{,*}$. Here ϕ' is an X-control on H' which exhibits it as a locally finite subspace of H, and $U' : H' \to H$ is the isometric embedding. We have a commuting diagram of non-unital C^*-categories

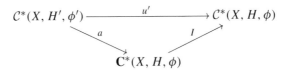

where a sends the unique object of its domain to (H', ϕ', U) and is the obvious map on morphisms. This factorization implies that x is also in the image of $K_*(I)$.

Injectivity Assume that x in $K_*(\mathbf{C}^*(X, H, \phi))$ is such that $K_*(I)(x) = 0$. There is a subcategory $\mathbf{D}$ of $\mathbf{C}^*(X, H, \phi)$ with finitely many objects and a class y in $K_*(\mathbf{D})$ with $x = K_*(i)(y)$, where $i : \mathbf{D} \to \mathbf{C}^*(X, H, \phi)$ is the inclusion. This follows from Proposition 8.57.

We define

$$\tilde{H} := \sum_{(H', \phi', U') \in \mathbf{D}} H' + H_1$$

with the sum taken in H, where H_1 is a locally finite subspace which will be chosen below. By Lemma 8.90 we know that $\tilde{H}$ is a locally finite subspace of H. Hence we can choose a control function $\tilde{\phi}$ on $\tilde{H}$ exhibiting $\tilde{H}$ as a locally finite subspace and let $\tilde{U} : \tilde{H} \to H$ denote the inclusion. We form the full subcategory category $\tilde{\mathbf{D}}$ of $\mathbf{C}^*(X, H, \phi)$ with set of objects $\mathrm{Ob}(\mathbf{D}) \cup \{(\tilde{H}, \tilde{\phi}, \tilde{U})\}$. We have a functor $\tilde{I} : \mathbf{D} \to C^*(X, \tilde{H}, \tilde{\phi})$ defined similarly as I and a diagram

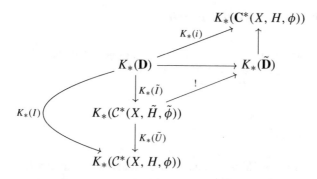

The upper right triangle is given by the obvious inclusions of C^*-categories. Also the left triangle commutes on the level of C^*-categories. The middle triangle commutes by an argument which is similar to the argument for the upper triangle in (8.32), where the marked morphism is induced by the inclusion $C^*(X, \tilde{H}, \tilde{\phi}) \to \tilde{\mathbf{D}}$ of C^*-categories.

Now by Lemma 8.92, since $K_*(I)(y) = 0$, we can choose H_1 so large such that $K_*(\tilde{I})(y) = 0$. This implies $x = K_*(i)(y) = 0$. $\square$

By definition $K\mathcal{X}(X) \simeq \mathrm{K}^{C^*Cat}(\mathbf{C}^*(X))$ and so the composition of an inverse $\mathrm{K}^{C^*Cat}(I)^{-1}$ with $\mathrm{K}^{C^*Cat}(F)$ provides the asserted equivalence

$$\kappa_{(X, H, \phi)} : K^{C^*}(C^*(X, H, \phi)) \to K\mathcal{X}(X).$$

We now show the second assertion of Theorem 8.99. We have the induced morphism of C^*-algebras

$$v : C^*(X', H', \phi') \to C^*(X, H, \phi), \quad v(A) := VAV^*.$$

The square (8.34) is induced from a commuting (in the one-categorical sense) diagram of C^*-categories

$$
\begin{array}{ccccc}
\mathcal{C}^*(X', H', \phi') & \xleftarrow{\ I'\ } & \mathbf{C}^*(X', H', \phi') & \xrightarrow{\ F'\ } & \mathbf{C}(X') \\
\downarrow{\scriptstyle v} & & \downarrow & & \downarrow{\scriptstyle f_*} \\
\mathcal{C}^*(X, H, \phi) & \xleftarrow{\ I\ } & \mathbf{C}^*(X, H, \phi) & \xrightarrow{\ F\ } & \mathbf{C}(X)
\end{array}
$$

where all horizontal arrows induce equivalences in K-theory, and the middle arrow is the functor which sends the object (H'', ϕ'', U'') to $(H'', \phi'' \circ f^*, V \circ U'')$ and is the identity on morphisms. It is then clear that the right square commutes. One also checks directly that the left square commutes.

This finishes the proof of Theorem 8.99.   □

8.8 Additivity and Coproducts

In this section we investigate how coarse K-homology interacts with coproducts and free unions. We refer to Sect. 6.2 for definitions.

8.8.1 Additivity

In order to approach additivity and strong additivity for coarse K-homology we need results stating that the K-theory for C^*-algebras preserves infinite products at least under certain additional conditions.

Let $(A_i)_{i \in I}$ be a family of C^*-algebras. The projections $p_j : \prod_{i \in I} A_i \to A_j$ for all j in I give a morphism of spectra

$$
(K^{C^*}(p_i))_{i \in I} : K^{C^*}\left(\prod_{i \in I} A_i\right) \to \prod_{i \in I} K^{C^*}(A_i). \tag{8.35}
$$

Then we ask for conditions on the factors ensuring that (8.35) is an equivalence.

Remark 8.102 In order to understand one of the difficulties note that K-theory classes are represented by projections or unitaries in matrix algebras. Matrices with coefficients in $\prod_{i \in I} A_i$ correspond to families of matrices in the factors of uniformly bounded size. On the other hand, a class in $\prod_{i \in I} K^{C^*}(A_i)$ is a family of K-theory classes for the factors. If the index set I is infinite, the members of this family might not be representable by a family of matrices of uniformly bounded size, see Remark 8.111 and the references give there.

Therefore conditions which help to avoid matrices are useful. $\mathbb{K}$-stability is such a condition. Recall that a C^*-algebra is called $\mathbb{K}$-stable if $A \cong A \otimes \mathbb{K}$, where $\mathbb{K} :=$

$\mathbb{K}(\ell^2)$. But note that an infinite product of $\mathbb{K}$-stable C^*-algebras might not be $\mathbb{K}$-stable again. The notion of quasi-$\mathbb{K}$-stability introduced in Definition 8.103 is better behaved in this respect by Lemma 8.107.   □

Let A be a C^*-algebra. We let $M_n(A)$ denote the $n \times n$-matrix algebra over A. Furthermore, for a C^*-algebra B we let $\mathcal{M}(B)$ denote the multiplier algebra of B.

Definition 8.103 ([WY20, Defn. 2.7.11]) *A* is called *quasi-$\mathbb{K}$-stable* if for every n in $\mathbb{N}$ there exists an isometry v in $\mathcal{M}(M_n(A))$ such that vv^* is the elementary matrix e_{11}.

Remark 8.104 Let v be an isometry as in Definition 8.103. Because of the relation $e_{11}v = vv^*v = v$ we know that v is of the form

$$v = \begin{pmatrix} v_{11} & v_{12} & \cdots & v_{1n} \\ 0 & 0 & \cdots & 0 \\ \vdots & \vdots & \ddots & \vdots \\ 0 & 0 & \cdots & 0 \end{pmatrix},$$

where v_{1i} belongs to $\mathcal{M}(A)$ for every i in $\{1, \ldots, n\}$. Here we use the compatibility of multiplier algebras with forming matrices: $\mathcal{M}(M_n(A)) = M_n(\mathcal{M}(A))$. Because v is an isometry, i.e., $v^*v = \mathrm{id}_{\mathcal{M}(M_n(A))}$, and since $vv^* = e_{11}$, we furthermore have the relations

$$v_{1i}^* v_{1j} = \delta_{ij} 1_{\mathcal{M}(A)}, \quad \sum_{i=1}^{n} v_{1i} v_{1i}^* = 1_{\mathcal{M}(A)}.$$

In particular, v_{1i} is an isometry for every i in $\{1, \ldots, n\}$.

Let x be any element of $M_n(A)$. Then vxv^* is supported only in the upper left corner of $M_n(A)$ and hence may be identified canonically with an element of A. This follows from the equality

$$e_{11}vxv^*e_{11} = vv^*vxv^*vv^* = vxv^*.$$

where we use that v is an isometry, i.e., $v^*v = \mathrm{id}_{\mathcal{M}(M_n(A))}$, and that $vv^* = e_{11}$.

Let $\iota_n \colon A \to M_n(A)$ denote the inclusion as the upper left corner. Let furthermore $\pi_{11} \colon M_n(A) \to A$ be the $\mathbb{C}$-linear projection onto this corner. The $\mathbb{C}$-linear map $A \to A$ given by $x \mapsto \pi_{11}(v\iota_n(x)v^*)$ is a homomorphism of C^*-algebras. It is equal to the conjugation map

$$x \mapsto v_{11}xv_{11}^*, \tag{8.36}$$

by the isometry v_{11} in $\mathcal{M}(A)$.   □

Example 8.105 Let H be an infinite-dimensional Hilbert space. The C^*-algebra $\mathbb{K}(H)$ of all compact operators on H is quasi-$\mathbb{K}$-stable. In order to see this we identify $M_n(\mathbb{K}(H))$ with $\mathbb{B}(H^{\oplus n})$. Since H is infinite-dimensional we can choose a unitary $u \colon H^{\oplus n} \to H$. We let $\iota \colon H \to H^{\oplus n}$ be the inclusion as the first summand. Then the isometry $v := \iota \circ u$ has the required properties.

The same isometries also exhibit $\mathbb{B}(H)$ as a quasi-$\mathbb{K}$-stable C^*-algebra.   □

Remark 8.106 An elaboration of Example 8.105 shows that every $\mathbb{K}$-stable C^*-algebra is quasi-$\mathbb{K}$-stable.

Furthermore, since the C^*-algebra $\mathbb{B}(H)$ of all bounded linear operators on an infinite-dimensional Hilbert space H is not $\mathbb{K}$-stable, we conclude that the notion of quasi-$\mathbb{K}$-stability is strictly more general than the notion of $\mathbb{K}$-stability.   □

Let $(A_i)_{i \in I}$ be a family of C^*-algebras.

Lemma 8.107 *If A_i is quasi-$\mathbb{K}$-stable for every i in I, then $\prod_{i \in I} A_i$ is quasi-K-stable.*

Proof Let n be in $\mathbb{N}$. For every i in I we choose an isometry v_i in $\mathcal{M}(M_n(A_i))$ as in the Definition 8.103. The family $(v_i)_{i \in I}$ gives rise to an element v in $\mathcal{M}(M_n(\prod_{i \in I} A_i))$ satisfying the condition of Definition 8.103.   □

Let $(A_i)_{i \in I}$ be a family of C^*-algebras.

Proposition 8.108 *If A_i is quasi-$\mathbb{K}$-stable for every i in I, then*

$$(K^{C^*}(p_i))_{i \in I} : K^{C^*}(\prod_{i \in I} A_i) \to \prod_{i \in I} K^{C^*}(A_i). \tag{8.37}$$

is an equivalence.

Proof We use the notation $K_* = \pi_* K^{C^*}$ for the group-valued K-theory functor. By Bott periodicity it suffices to show that

$$(K_*(p_i))_{i \in I} : K_*(\prod_{i \in I} A_i) \to \prod_{i \in I} K_*(A_i) \tag{8.38}$$

is an isomorphism for $* = 0, 1$.

In the following we identify a C^*-algebra A with a subalgebra of $M_n(A)$ using the left-upper corner inclusion.

We let A^+ denote the unitalization of A. We let further $e : A^+ \to \mathbb{C}$ be the canonical projection such that $\ker(e) = A$. We use the same symbol for the extension of e to matrix algebras. For simplicity we will not add a subscript indicating that e is associated with A.

In general, an element x in $K_0(A)$ can represented by a pair of projections (p, q) in $M_n(A^+)$ for some n in $\mathbb{N}$ such that $e(p) = e(q)$. We write $x = [p, q]$. If A is quasi-$\mathbb{K}$-stable and v is the isometry as in Definition 8.103, then using Property 9 of K^{C^*} we see that $x = [vpv^*, vqv^*]$. Note that vpv^* and vqv^* are projections in A^+ such that $e(vpv^*) = e(vqv^*)$. Therefore, if A is quasi-$\mathbb{K}$-stable, then every element in $K_0(A)$ can actually be represented by a pair (p, q) of projections p, q in A^+ with $e(p) = e(q)$.

Let p, q be projections in A^+ such that $e(p) = e(q)$. If $[p, q] = 0$, then after increasing n (and filling matrices up by zero) if necessary, there exists a partial isometry u in $M_n(A^+)$ such that $p = uu^*$ and $q = u^*u$. We define the partial isometry $w := puq$ in A^+. Then $p = ww^*$ and $q = w^*w$.

An element y in $K_1(A)$ can be represented by a unitary u in $M_n(A^+)$ for some n in $\mathbb{N}$ such that $e(u) = 1_{M_n(\mathbb{C})}$. If A is quasi-$\mathbb{K}$-stable and v is the isometry as in Definition 8.103, then $u' := v(u - 1_{M_n(A^+)})v^* + 1_{A^+}$ is a unitary in A^+ (where we use that the matrix $v(u - 1_{M_n(A^+)})v^*$ is supported in the left upper corner and is thus considered as an element in A, see Remark 8.104) with $e(u') = 1$. By Property 9 of K^{C^*} we have $y = [u']$. Hence, if A is quasi-$\mathbb{K}$-stable, then every element in $K_1(A)$ can be represented by a unitary u in A^+ with $e(u) = 1$.

After these preparations we now turn to the actual proof of Proposition 8.37.

Surjectivity on K_0

An element $x = (x_i)_{i \in I}$ of $\prod_{i \in I} K_0(A_i)$ may be represented by a family of pairs $((p_i, q_i))_{i \in I}$ of projections p_i, q_i in A_i^+ such that $e(p_i) = e(q_i)$ for every i in I. Since $e(p_i)$ is a projection we have $e(p_i) \in \{0, 1\}$. For k in $\{0, 1\}$ we define the subsets $I_k := \{i \in I \mid e(p_i) = k\}$ of I. Then (I_0, I_1) is a partition of I. For k in $\{0, 1\}$ we form the families $(p_{k,i})_{i \in I}$ and $(q_{k,i})_{i \in I}$

$$p_{k,i} := \begin{cases} p_i - k1_{A^+} & i \in I_k \\ 0 & i \notin I_k \end{cases}, \quad q_{k,i} := \begin{cases} q_i - k1_{A^+} & i \in I_k \\ 0 & i \notin I_k \end{cases}$$

Note that $p_{k,i}, q_{k,i} \in A_i$ for all i in I. Then we define the elements

$$P_k := (p_{k,i})_{i \in I} + k1_{(\prod_{i \in I} A_i)^+}, \quad Q_k := (q_{k,i})_{i \in I} + k1_{(\prod_{i \in I} A_i)^+}$$

in $(\prod_{i \in I} A_i)^+$. One checks that P_k and Q_k are projections, and that $e(P_k) = e(Q_k)$. Hence we get classes $[P_k, Q_k]$ in $K_0(\prod_{i \in I} A_i)$ for k in $\{0, 1\}$. We claim that $y := [P_0, Q_0] + [P_1, Q_1]$ in $K_0(\prod_{i \in I} A_i)$ is a preimage of x. Indeed, the projection of y to the i'th factor is given by

$$[p_{0,i}, q_{0,1}] + [p_{1,i} + 1_{A_i^+}, q_{1,i} + 1_{A_i^+}].$$

If $i \in I_0$, then this is equal to

$$[p_i, q_i] + [1_{A_i^+}, 1_{A_i^+}] = [p_i, q_i] = x_i.$$

If $i \in I_1$, then this class is equal to

$$[0, 0] + [p_i, q_i] = [p_i, q_i] = x_i$$

again.

Injectivity on K_0

We consider a class $[P, Q]$ in $K_0(\prod_{i \in I} A_i)$, where P, Q are projections in $(\prod_{i \in I} A_i)^+$ such that $e(P) = e(Q)$. Let $k := e(P)$. Then we have

$$P = (\tilde{p}_i)_{i \in I} + k 1_{(\prod_{i \in I} A_i)^+} \qquad \text{and} \qquad Q = (\tilde{q}_i)_{i \in I} + k 1_{(\prod_{i \in I} A_i)^+}$$

such that $\tilde{p}_i, \tilde{q}_i$ belong to A_i and $p_i := \tilde{p}_i + k 1_{A_i^+}$ and $q_i := \tilde{q}_i + k 1_{A_i^+}$ are projections for every i in I.

Assume that $[P, Q]$ is sent to zero. Then for every i in I there exists a partial isometry u_i in A_i^+ such that $u_i^* u_i = p_i$ and $u_i u_i^* = q_i$. If $k = 0$, then $e(u_i) = 0$ for all i in I. If $k = 1$, then we can replace u_i by $e(u_i)^{-1} u_i$. After this modification we have $e(u_i) = k 1_{A_i^+}$ for all i in I. Then $U := (u_i - k 1_{A_i^+}) + k 1_{(\prod_{i \in I} A_i)^+}$ is a partial isometry in $(\prod_{i \in I} A_i)^+$ such that $U^* U = P$ and $U U^* = Q$. We conclude that $[P, Q] = 0$.

Surjectivity on K_1

We consider a class in $\prod_{i \in I} K_1(A_i)$ which may be represented by a family $(u_i)_{i \in I}$ of unitaries u_i in A_i^+ such that $e(u_i) = 1_{A_i^+}$. Then

$$U := (u_i - 1_{A_i^+})_{i \in I} + 1_{(\prod_{i \in I} A_i)^+}$$

is a unitary in $(\prod_{i \in I} A_i)^+$ with $e(U) = 1$. It represents the desired preimage of our class in $K_1(\prod_{i \in I} A_i)$.

Injectivity on K_1

This is different from the above cases since the relation defining K_1 is not algebraic. A unitary represents the trivial element in K-theory if it can be connected by a path with the identity. The problem is that the product of an infinite family of paths is not necessarily continuous. This is true only if the family is equi-continuous.

We will solve the problem by showing that by passing to large matrix algebras one can improve the Lipschitz constants of homotopies. This argument is inspired by [WY20, Proof of Prop. 12.6.3].

Let U be a unitary in $(\prod_{i \in I} A_i)^+$ with $e(U) = 1$. Then

$$U = (u_i - 1_{A_i^+})_{i \in I} + 1_{(\prod_{i \in I} A_i)^+}$$

for unitaries u_i in A_i^+ with $e(u_i) = 1_{A_i^+}$. Assume that the image of U in $\prod_{i \in I} K_1(A_i)$ vanishes. Then for each i in I there is an n_i in $\mathbb{N}$ and a homotopy $\tilde{u}_{i,t}$ parametrized by $[0, 1]$ in the unitary group of $M_{n_i}(A_i)^+$ from $\tilde{u}_{i,0} := \operatorname{diag}(u_i, 1_{A_i^+}, \dots, 1_{A_i^+})$ to $\tilde{u}_{i,1} = 1_{M_{n_i}(A_i)^+}$ satisfying $e(\tilde{u}_{i,t}) = 1_{M_{n_i}(\mathbb{C})}$ for all t in $[0, 1]$.

We can replace the homotopies $\tilde{u}_{i,t}$ by Lipschitz paths. The argument is as follows. For any unital C^*-algebra B the group $\mathrm{Gl}_n(B)$ of invertible matrices with entries in B is an open subgroup of $M_n(B)$. If we have a continuous path parametrized by $[0, 1]$ in $\mathrm{Gl}_n(B)$, then we get a Lipschitz continuous path in $\mathrm{Gl}_n(B)$ between the same endpoints by subdividing the interval $[0, 1]$ sufficiently fine and using linear interpolation. The map $(z, s) \mapsto (1 - s)z + s(zz^*)^{-1/2}z$ defines a deformation retraction of $\mathrm{Gl}_n(B)$ onto $U_n(B)$ with the property that it maps Lipschitz continuous path in $\mathrm{Gl}_n(B)$ to Lipschitz continuous path in $U_n(B)$. Therefore, if we have a continuous path in $U_n(B)$, then there exists also a Lipschitz continuous path in $U_n(B)$ between the same endpoints. Finally, if A is a non-unital C^*-algebra and we have a continuous path u_t in $U_n(A^+)$ with $e(u_t) = 1_{M_n(A^+)}$, then the construction above gives a Lipschitz continuous path u'_t in $U_n(A^+)$ from u_0 to u_1 with $e(u'_t) = 1_{M_n(A^+)}$.

For i in I and a natural number N_i we write $\mathrm{diag}(\tilde{u}_{i,0}, 1, \ldots, 1)$ as the product

$$
\begin{pmatrix}
\tilde{u}_{i,0} & 0 & \cdots & 0 \\
0 & \tilde{u}_{i,1/N_i} & \cdots & 0 \\
\vdots & \vdots & \ddots & \vdots \\
0 & 0 & \ldots & \tilde{u}_{i,1}
\end{pmatrix}
\cdot
\begin{pmatrix}
1_{M_{n_i}(A_i)^+} & 0 & \cdots & 0 \\
0 & (\tilde{u}_{i,1/N_i})^* & \cdots & 0 \\
\vdots & \vdots & \ddots & \vdots \\
0 & 0 & \ldots & (\tilde{u}_{i,1})^*
\end{pmatrix}
\tag{8.39}
$$

We now define a homotopy $\hat{u}_{i,t}$ from $\mathrm{diag}(\tilde{u}_{i,0}, 1, \ldots, 1)$ to $1_{M_{N_i n_i}(A_i)^+}$ inside the unitary group of $M_{N_i n_i}(A_i)^+$ with $e(\hat{u}_{i,t}) = 1_{M_{N_i n_i}(A_i)^+}$. The first part of this homotopy is parametrized by the interval $[0, 1/2]$ and given by the paths from $(\tilde{u}_{i,k/N_i})^*$ to $(\tilde{u}_{i,(k-1)/N_i})^*$ of the matrix entries of the second factor for each k in $\{1, \ldots, N_i\}$. For the second part, we let $r_{i,t}$ denote the rotation homotopy between

$$
\begin{pmatrix}
1_{M_{n_i}(A_i)^+} & 0 & 0 & \cdots & 0 \\
0 & (\tilde{u}_{i,0/N_i})^* & 0 & \cdots & 0 \\
0 & 0 & (\tilde{u}_{i,1/N_i})^* & \cdots & 0 \\
\vdots & \vdots & \vdots & \ddots & \vdots \\
0 & 0 & 0 & \ldots & (\tilde{u}_{i,(N_i-1)/N_i})^*
\end{pmatrix}
$$

and

$$
\begin{pmatrix}
(\tilde{u}_{i,0/N_i})^* & 0 & \cdots & 0 & 0 \\
0 & (\tilde{u}_{i,1/N_i})^* & \cdots & 0 & 0 \\
\vdots & \vdots & \ddots & \vdots & \vdots \\
0 & 0 & \cdots & (\tilde{u}_{i,(N_i-1)/N_i})^* & 0 \\
0 & 0 & \ldots & 0 & 1_{M_{n_i}(A_i)^+}
\end{pmatrix}
$$

parametrized by $[1/2, 1]$. The second part of the homotopy consists of applying the homotopy $e(r_{i,t})^* r_{i,t}$ to the second factor of the product in (8.39). We can arrange that the Lipschitz constant $e(r_{i,t})^* r_{i,t}$ is bounded by 8π independently of i.

If the Lipschitz constant of $\tilde{u}_{i,t}$ was bounded from above by L_i, then the Lipschitz constant of $\hat{u}_{i,t}$ is bounded from above by $\max\{2L_i/N_i, 8\pi\}$.

We choose for each i in I a number N_i such that the resulting family of homotopies $(\hat{u}_{i,t})_{i \in I}$ has a uniform upper bound on their Lipschitz constants. For i in I we let v_i in $\mathcal{M}(M_{N_i n_i}(A))$ be the isometry from Definition 8.103. Then $\bar{u}_{i,t} := v_i(\hat{u}_{i,t} - 1_{M_{N_i n_i}(A_i)^+})v_i^* + 1_{A_i^+}$ defines a homotopy in the unitary group of A_i^+ connecting $v_{i,11}(u_i - 1_{A_i^+})v_{i,11}^* + 1_{A_i^+}$ to $1_{A_i^+}$. Here we use (8.36) and $v_{i,11}$ denotes the upper left corner of v_i. The family $(\bar{u}_{i,t})_{i \in I}$ of Lipschitz paths has a uniform upper bound on their Lipschitz constants. Then

$$\bar{U}_t := (\bar{u}_{i,t} - 1_{A_i^+})_{i \in I} + 1_{(\prod_{i \in I} A_i)^+}$$

is a homotopy in the unitary group of $(\prod_{i \in I} A_i)^+$ connecting

$$(v_{i,11})_{i \in I}(U - 1_{(\prod_{i \in I} A_i)^+})(v_{i,11})_{i \in I}^* + 1_{(\prod_{i \in I} A_i)^+}$$

to the identity $1_{(\prod_{i \in I} A_i)^+}$. Since $(v_{i,11})_{i \in I}$ is an isometry in the C^*-algebra $\prod_{i \in I} \mathcal{M}(A_i)$ which acts via multipliers on $\prod_{i \in I} A_i$ we have the first equality in

$$[U] = [(v_{i,11})_{i \in I}(U - 1_{(\prod_{i \in I} A_i)^+})(v_{i,11})_{i \in I}^* + 1_{(\prod_{i \in I} A_i)^+}] = 0. \qquad \square$$

Remark 8.109 In [WY20, Thm. 2.7.12] it is claimed that (8.38) is an isomorphism for $* \in \{0, 1\}$. The argument for surjectivity in the case $* = 0$ is sketched in the reference, while rest of the argument was left to the reader as a straightforward exercise. So Proposition 8.108 is a solution to this exercise in [WY20], in which the argument for the injectivity on K_1 turned out to be more complicated than expected. $\qquad \square$

For later use let us record the following special case where $A_i = \mathbb{K}$ for every i in I. Let I be any set.

Corollary 8.110 *The map* $(K^{C^*}(p_i))_{i \in I} : K^{C^*}(\prod_{i \in I} \mathbb{K}) \to \prod_{i \in I} K^{C^*}(\mathbb{K})$ *is an equivalence.*

Proof We note that $\mathbb{K}$ is quasi-$\mathbb{K}$-stable, see Example 8.105. $\qquad \square$

Remark 8.111 If one drops the assumption of quasi-$\mathbb{K}$-stability on the factors, then in general the morphism (8.35) is not an equivalence. An explicit counter-example is the case where $I := \mathbb{N}$ and $A_i := \mathbb{C}$ for each i in I (see [WY20, Ex. 2.11.15] or [HR00b, Ex. 7.7.3]). $\qquad \square$

Let X be a bornological coarse space, and let (H, ϕ) be an ample X-controlled Hilbert space.

Lemma 8.112 *The Roe algebra* $C^*(X, H, \phi)$ *is quasi-$\mathbb{K}$-stable.*

Proof By Lemma 8.21 we can assume that (H, ϕ) is of the form $(H' \otimes \ell^2, \phi' \otimes \mathrm{id}_{\ell^2})$ for some X-controlled Hilbert space (H', ϕ'). Then we can identify $M_n(C^*(X, H' \otimes \ell^2, \phi' \otimes \mathrm{id}_{\ell^2}))$ with $C^*(X, H' \otimes (\ell^2 \otimes \mathbb{C}^n), \phi' \otimes \mathrm{id}_{\ell^2 \otimes \mathbb{C}^n})$. We choose a unitary

$u : \ell^2 \otimes \mathbb{C}^n \to \ell^2$ and let $\iota : \ell^2 \to \ell^2 \otimes \mathbb{C}^n$ be the embedding induced by the first basis vector of $\mathbb{C}^n$. Then $v := \mathrm{id}_{H_D} \otimes (\iota \circ u)$ is an isometry which acts as a multiplier on $M_n(C^*(X, H' \otimes \ell^2, \phi' \otimes \mathrm{id}_{\ell^2}))$ and satisfies $vv^* = e_{1,1}$ as required. $\qquad \square$

Recall Proposition 8.20 which states that a bornological coarse space X admits an ample X-controlled Hilbert space if and only if it is locally countable. Let $(X_i)_{i \in I}$ be a family of locally countable bornological coarse spaces. Recall the Definition 2.27 of a free union of such a family. Note that $\bigsqcup_{i \in I}^{\mathrm{free}} X_i$ is again locally countable and therefore admits an ample $\bigsqcup_{i \in I}^{\mathrm{free}} X_i$-controlled Hilbert space (H, ϕ). Then for every i in I the X_i-controlled Hilbert space $(H(X_i), \phi_{X_i})$ is also ample, where $H(X_i)$ is the image of the projection $\phi(X_i) := \phi(\chi_{X_i})$, and ϕ_{X_i} is the control obtained from ϕ by restriction (see the proof of Proposition 8.82 for a similar construction).

Proposition 8.113 *We have an isomorphism of C^*-algebras*

$$C^*\left(\bigsqcup_{i \in I}^{\mathrm{free}} X_i, H, \phi\right) \cong \prod_{i \in I} C^*(X_i, H(X_i), \phi_{X_i}). \tag{8.40}$$

Proof Let $e_i : H(X_i) \to H$ denote the inclusion. We define a homomorphism

$$C^*\left(\bigsqcup_{i \in I}^{\mathrm{free}} X_i, H, \phi\right) \to \prod_{i \in I} C^*(X_i, H(X_i), \phi_{X_i})$$

such that it sends an operator A in $C^*\left(\bigsqcup_{i \in I}^{\mathrm{free}} X_i, H, \phi\right)$ to the family $(e_i^* A e_i)_{i \in I}$. Since A is bounded, this family is uniformly bounded. Since the components $(X_i)_{i \in I}$ are mutually coarsely disjoint in $\bigsqcup_{i \in I}^{\mathrm{free}} X_i$ we have $e_i^* A e_{i'} = 0$ if i, i' are in I with $i \neq i'$, and $e_i e_i^* A e_i = A e_i$ for every i in I. This implies that $A \mapsto (e_i^* A e_i)_{i \in I}$ is compatible with the product. It is also immediate from the formula that this map is compatible with the involution. If A is locally finite and has controlled propagation, then the operators $e_i^* A e_i$ on $(H(X_i), \phi_{X_i})$ are locally finite and have controlled propagation for all i in I. This implies that the map $A \mapsto (e_i^* A e_i)_{i \in I}$ has values in the product of Roe algebra.

The inverse sends a family $(A_i)_{i \in I}$ in $\prod_{i \in I} C^*(X_i, H(X_i), \phi_{X_i})$ to $\sum_{i \in I} e_i A_i e_i^*$, where the sum converges strongly. Since the family $(A_i)_{i \in I}$ is uniformly bounded and the family of inclusions $(e_i)_{i \in I}$ is mutually orthogonal the sum defines a bounded operator A, and the described map $(A_i)_{i \in I} \mapsto A$ is compatible with the composition and the involution. Again, if A_i has controlled propagation and is locally finite for every i, then A has controlled propagation and is locally finite. So the inverse also has values in the Roe algebra. $\qquad \square$

We will need the following special case of Proposition 8.113. Let I be a set, and let (H, ϕ) be an ample Hilbert space on $\bigsqcup_{i \in I}^{\mathrm{free}} *$.

Corollary 8.114 *We have an isomorphism* $C^*\left(\bigsqcup_{i\in I}^{\text{free}} *, H, \phi\right) \cong \prod_{i\in I} \mathbb{K}$.

Proof We let $*_i$ denote the point in the component with index i. We note that $H(*_i) \cong \ell^2$ for every i in I, and that

$$C^*(*_i, H(*_i), \phi_{*_i}) \cong \mathbb{K}. \tag{8.41}$$

$\square$

Corollary 8.115 *The functors* $K\mathcal{X}$ *and* $K\mathcal{X}_{\text{ql}}$ *are additive.*

Proof Since $K\mathcal{X} \to K\mathcal{X}_{\text{ql}}$ is an equivalence on discrete bornological coarse spaces by Lemma 8.87 it suffices to consider the case of $K\mathcal{X}$. We choose an ample Hilbert space (H, ϕ) on $\bigsqcup_{i\in I}^{\text{free}} *$. The chain of equivalences (which is a factorization of the map to be considered)

$$K\mathcal{X}\left(\bigsqcup_{i\in I}^{\text{free}} *\right) \overset{\text{Thm. 8.99}}{\simeq} K^{C^*}\left(C^*\left(\bigsqcup_{i\in I}^{\text{free}} *, H, \phi\right)\right)$$

$$\overset{\text{Cor. 8.114}}{\simeq} K^{C^*}\left(\prod_{i\in I} \mathbb{K}\right)$$

$$\overset{\text{Cor. 8.110}}{\simeq} \prod_{i\in I} K^{C^*}(\mathbb{K})$$

$$\overset{(8.41)}{\simeq} \prod_{i\in I} K^{C^*}(C^*(*_i, H(*_i), \phi_{*_i}))$$

$$\overset{\text{Thm. 8.99}}{\simeq} \prod_{i\in I} K\mathcal{X}(*_i)$$

proves the claim. $\square$

The following is a step towards strong additivity. Let $(X_i)_{i\in I}$ be a family of bornological coarse spaces.

Corollary 8.116 *If* X_i *is locally countable for every* i *in* I, *then the canonical morphism*

$$K\mathcal{X}\left(\bigsqcup_{i\in I}^{\text{free}} X_i\right) \to \prod_{i\in I} K\mathcal{X}(X_i) \tag{8.42}$$

is an equivalence.

Proof This follows from Proposition 8.113, Theorem 8.99, Proposition 8.108, and Lemma 8.112. $\square$

Remark 8.117 In [BEe] we will show that the coarse K-homology functor $K\mathcal{X}$ is actually strongly additive. Therefore the homomorphism (8.42) is an isomorphism for every family $(X_i)_{i \in I}$ of bornological coarse spaces. The proof does not use Roe algebras but rather works directly with the Roe categories and uses the fact shown in [BEa] that the K-theory functor for C^*-categories preserves products of additive C^*-categories. □

Example 8.118 In this example we show that for an infinite set I the product $\prod_{i \in I} \mathbb{K}$ is not $\mathbb{K}$-stable. In view of Corollary 8.114 this implies that the Roe algebra $C^*(X, H, \phi)$ associated to a bornological coarse space X and an ample X-controlled Hilbert space (H, ϕ) is in general not $\mathbb{K}$-stable.

Let I be an infinite set. In order to show that $\prod_{i \in I} \mathbb{K}$ is not $\mathbb{K}$-stable we use the following property of $\mathbb{K}$-stable C^*-algebras.

If A is $\mathbb{K}$-stable, then there is a sequence $\{E_n\}_{n \in \mathbb{N}}$ of mutually orthogonal, mutually equivalent projections in the multiplier algebra $\mathcal{M}(A)$ of A such that $\sum_{n=0}^{\infty} E_n = 1$, where the sum converges in the strict topology.[7]

We now show that $\prod_{i \in I} \mathbb{K}$ does not admit such a sequence. Assume that $\{E_n\}_{n \in \mathbb{N}}$ is such a family of projections. Then $E_n = (E_{i,n})_{i \in I}$ is a family of projections in $\mathbb{K}$ for every n in $\mathbb{N}$. Let $\kappa : \mathbb{N} \to I$ be an injective map. It exists since we assume that I is infinite. For every n in $\mathbb{N}$ we choose a non-zero projection P_n in $\mathbb{K}$ which is orthogonal to $E_{\kappa(n),m}$ for all m in $\mathbb{N}$ with $m \leq n$. We now define a projection $Q := (Q_i)_{i \in I}$ in $\prod_{i \in I} \mathbb{K}$, where $Q_i := P_n$ if $i = \kappa(n)$ for some (uniquely determined) n in $\mathbb{N}$, and $Q_i := 0$ else. We have $E_m Q = (E_{i,m} Q_i)_{i \in I}$. If $i = \kappa(n)$ for n in $\mathbb{N}$, then $E_{i,m} Q_i = 0$ for all m in $\mathbb{N}$ with $m \leq n$. Hence $\| \sum_{m=0}^{n} E_m Q - Q \| = 1$ for all n in $\mathbb{N}$. This is a contradiction to the condition $\sum_{n=0}^{\infty} E_n = 1$ in the strict topology. □

8.8.2 Coproducts

The goal of this subsection is to show that the functors $K\mathcal{X}$ and $K\mathcal{X}_{ql}$ preserve coproducts. We start with an explicit description of the coproduct $\mathbf{C} := \coprod_{i \in I} \mathbf{C}_i$ of a family $(\mathbf{C}_i)_{i \in I}$ of C^*-categories. The set of objects of $\mathbf{C}$ is given by $\bigsqcup_{i \in I} \mathrm{Ob}(\mathbf{C}_i)$. We write (c, i) for the object given by c in $\mathbf{C}_i$. The morphisms of $\mathbf{C}$ are given by

$$\mathrm{Hom}_{\mathbf{C}}((c, i), (c', i')) := \begin{cases} \mathrm{Hom}_{\mathbf{C}_i}(c, c') & i = i' \\ 0 & i \neq i' \end{cases}.$$

The composition and the involution are given in the obvious way.

---

[7] If A is σ-unital, then the converse also holds [Rør04, Thm. 2.2].

By an inspection of the definition of A (see e.g. (8.9)) we observe that have a canonical isomorphism

$$\bigoplus_{i \in I} A(\mathbf{C}_i) \cong A(\mathbf{C}) \tag{8.43}$$

of C^*-algebras.

By Definition 8.56, Proposition 8.55 and since K-theory of C^*-algebras sends direct sums to coproducts (Property 4 of K^{C^*}) we have a canonical equivalence

$$\bigoplus_{i \in I} \mathrm{K}^{C^*Cat}(\mathbf{C}_i) \simeq \mathrm{K}^{C^*Cat}(\mathbf{C}) . \tag{8.44}$$

Remark 8.119 Since A^f is a left adjoint it preserves colimits. The coproduct in C^*-algebras is the free product. Hence $A^f(\mathbf{C})$ is the free product of the C^*-algebras $A^f(\mathbf{C}_i)$ for i in I. Since K^{C^*} sends free products to sums this would also lead to the equivalence (8.44) avoiding the use of Proposition 8.55.   □

Proposition 8.120 *The functors $K\mathcal{X}$ and $K\mathcal{X}_{\mathrm{ql}}$ preserve coproducts.*

Proof We will only discuss the case of $K\mathcal{X}$ since the quasi-local case $K\mathcal{X}_{\mathrm{ql}}$ is analogous.

Let $(X_i)_{i \in I}$ be a family of bornological coarse spaces and set $X := \coprod_{i \in I} X_i$. We must show that the canonical morphism

$$\bigoplus_{i \in I} K\mathcal{X}(X_i) \to K\mathcal{X}(X)$$

is an equivalence. This morphism has the following factorization

$$\bigoplus_{i \in I} K\mathcal{X}(X_i) \quad \simeq \quad \bigoplus_{i \in I} \mathrm{K}^{C^*Cat}(\mathbf{C}^*(X_i))$$

$$\overset{(8.44)}{\simeq} \quad \mathrm{K}^{C^*Cat}(\coprod_{i \in I} \mathbf{C}^*(X_i))$$

$$\overset{(1)}{\simeq} \quad \mathrm{K}^{C^*Cat}(\coprod_{i \in I} \mathbf{C}^*(X_i)_\oplus)$$

$$\overset{(2)}{\simeq} \quad \mathrm{K}^{C^*Cat}(\mathbf{C}^*(X))$$

$$\simeq \quad K\mathcal{X}(X)$$

This first and the last equivalence follow from the definition of $K\mathcal{X}$. In order to finish the proof we must explain the equivalences marked by (1) and (2).

From the universal property of the coproduct of C^*-categories we get a functor

$$\coprod_{i \in I} \mathbf{C}^*(X_i) \to \mathbf{C}^*(X) \tag{8.45}$$

which sends for every i in I the X_i-controlled Hilbert space (H, ϕ) to the X-controlled Hilbert space $\iota_{i,*}(H, \phi)$, where $\iota_i : X_i \to X$ is the inclusion. This functor induces the composition $(2) \circ (1)$. It is not yet essentially surjective since its image contains only those X-controlled Hilbert spaces which are supported on a single component X_i. A general X-controlled Hilbert space (H, ϕ) is locally finite, i.e., $\mathrm{supp}(H, \phi)$ is a locally finite subset of X. Note that a subset B of X is bounded if and only if $B \cap X_i$ is bounded for every i in I. This implies that the set $\{i \in I : H(X_i) \not\cong 0\}$ is finite. In other words, (H, ϕ) is supported on finitely many of the components X_i.

In order to obtain these objects as well we extend the functor (8.45) to the additive completion of the coproduct (we use the model introduced by Davis–Lück [DL98, Sec. 2])

$$\left(\coprod_{i \in I} \mathbf{C}^*(X_i) \right)_{\oplus} \to \mathbf{C}^*(X). \tag{8.46}$$

This extension exists since $\mathbf{C}^*(X)$ admits finite sums. This extension (8.46) sends the finite family $((H_1, \phi_1), \ldots, (H_n, \phi_n))$ in $(\coprod_{i \in I} \mathbf{C}^*(X_i))_{\oplus}$ to the sum $(\bigoplus_{k=1}^{n} H_k, \bigoplus_{k=1}^{n} \phi_k)$ in $\mathbf{C}^*(X)$.

In order to define an inverse functor we choose an ordering on the index set I. Then we can define a functor

$$\mathbf{C}^*(X) \to \left(\coprod_{i \in I} \mathbf{C}^*(X_i) \right)_{\oplus} \tag{8.47}$$

which sends (H, ϕ) to the tuple $(H(X_i), \phi_{X_i})'$. Here the index $'$ indicates that the tuple is obtained from the infinite family $(H(X_i), \phi_{X_i})_{i \in I}$ by deleting all zero spaces and listing the remaining (finitely many members) in the order determined by the order on I. On morphisms these functors are defined in the obvious way. One now checks that (8.46) and (8.47) are inverse to each other equivalences of categories. Indeed, both are fully faithful and essentially surjective.

The equivalence of C^*-categories (8.46) induces the functor (2) which is then an equivalence by Corollary 8.63.

The functor (1) is induced by the canonical inclusion

$$\coprod_{i \in I} \mathbf{C}^*(X_i) \to \left(\coprod_{i \in I} \mathbf{C}^*(X_i) \right)_{\oplus}.$$

Now in general, for a C^*-category $\mathbf{C}$ we have an isomorphism $A(\mathbf{C}_\oplus) \simeq A(\mathbf{C}) \otimes \mathbb{K}$, and that the inclusion $\mathbf{C} \to \mathbf{C}_\oplus$ induces the homomorphism

$$A(\mathbf{C}) \to A(\mathbf{C}_\oplus) \cong A(\mathbf{C}) \otimes \mathbb{K}$$

given by the left-upper corner inclusion. In particular, it induces an equivalence in K-theory by the stability Property 6 of K^{C^*}. Using Proposition 8.55 we see that

$$\mathrm{K}^{C^*Cat}(\mathbf{C}) \to \mathrm{K}^{C^*Cat}(\mathbf{C}_\oplus) \tag{8.48}$$

is an equivalence. In particular, the morphism (1) is an equivalence.   □

Remark 8.121 The functors $K\mathcal{X}$ and $K\mathcal{X}_{\mathrm{ql}}$ have the stronger property called continuity discussed in [BEKW20, Sec. 5], see Remark 6.15. In order to verify this property one uses that the objects of $\mathbf{C}^*(X)$ and $\mathbf{C}^*_{\mathrm{ql}}(X)$ are supported on locally finite subsets of X. Then the argument is similar to the proof of [BEKW20, Prop. 8.17]. By [BEKW20, Lem. 5.17] a continuous coarse homology theory preserves coproducts. This would give an alternative proof of Proposition 8.120.
                                                                                      □

Remark 8.122 In [BEa] we will show that K^{C^*Cat} preserves Morita equivalences of C^*-categories. Because $\mathbf{C} \to \mathbf{C}_\oplus$ is a Morita equivalence, this immediately implies that $\mathrm{K}^{C^*Cat}(\mathbf{C}) \to \mathrm{K}^{C^*Cat}(\mathbf{C}_\oplus)$ is an equivalence.   □

8.9   Dirac Operators

The goal of this subsection is to explain how the coarse index class with support of a Dirac operator defined in [Roe16] is captured by the coarse K-homology theory $K\mathcal{X}$. It catches a glimpse of the theory developed in [Buna, BEd].

Let (M, g) be a complete Riemannian manifold. The Riemannian metric induces a metric on the underlying set of M and therefore (Example 2.18) a coarse and a bornological structure $\mathcal{C}_g$ and $\mathcal{B}_g$. We thus get a bornological coarse space M_g in **BornCoarse**. Let $S \to M$ be a Dirac bundle (Gromov–Lawson [GL83]) and denote the associated Dirac operator by $\slashed{D}$. In the following we discuss the index class of the Dirac operator in the K-theory of the Roe algebra.

The Weitzenböck formula

$$\slashed{D}^2 = \Delta + \mathcal{R} \tag{8.49}$$

expresses the square of the Dirac operator in terms of the connection Laplacian $\Delta := \nabla^*\nabla$ on S and a bundle endomorphism $\mathcal{R}$ in $C^\infty(M, \mathrm{End}(S))$.

For every m in M the value $\mathcal{R}(m)$ in $\mathrm{End}(S_m)$ is selfadjoint. For c in $\mathbb{R}$ we define the subset

$$M_c := \{m \in M : \mathcal{R}(m) \geq c\}$$

of points in M on which $\mathcal{R}(m)$ is bounded below by c.

Example 8.123 For example, if S is the spinor bundle associated to some spin structure on M, then

$$\mathcal{R} = \frac{s}{4}\mathrm{id}_S,$$

where s in $C^\infty(M)$ is the scalar curvature function, see Lawson–Michelsohn [LM89, Equation (8.18)], Lichnerowicz [Lic63] or Schrödinger [Sch32]. In this case M_c is the subset of M on which the scalar curvature is bounded below by $4c$. □

Since M is complete the Dirac operator $\slashed{D}$ and the connection Laplacian Δ are essentially selfadjoint unbounded operators on the Hilbert space $H := L^2(M, S)$ defined on the dense domain $C_c^\infty(M, S)$.

The Laplacian Δ is non-negative. So morally the formula (8.49) shows that $\slashed{D}^2$ is lower bounded by c on the submanifold M_c, and the restriction of $\slashed{D}$ to this subset is invertible and therefore has vanishing index. Roe [Roe16, Thm. 2.4] constructs a large scale index class of the Dirac operator which takes this positivity into account (the original construction without taking the positivity into account may be found in [Roe96]). Our goal in this section is to interpret the construction of Roe in the language of bornological coarse spaces.

We fix c in $\mathbb{R}$ with $c > 0$.

Definition 8.124 We define the big family $\mathcal{Y}_c := \{M \setminus M_c\}$ on M_g.

We use the construction explained in Remark 8.23 in order to turn $H := L^2(M, S)$ into an M-controlled Hilbert space (H, ϕ).

We equip every member Y of $\mathcal{Y}_c$ with the induced bornological coarse structure. For every member Y of $\mathcal{Y}_c$ the inclusion $(H(Y), \phi_Y) \hookrightarrow (H, \phi)$ provides an inclusion of Roe algebras (Definition 8.34) $C_{\mathrm{lc}}^*(Y, H(Y), \phi_Y) \hookrightarrow C_{\mathrm{lc}}(M, H, \phi)$. We define the C^*-algebra

$$C_{\mathrm{lc}}^*(\mathcal{Y}_c, H, \phi) := \operatorname*{colim}_{Y \in \mathcal{Y}_c} C_{\mathrm{lc}}^*(Y, H(Y), \phi_Y).$$

We will interpret $C_{\mathrm{lc}}^*(\mathcal{Y}_c, H, \phi)$ as a closed subalgebra of $C_{\mathrm{lc}}(M, H, \phi)$ obtained by forming the closure of the union of subalgebras $C_{\mathrm{lc}}^*(Y, H(Y), \phi_Y)$.

Remark 8.125 The C^*-algebra $C_{\mathrm{lc}}^*(\mathcal{Y}_c, H, \phi)$ has first been considered by Roe [Roe16]. In his notation it would have the symbol $C^*((M \setminus M_c) \subseteq M)$. □

We now recall the main steps of Roe's construction of the large scale index classes $\mathrm{index}(\not{D}, c)$ in $K_*(\mathcal{C}^*_{\mathrm{lc}}(\mathcal{Y}_c, H, \phi))$. To this end we consider the C^*-algebra $\mathcal{D}^*(M, H, \rho)$. Its definition uses the notion of pseudo-locality and depends on the representation by multiplication operators $\rho : C_0(M) \to B(H)$ of the C^*-algebra of continuous functions on M vanishing at ∞.

Definition 8.126 An operator A in $B(H)$ is pseudo-local, if $[A, \rho(f)]$ is a compact operator for every function f in $C_0(M)$.

Note that in this definition we can not replace ρ by ϕ. Pseudo-locality is a topological and not a bornological coarse concept. In order to control propagation and to define local compactness we could use ρ instead of ϕ, see Remark 8.43.

Definition 8.127 The C^*-algebra $\mathcal{D}^*(M, H, \rho)$ is defined as the closed subalgebra of $B(H)$ generated by all pseudo-local operators of controlled propagation.

Lemma 8.128 *We have an inclusion*

$$\mathcal{C}^*_{\mathrm{lc}}(\mathcal{Y}_c, H, \phi) \subseteq \mathcal{D}^*(M, H, \rho)$$

*as a closed two-sided *-ideal.*

Proof Recall that $\mathcal{C}^*_{\mathrm{lc}}(\mathcal{Y}_c, H, \phi)$ is generated by locally compact, bounded operators of controlled propagation which belong to $\mathcal{C}^*_{\mathrm{lc}}(Y, H(Y), \phi_Y)$ some Y in $\mathcal{Y}_c$. For short we say that such an operator is supported on Y.

A locally compact operator A is pseudo-local. Indeed, for f in $C_0(M)$ we can find a bounded subset B in $\mathcal{B}_g$ such that $\rho(f)\phi(B) = \rho(f) = \phi(B)\rho(f)$. But then the commutator

$$[A, \rho(f)] = A\phi(B)\rho(f) - \rho(f)\phi(B)A$$

is compact.

In order to show that $\mathcal{C}^*_{\mathrm{lc}}(\mathcal{Y}_c, H, \phi)$ is an ideal we consider generators A of $\mathcal{C}^*_{\mathrm{lc}}(\mathcal{Y}_c, H, \phi)$ supported on Y in $\mathcal{Y}_c$ and Q in $\mathcal{D}^*(M, H, \rho)$. In particular, both A and Q have controlled propagation.

We argue that $QA \in \mathcal{C}^*_{\mathrm{lc}}(\mathcal{Y}_c, H, \phi)$. The argument for AQ is analogous. It is clear that QA again has controlled propagation. Let $U := \mathrm{supp}(Q)$ be the propagation of Q (measured with the control ϕ). Then QA is supported on $U[Y]$ which is also a member of $\mathcal{Y}_c$.

For every bounded subset B of M_g we furthermore have the identity

$$\phi(B)QA = \phi(B)Q\phi(U^{-1}[B])A \, .$$

Since $U^{-1}[B]$ is bounded and A is locally compact we conclude that $\phi(B)QA$ is compact. Clearly also $QA\phi(B)$ is compact. This shows that QA is locally compact.

□

In order to define the index class we use the boundary operator in K-theory

$$\partial : K_{*+1}(\mathcal{D}^*(M, H, \rho)/\mathcal{C}^*_{lc}(\mathcal{Y}_c, H, \phi)) \to K_*(\mathcal{C}^*_{lc}(\mathcal{Y}_c, H, \phi)) .$$

associated to the short exact sequence of C^*-algebras

$$0 \to \mathcal{C}^*_{lc}(\mathcal{Y}_c, H, \phi) \to \mathcal{D}^*(M, H, \rho) \to \mathcal{D}^*(M, H, \rho)/\mathcal{C}^*_{lc}(\mathcal{Y}_c, H, \phi) \to 0 .$$

The value of $*$ in $\{0, 1\}$ will depend on whether the operator $\slashed{D}$ is graded or not.

We choose a function χ_c in $C^\infty(\mathbb{R})$ with the following properties:

1. $\mathrm{supp}(\chi_c^2 - 1) \subseteq [-c, c]$
2. $\lim_{t \to \pm\infty} \chi_c(t) = \pm 1$.

If $\psi : \mathbb{R} \to \mathbb{R}$ is a bounded measurable function on $\mathbb{R}$, then we can define the bounded operator $\psi(\slashed{D})$ using functional calculus. The construction of the large scale index class of $\slashed{D}$ depends on the following key result of Roe:

Lemma 8.129 ([Roe16, Lem. 2.3]) *If ψ in $C^\infty(\mathbb{R})$ is supported in $[-c, c]$, then we have $\psi(\slashed{D}) \in \mathcal{C}^*_{lc}(\mathcal{Y}_c, H, \phi)$.*

Note that $\chi_c^2 - 1$ is supported in $[-c, c]$. The lemma implies therefore that

$$[(1 + \chi_c(\slashed{D}))/2] \in K_0(\mathcal{D}^*(M, H, \rho)/\mathcal{C}^*_{lc}(\mathcal{Y}_c, H, \phi)) .$$

We then define the class

$$\partial[(1 + \chi_c(\slashed{D}))/2] \in K_1(\mathcal{C}^*_{lc}(\mathcal{Y}_c, H, \phi)) . \tag{8.50}$$

This class is independent of the choice of χ_c. Indeed, if $\tilde{\chi}_c$ is a different choice, then again by the lemma $(1 + \chi_c(\slashed{D}))/2$ and $(1 + \tilde{\chi}_c(\slashed{D}))/2$ represent the same class in the quotient $\mathcal{D}^*(M, H, \rho)/\mathcal{C}^*_{lc}(\mathcal{Y}_c, H, \phi)$.

The class (8.50) is the coarse index class in the case the Dirac operator $\slashed{D}$ is ungraded. In the case that S is graded, i.e., $S = S^+ \oplus S^-$, and the Dirac operator $\slashed{D}$ is an odd operator, i.e., $\slashed{D}^{\pm} : S^{\pm} \to S^{\mp}$, we choose a unitary $U : S^- \to S^+$ of controlled propagation. At this point, for simplicity we assume that M has no zero-dimensional components. Then the relevant controlled Hilbert spaces are ample and we can find such an operator U by Lemma 8.21. The above lemma implies that

$$[U \chi_c(\slashed{D}^+)] \in K_1(\mathcal{D}^*(M, H^+, \rho^+)/\mathcal{C}^*_{lc}(\mathcal{Y}_c, H^+, \phi^+)) ,$$

where $H^+ := L^2(M, S^+)$ and ρ^+, ϕ^+ are the restrictions of ρ, ϕ to H^+. We then define the class

$$\partial[U \chi_c(\slashed{D}^+)] \in K_0(\mathcal{C}^*_{lc}(\mathcal{Y}_c, H^+, \phi^+)) . \tag{8.51}$$

Again by the lemma, this class is independent of the choice of χ_c. It is also independent of the choice of U since two different choices will result in operators $U\chi_c(\not{D}^+)$ and $\tilde{U}\chi_c(\not{D}^+)$ such that their difference in the quotient algebra $U\chi_c(\not{D}^+)(\tilde{U}\chi_c(\not{D}^+))^* \sim U\tilde{U}^*$ comes from $\mathcal{D}^*(M, H^+, \rho^+)$. Hence they map to the same element under the boundary operator.

In the following we argue that the coarse index classes (8.50) and (8.51) can naturally be interpreted as coarse K-homology classes in $K\mathcal{X}_*(\mathcal{Y}_c)$. We also assume that M has no zero-dimensional components.

By an inspection of the construction in Example 8.23 we see that there is a cofinal subfamily of members Y of $\mathcal{Y}_c$ with the property that $(H(Y), \phi_Y)$ is ample. We will call such Y good. An arbitrary member may not be not good, since it may not contain enough of the evaluation points denoted by d_α in Example 8.23.

Every member of $\mathcal{Y}_c$ is a separable bornological coarse space. If the member Y is good, then by Proposition 8.40 the canonical inclusion is an equality

$$C^*(Y, H(Y), \phi_Y) = C^*_{\mathrm{lc}}(Y, H(Y), \phi_Y) \,.$$

In view of Theorem 8.88 we have a canonical isomorphism

$$K_*(C^*_{\mathrm{lc}}(Y, H(Y), \phi_Y)) \cong K\mathcal{X}_*(Y) \,.$$

Recall that we define

$$K\mathcal{X}(\mathcal{Y}_c) := \operatorname*{colim}_{Y \in \mathcal{Y}_c} K\mathcal{X}(Y) \,.$$

We can restrict the colimit to the cofinal subfamily of good members. Since taking C^*-algebra K-theory and homotopy groups commutes with filtered colimits of C^*-algebras we get the canonical isomorphism

$$K_*(C^*_{\mathrm{lc}}(\mathcal{Y}_c, H, \phi)) \cong K\mathcal{X}_*(\mathcal{Y}_c) \,. \tag{8.52}$$

Definition 8.130 The large scale index $\mathtt{index}_c(\not{D})$ in $K\mathcal{X}_*(\mathcal{Y}_c)$ of $\not{D}$ is defined by (8.50) in the ungraded case, resp. by (8.51) in the graded case, under the identification (8.52).

Remark 8.131 A much more detailed construction of this index class (even in the equivariant case) is discussed in [BEd]. In this paper we furthermore prove a relative index theorem and a compatibility with suspension. One of the goals of that paper is to connect the analytical constructions in [Zei16] with the coarse homotopy theory as developed in the present book. As the discussion above shows this is not completely trivial since the analytic representative of index class lives in the K-theory of a Roe algebra associated to an ample M-controlled Hilbert space, while the K-theory $K\mathcal{X}(M)$ is build from Roe algebras of objects in $\mathbf{C}^*(M)$ which are never ample by local finiteness. The bridge is provided by the comparison Theorem 8.88 and its functorial version Theorem 8.99. $\square$

These index classes are compatible for different choices of c. If c, c' are in $\mathbb{R}$ such that $0 < c < c'$, then $M_{c'} \subseteq M_c$. Consequently every member of $\mathcal{Y}_c$ is contained in some member of $\mathcal{Y}_{c'}$. We thus get a map

$$\iota : K\mathcal{X}(\mathcal{Y}_c) \to K\mathcal{X}(\mathcal{Y}_{c'}).$$

An inspection of the construction using the independence of the choice of χ_c discussed above one checks the relation

$$\iota_* \mathrm{index}_c(\not{D}) = \mathrm{index}_{c'}(\not{D}).$$

At the cost of losing some information we can also encode the positivity of $\not{D}$ in a bornology $\mathcal{B}_c$ on M.

Definition 8.132 We define $\mathcal{B}_c$ to be the family of subsets B of X such that $B \cap Y \in \mathcal{B}_g$ for every Y in $\mathcal{Y}_c$.

It is clear that $\mathcal{B}_g \subseteq \mathcal{B}_c$.

Lemma 8.133 $\mathcal{B}_c$ is a bornology on M which is compatible with the coarse structure $\mathcal{C}_g$.

Proof It is clear that $\mathcal{B}_c$ covers M, is closed under subsets and finite unions. Let B be in $\mathcal{B}_c$. For every entourage U of M and Y a member of $\mathcal{Y}_c$ we have

$$U[B] \cap Y \subseteq U[B \cap U^{-1}[Y]].$$

This subset of M belongs to $\mathcal{B}_g$, since $U^{-1}[Y] \in \mathcal{Y}_c$ and hence $B \cap U^{-1}[Y] \in \mathcal{B}_g$.
□

We consider the bornological coarse space $M_{g,c} := (M, \mathcal{C}_g, \mathcal{B}_c)$ in **BornCoarse**. If Y is a member of $\mathcal{Y}_c$ then it is considered as a bornological coarse space with the structures induced from M_g. By construction of $\mathcal{B}_c$ the inclusion $Y \to M_{g,c}$ is also proper and hence a morphism. Taking the colimit we get a morphism of spectra

$$\kappa : K\mathcal{X}(\mathcal{Y}_c) \to K\mathcal{X}(M_{g,c}).$$

We can consider the class

$$\mathrm{index}_M(\not{D}, c) := \kappa(\mathrm{index}(\not{D}, c))$$

in $K\mathcal{X}_*(M_{g,c})$.

Example 8.134 The Dirac operator $\not{D}$ is invertible at ∞, if there exists $c > 0$ such that $M \setminus M_c$ is compact. In this case $\mathcal{Y}_c = \mathcal{B}_g$ is the family of relatively compact subsets of M and $\mathcal{B}_c$ is the maximal bornology. Consequently, the map

$$p : M_{g,c} \to *$$

is a morphism in **BornCoarse** and therefore we can consider $p_* \mathrm{index}_M(\slashed{D}, c)$ in $K\mathcal{X}_*(*)$. If $\slashed{D}$ is invertible at ∞, then it is Fredholm and $p_* \mathrm{index}_M(\slashed{D}, c)$ is its Fredholm index.

Note that we recover the Fredholm index of $\slashed{D}$ by applying a morphism. This shows the usefulness of the category **BornCoarse**, where we can change the bornology and coarse structure quite independently from each other.   □

Remark 8.135 In [Buna] we discuss more aspects of index theory in realm of coarse homotopy theory including the construction of secondary invariants in [Buna] and boundary value problems.   □

8.10 K-Theoretic Coarse Assembly Map

In this section we discuss the construction of the coarse assembly map following Higson–Roe [HR95] and Roe–Siegel [RS12]. We will just explain its definition using the language developed in this book. We will explain why the comparison result Corollary 6.43 is not directly applicable in order to prove that it is an isomorphism under a finite asymptotic dimension assumption, i.e., the coarse Baum–Connes conjecture. In [BEb] we will provide a better construction of the assembly map which indeed allows to apply Corollary 6.43 in order to deduce the coarse Baum–Connes conjecture.

Let (Y, d) be a separable proper metric space, and let Y_d be the associated coarse bornological space. We let Y_t denote the underlying locally compact topological space of Y which we consider as a topological bornological space. Let furthermore (H, ϕ) be an ample Y_d-controlled Hilbert space, and let $\rho : C_0(Y_t) \to B(H)$ be a *-representation satisfying the three conditions mentioned in Remark 8.43. The notions of local compactness and controlled propagation for operators on H do not depend on using ρ or ϕ. We refer to the Example 8.23 which explains why we need these requirements and can not just assume that ϕ extends ρ.

We get an exact sequence of C^*-algebras

$$0 \to \mathcal{C}^*_{\mathrm{lc}}(Y_d, H, \phi) \to \mathcal{D}^*(Y_d, H, \rho) \to Q^*(Y_d, H, \rho) \to 0 \qquad (8.53)$$

defining the quotient $Q^*(Y_d, H, \rho)$, where the C^*-algebra in the middle is characterized in Definition 8.127.

Since (H, ρ) is ample, we have the Paschke duality isomorphism (marked by P)

$$K_{*+1}(Q^*(Y_d, H, \rho)) \overset{P}{\cong} KK_*(C_0(Y_t), \mathbb{C}) \cong K^{\mathrm{an,lf}}_*(Y_t), \qquad (8.54)$$

see [Pas81, HR00b, HR05, Hig95]. The analytic locally finite K-homology $K^{\mathrm{an,lf}}(Y_t)$ was introduced in Definition 7.66 (with the notation $K^{\mathrm{an,lf}}_{\mathbb{C}}$, but here we drop the subscript $\mathbb{C}$), and we use Lemma 7.57 and Proposition 7.62 (the fact that K^{an} is locally finite) for the second isomorphism.

The boundary map in K-theory associated to the exact sequence of C^*-algebras (8.53) therefore gives rise to a homomorphism

$$A : K_*^{an,lf}(Y_t) \overset{(8.54)}{\cong} K_{*+1}(Q^*(Y_d, H, \rho)) \overset{\partial}{\to} K_*(C_{lc}^*(Y_d, H, \phi))$$
$$\cong K\mathcal{X}_{lc,*}(Y_d) \cong K\mathcal{X}_*(Y_d),$$

where the last two isomorphisms are due to Theorem 8.88 and Corollary 8.96. It is further known that the homomorphism A is independent of the choice of the H, ϕ and ρ above [HRY93, Sec. 4]. Following [HR05, Def. 1.5] we adopt the following definition.

Let (Y, d) be a separable proper metric space.

Definition 8.136 The homomorphism

$$A : K_*^{an,lf}(Y_t) \to K\mathcal{X}_*(Y_d)$$

described above is called the analytic assembly map.

We now apply the above construction to the space of controlled probability measures $P_U(X)$ on a bornological coarse space X, see (6.6).

Let X be a bornological coarse space.

Definition 8.137 X has strongly locally bounded geometry if it has the minimal compatible bornology and for every entourage U of X every U-bounded subset of X is finite.

In contrast to the notion of "strongly bounded geometry" (Definition 7.75) we drop the condition of uniformity of the bound on the cardinalities of U-bounded subsets.

Assume that X is a bornological coarse space which has strongly locally bounded geometry. For every entourage U of X containing the diagonal we consider the space $P_U(X)$. Since X has strongly locally bounded geometry $P_U(X)$ is a locally finite simplicial complex. We equip the simplices of $P_U(X)$ with the spherical metric and $P_U(X)$ itself with the induced path-metric d (recall that points in different components have infinite distance). With this metric $(P_U(X), d)$ is a proper metric space and we denote by $P_U(X)_d$ the underlying bornological coarse space, and by $P_U(X)_t$ the underlying topological bornological space. The Dirac measures provide an embedding $X \to P_U(X)$. This map is an equivalence $X_U \to P_U(X)_d$ in **BornCoarse**, see Definition 3.14. We therefore get a map

$$A_U : K_*^{an,lf}(P_U(X)_t) \overset{A}{\to} K\mathcal{X}_*(P_U(X)_d) \overset{8.80}{\cong} K\mathcal{X}_*(X_U).$$

These maps are compatible with the comparison maps on domain and target associated to inclusions $U \subseteq U'$ of entourages. We now form the colimit over the entourages $\mathcal{C}$ of X and get the homomorphism

$$\mu : QK_*^{\mathrm{an,lf}}(X) \cong \operatorname*{colim}_{U \in \mathcal{C}} K_*^{\mathrm{an,lf}}(P_U(X)_t) \to \operatorname*{colim}_{U \in \mathcal{C}} K\mathcal{X}_*(X_U) \cong K\mathcal{X}_*(X),$$

where $QK^{\mathrm{an,lf}}$ is defined in Definition 7.44, the first isomorphism explained in Remark 7.45 and the second isomorphism follows from Proposition 8.86.

In the construction above we assumed that X is a bornological coarse space with strongly locally bounded geometry. The condition of strongly locally bounded geometry is not invariant under equivalences of bornological coarse spaces. For example, the inclusion $\mathbb{Z} \to \mathbb{R}$ is an equivalence, but $\mathbb{Z}$ has strongly locally bounded geometry, while $\mathbb{R}$ has not.

Let X be a bornological coarse space.

Definition 8.138 X has locally bounded geometry if it is equivalent to a bornological coarse space of strongly locally bounded geometry.

Note that bounded geometry (Definition 7.77) implies locally bounded geometry.

We now observe that the domain (see Proposition 7.46) and target of the homomorphism μ are coarsely invariant. By naturality we can therefore define the homomorphism μ for all X of locally bounded geometry using a locally finite approximation $X' \to X$

$$\mu : QK_*^{\mathrm{an,lf}}(X) \cong QK_*^{\mathrm{an,lf}}(X') \to K\mathcal{X}_*(X') \cong K\mathcal{X}_*(X).$$

Let X be a bornological coarse space of locally bounded geometry.

Definition 8.139 The homomorphism

$$\mu : QK_*^{\mathrm{an,lf}}(X) \to K\mathcal{X}_*(X)$$

described above is called the K-theoretic coarse assembly map.

Remark 8.140 The notation $QK_*^{\mathrm{an,lf}}(X)$ for the domain of the assembly map looks complicated and this is not an accident. This coarse homology group is defined using a complicated procedure starting with functional analytic data and using the homotopy theoretic machine of locally finite homology theories in order to produce the functor $K^{\mathrm{an,lf}}$ which we then feed into the coarsification machine introduced in Definition 7.44. The construction of the assembly map heavily depends on the analytic picture, in particular on Paschke duality and the boundary operator in K-theory.

The spaces $P_U(X)$ which occur during the construction of the assembly map are all locally finite-dimensional simplicial complexes. So by Corollary 7.65 we could replace $QK^{\mathrm{an,lf}}(X)$ by $Q(KU \wedge \Sigma_+^\infty)^{\mathrm{lf}}$ if we wished. $\square$

Remark 8.141 One could ask wether the comparison result Corollary 6.43 implies the coarse Baum–Connes conjecture stating that the coarse assembly map μ is an isomorphism, if X is a bornological coarse space of bounded geometry of weakly finite asymptotic dimension.

Unfortunately, Corollary 6.43 does not directly apply to the coarse assembly map μ.

First of all μ is not defined for all bornological coarse spaces. Furthermore it is not a transformation between spectrum-valued functors.

The first problem can easily be circumvented by replacing the category **BornCoarse** by its full subcategory of bornological coarse spaces of locally bounded geometry everywhere and applying a corresponding version of Corollary 6.43.

The second problem is more serious. One would need to refine the constructions leading to the map A (Definition 8.136) to the spectrum level.

In [BEb] we will study the construction of spectrum-valued assembly maps in a systematic manner. It will turn out that in general we must modify the construction of the domain of the assembly map. □

Remark 8.142 There are examples of spaces X with bounded geometry such that the K-theoretic coarse assembly map $\mu : QK_*^{\mathrm{an,lf}}(X) \to K\mathcal{X}_*(X)$ is not surjective, i.e., these spaces are counter-examples to the coarse Baum–Connes conjecture. For the construction of these examples see Higson [Hig99] and Higson–Lafforgue–Skandalis [HLS02].

Composing the K-theoretic coarse assembly map with the natural transformation $K\mathcal{X} \to K\mathcal{X}_{\mathrm{ql}}$ we get the quasi-local version

$$\mu_{\mathrm{ql}} : QK_*^{\mathrm{an,lf}}(X) \to K\mathcal{X}_{ql,*}(X)$$

of the coarse assembly map. The natural question is now whether the above surjectivity counter-examples still persist in the quasi-local case.

We were not able to adapt Higson's arguments to the quasi-local case, i.e., the question whether we do have surjectivity counter-examples to the quasi-local version of the coarse Baum–Connes conjecture is currently open. □

References

[ALR03] J. Adámek, F.W. Lawvere, J. Rosický, Continuous categories revisited. Theory Appl. Cat. **11**(11), 252–282 (2003)

[Bar16] A. Bartels, On proofs of the Farrell-Jones conjecture, in *Topology and Geometric Group Theory*, vol. 184 of *Springer Proc. Math. Stat.* (Springer, Cham, 2016), pp. 1–31

[BCa] U. Bunke, L. Caputi, Controlled objects as a symmetric monoidal functor. arXiv:1902.03053

[BCb] U. Bunke, L. Caputi, Localization for coarse homology theories. arXiv:1902.04947

[BC20] U. Bunke, D.-Ch. Cisinski, A universal coarse K-theory. New York J. Math. **26**, 1–27 (2020). arXiv:1705.05080

[BCKW] U. Bunke, D.-Ch. Cisinski, D. Kasprowski, Ch. Winges, Controlled objects in left-exact ∞-categories and the Novikov conjecture. arXiv:1911.02338

[BEa] U. Bunke, A. Engel, Additive C^*-categories and K-theory. In preparation

[BEb] U. Bunke, A. Engel, Coarse assembly maps. arXiv:1706.02164

[BEc] U. Bunke, A. Engel, Coarse cohomology theories. arXiv:1711.08599

[BEd] U. Bunke, A. Engel, The coarse index class with support. arXiv:1706.06959

[BEe] U. Bunke, A. Engel, Topological equivariant coarse K-homology and injectivity of assembly maps. In preparation

[BEKWa] U. Bunke, A. Engel, D. Kasprowski, Ch. Winges, Injectivity results for coarse homology theories. Proc. London Math. Soc. (to appear) arXiv:1809.11079

[BEKWb] U. Bunke, A. Engel, D. Kasprowski, Ch. Winges, Transfers in coarse homology. Münster J. Math. (to appear) arXiv:1809.08300

[BEKW19] U. Bunke, A. Engel, D. Kasprowski, Ch. Winges, Coarse homology theories and finite decomposition complexity. Alg. Geom. Topol. **19**(6), 3033–3074 (2019). arXiv:1712.06932

[BEKW20] U. Bunke, A. Engel, D. Kasprowski, Ch. Winges, Equivariant coarse homotopy theory and coarse algebraic K-homology. Contemp. Math. **749**, 13–194 (2020). arXiv:1710.04935

[BG] U. Bunke, D. Gepner, Differential function spectra, the differential Becker-Gottlieb transfer, and applications to differential algebraic K-theory. arXiv:1306.0247

[BGT13] A.J. Blumberg, D. Gepner, G. Tabuada, A universal characterization of higher algebraic K-theory. Geom. Topol. **17**(2), 733–838 (2013)

[BJM17] I. Barnea, M. Joachim, S. Mahanta, Model structure on projective systems of C^*-algebras and bivariant homology theories. New York J. Math. **23**, 383–439 (2017)

© The Editor(s) (if applicable) and The Author(s), under exclusive licence to Springer Nature Switzerland AG 2020
U. Bunke, A. Engel, *Homotopy Theory with Bornological Coarse Spaces*, Lecture Notes in Mathematics 2269, https://doi.org/10.1007/978-3-030-51335-1

[BKWa] U. Bunke, D. Kasprowski, Ch. Winges, Assembly maps and coarse homology theories. In preparation

[BKWb] U. Bunke, D. Kasprowski, Ch. Winges, Split injectivity of A-theoretic assembly maps. Int. Math. Res. Not. IMRN (to appear). arXiv:1811.11864

[BL11] A. Bartels, W. Lück, The Farrell-Hsiang method revisited. Math. Ann. **354**, 209–226 (2011). arXiv:1101.0466

[Bla98] B. Blackadar, *K-Theory for Operator Algebras*, 2nd edn. (Cambridge University Press, Cambridge, 1998)

[BLR08] A. Bartels, W. Lück, H. Reich, The *K*-theoretic Farrell–Jones conjecture for hyperbolic groups. Invent. math. **172**, 29–70 (2008)

[BNV16] U. Bunke, Th. Nikolaus, M. Völkl, Differential cohomology theories as sheaves of spectra. J. Homotopy Relat. Struct. **11**(1), 1–66 (2016)

[Bro82] K.S. Brown, *Cohomology of Groups*, vol. 87 of *Graduate Texts in Mathematics* (Springer, 1982)

[Buna] U. Bunke, . Coarse homotopy theory and boundary value problems. arXiv:1806.03669

[Bunb] U. Bunke, Non-unital C^*-categories, (co)limits, crossed products and exactness. In preparation

[Bun19] U. Bunke, Homotopy theory with *-categories. Theory Appl. Cat. **34**(27), 781–853 (2019)

[BW92] J. Block, S. Weinberger, Aperiodic tilings, positive scalar curvature, and amenability of spaces. J. Am. Math. Soc. **5**(4), 907–918 (1992)

[Cap] L. Caputi, Cyclic homology for bornological coarse spaces, PhD-thesis, University of Regensburg 2019. arXiv:1907.02849

[CDV14] M. Cencelj, J. Dydak, A. Vavpetič, Coarse amenability versus paracompactness. J. Topol. Anal. **6**(1), 125–152 (2014). arXiv:math/1208.2864v3

[Cis19] D.C. Cisinski, *Higher Categories and Homotopical Algebra*, vol. 180 of *Cambridge Studies in Advanced Mathematics* (Cambridge University Press, Cambridge, 2019). Available online under http://www.mathematik.uni-regensburg.de/cisinski/CatLR.pdf

[CP98] G. Carlsson, E.K. Pedersen, Čech homology and the Novikov conjectures for *K*- and *L*-theory. Math. Scand. **82**(1), 5–47 (1998)

[DG07] M. Dadarlat, E. Guentner, Uniform embeddability of relatively hyperbolic groups. J. Reine Angew. Math. **612**, 1–15 (2007)

[DL98] J.F. Davis, W. Lück, Spaces over a category and assembly maps in isomorphism conjectures in *K*- and *L*-theory. *K*-Theory **15**, 201–252 (1998)

[Dyd16] J. Dydak, Coarse amenability and discreteness. J. Aust. Math. Soc. **100**, 65–77 (2016). arXiv:math/1307.3943v2

[EM06] H. Emerson, R. Meyer, Dualizing the coarse assembly map. J. Inst. Math. Jussieu **5**(2), 161–186 (2006)

[Ger93] S.M. Gersten, Isoperimetric functions of groups and exotic cohomology, in *Combinatorial and Geometric Group Theory*, vol. 204 of *London Mathematical Society Lecture Note Series*, pp. 87–104, 1993

[GL83] M. Gromov, H.B. Lawson, Jr., Positive scalar curvature and the Dirac operator on complete Riemannian manifolds. Publ. Math. IHÉS **58**(1), 83–196 (1983)

[Gro93] M. Gromov, Asymptotic invariants of infinite groups, in G.A. Niblo, M.A. Roller (eds.), *Geometric Group Theory*, vol. 182 of *London Mathematical Society Lecture Note Series*, 1993

[GTY12] E. Guentner, R. Tessera, G. Yu, A notion of geometric complexity and its application to topological rigidity. Invent. Math. **189**(2), 315–357 (2012)

[Hei] D. Heiss, Generalized bornological coarse spaces and coarse motivic spectra. arXiv:1907.03923

[Hig95] N. Higson, C^*-algebra extension theory and duality. J. Funct. Anal. **129**, 349–363 (1995)

[Hig99] N. Higson, Counterexamples to the coarse Baum–Connes conjecture. Unpublished. http://www.personal.psu.edu/ndh2/math/Unpublished_files/, 1999

[HLS02] N. Higson, V. Lafforgue, G. Skandalis, Counterexamples to the Baum–Connes conjecture. Geom. Funct. Anal. **12**, 330–354 (2002)

[Hoy17] M. Hoyois, The six operations in equivariant motivic homotopy theory. Adv. Math. **305**, 197–279 (2017). arXiv:1509.02145

[HP04] I. Hambleton, E.K. Pedersen, Identifying assembly maps in K- and L-theory. Math. Ann. **328**, 27–57 (2004)

[HPR96] N. Higson, E.K. Pedersen, J. Roe, C^*-algebras and controlled topology. K-Theory **11**, 209–239 (1996)

[HR95] N. Higson, J. Roe, On the coarse Baum–Connes conjecture, in *Novikov conjectures, index theorems and rigidity, Vol. 2*, ed. by S.C. Ferry, A. Ranicki, J. Rosenberg. London Mathematical Society Lecture Notes, vol. 227 (Cambridge University Press, Cambridge, 1995)

[HR00a] N. Higson, J. Roe, Amenable group actions and the Novikov conjecture. J. Reine Angew. Math. **519**, 143–153 (2000)

[HR00b] N. Higson, J. Roe, *Analytic K-Homology* (Oxford University Press, Oxford, 2000)

[HR05] N. Higson, J. Roe, Mapping surgery to analysis. III. Exact sequences. K-Theory **33**(4), 325–346 (2005)

[HRY93] N. Higson, J. Roe, G. Yu, A coarse Mayer–Vietoris principle. Math. Proc. Camb. Phil. Soc. **114**, 85–97 (1993)

[JJ06] M. Joachim, M.W. Johnson, Realizing Kasparov's KK-theory groups as the homotopy classes of maps of a Quillen model category. Contemp. Math. **399**, 163–198 (2006). arXiv:0705.1971

[Joa03] M. Joachim, K-homology of C^*-categories and symmetric spectra representing K-homology. Math. Ann. **327**, 641–670 (2003)

[Joy08] A. Joyal, The Theory of Quasi-Categories and its Applications. Online available at http://mat.uab.cat/~kock/crm/hocat/advanced-course/Quadern45-2.pdf, 2008

[Kle01] J.R. Klein, The dualizing spectrum of a topological group. Math. Ann. **319**(3), 421–456 (2001)

[Lic63] A. Lichnerowicz, Spineurs harmoniques. C. R. Acad. Sci. Paris **257**, 7–9 (1963)

[LM89] H.B. Lawson, Jr., M.-L. Michelsohn, *Spin Geometry* (Princeton University Press, 1989)

[LN18] M. Land, Th. Nikolaus, On the relation between K- and L-theory of C^*-algebras. Math. Ann. **371**, 517–563 (2018)

[LR85] B.V. Lange, V.S. Rabinovich, Noether property for multidimensional discrete Convolution operators. Math. Notes Acad. Sci. USSR **37**(3), 228–237 (1985). Translated from Matematicheskie Zametki **37**(3), 407–421 (1985)

[Lur09] J. Lurie, *Higher Topos Theory*, vol. 170 of *Annals of Mathematics Studies* (Princeton University Press, Princeton, NJ, 2009)

[Lur17] J. Lurie, Higher algebra. Available at www.math.harvard.edu/lurie, 2017

[Mah15] S. Mahanta, Noncommutative stable homotopy and stable infinity categories. J. Topol. Anal. **7**(1), 135–165 (2015)

[Mit01] P.D. Mitchener, Coarse homology theories. Alg. Geom. Topol. **1**, 271–297 (2001)

[Mit02] P.D. Mitchener, C^*-categories. Proc. Lond. Math. Soc. **84**, 375–404 (2002)

[Mit10] P.D. Mitchener, The general notion of descent in coarse geometry. Alg. Geom. Topol. **10**, 2419–2450 (2010)

[MV99] F. Morel, V. Voevodsky, $\mathbb{A}^1$-homotopy theory of schemes. Publications Mathématiques de l'Institut des Hautes Études Scientifiques **90**, 45–143 (1999)

[NY12] P.W. Nowak, G. Yu, *Large Scale Geometry* (EMS, 2012)

[Pas81] W.L. Paschke, K-Theory for Commutants in the Calkin Algebra. Pacific J. Math. **95**(2), 427–434 (1981)

[Roe93a] J. Roe, Coarse cohomology and index theory on complete Riemannian manifolds. Mem. Am. Math. Soc. **104**(497), x+90 (1993)

[Roe93b] J. Roe, Coarse cohomology and index theory on complete Riemannian manifolds. Memoirs Am. Math. Soc. **104**(497), 1–90 (1993)

[Roe96] J. Roe, *Index Theory, Coarse Geometry, and Topology of Manifolds*, vol. 90 of *CBMS Regional Conference Series in Mathematics* (AMS, 1996)

[Roe03] J. Roe, *Lectures on Coarse Geometry*, vol. 31 of *University Lecture Series* (American Mathematical Society, 2003)

[Roe16] J. Roe, Positive curvature, partial vanishing theorems and coarse indices. Proc. Edinburgh Math. Soc. **59**, 223–233 (2016)

[Rør04] M. Rørdam, Stable C^*-algebras, in *Operator Algebras and Applications*, ed. by H. Kosaki. Adv. Stud. Pure Math., vol. 38 (Math. Soc. Japan, 2004), pp. 177–199

[RS12] J. Roe, P. Siegel, Sheaf theory and Paschke duality. arXiv:math/1210.6420, 2012

[RTY14] D.A. Ramras, R. Tessera, G. Yu, Finite decomposition complexity and the integral Novikov conjecture for higher algebraic K-theory. J. Reine Angew. Math. **694**, 129–178 (2014)

[Sch32] E. Schrödinger, Reflections and spinors on manifolds. Sitzungsber. Preuss. Akad. Wissen. Phys.–Math. **11**, 105–128 (1932)

[ŠT19] J. Špakula, A. Tikuisis, Relative commutant pictures of Roe algebras. Commun. Math. Phys. **365**, 1019–1048 (2019)

[ŠŽ20] J. Špakula, J. Zhang, Quasi-locality and property A. J. Funct. Anal. **278**(1), (2020)

[Tu01] J.-L. Tu, Remarks on Yu's 'Property A' for discrete metric spaces and groups. Bull. Soc. Math. France **129**(1), 115–139 (2001)

[Wil09] R. Willett, Some notes on property A, in *Limits of Graphs in Group Theory and Computer Science*, ed. by G. Arzhantseva, A. Valette (EPFL Press, 2009), pp. 191–281

[Wri05] N.J. Wright, The coarse Baum-Connes conjecture via C_0 coarse geometry. J. Funct. Anal. **220**(2), 265–303 (2005)

[WW95] M. Weiss, B. Williams, Pro-excisive functors, in *Novikov Conjectures, Index Theorems and Rigidity, Vol. 2*, ed. by S.C. Ferry, A. Ranicki, J. Rosenberg. LMS Lecture Notes, vol. 227 (Cambridge University Press, 1995)

[WY20] R. Willett, G. Yu, *Higher Index Theory*. Cambridge Studies in Advanced Mathematics, vol. 189 (Cambridge University Press, 2020)

[Yu95] G. Yu, Baum–Connes conjecture and coarse geometry. *K*-Theory **9**, 223–231 (1995)

[Yu98] G. Yu, The Novikov conjecture for groups with finite asymptotic dimension. Ann. Math. **147**, 325–355 (1998)

[Yu00] G. Yu, The coarse Baum–Connes conjecture for spaces which admit a uniform embedding into Hilbert space. Invent. Math. **139**(1), 201–240 (2000)

[Zei16] R. Zeidler, Positive scalar curvature and product formulas for secondary index invariants. J. Topol. **9**(3), 687–724 (2016)

Index

U. Bunke, A. Engel, *Homotopy Theory with Bornological Coarse Spaces*,
Lecture Notes in Mathematics 2269, https://doi.org/10.1007/978-3-030-51335-1

LECTURE NOTES IN MATHEMATICS   🐎 Springer

Editors in Chief: J.-M. Morel, B. Teissier;

Editorial Policy

1. Lecture Notes aim to report new developments in all areas of mathematics and their applications – quickly, informally and at a high level. Mathematical texts analysing new developments in modelling and numerical simulation are welcome.

   Manuscripts should be reasonably self-contained and rounded off. Thus they may, and often will, present not only results of the author but also related work by other people. They may be based on specialised lecture courses. Furthermore, the manuscripts should provide sufficient motivation, examples and applications. This clearly distinguishes Lecture Notes from journal articles or technical reports which normally are very concise. Articles intended for a journal but too long to be accepted by most journals, usually do not have this "lecture notes" character. For similar reasons it is unusual for doctoral theses to be accepted for the Lecture Notes series, though habilitation theses may be appropriate.

2. Besides monographs, multi-author manuscripts resulting from SUMMER SCHOOLS or similar INTENSIVE COURSES are welcome, provided their objective was held to present an active mathematical topic to an audience at the beginning or intermediate graduate level (a list of participants should be provided).

   The resulting manuscript should not be just a collection of course notes, but should require advance planning and coordination among the main lecturers. The subject matter should dictate the structure of the book. This structure should be motivated and explained in a scientific introduction, and the notation, references, index and formulation of results should be, if possible, unified by the editors. Each contribution should have an abstract and an introduction referring to the other contributions. In other words, more preparatory work must go into a multi-authored volume than simply assembling a disparate collection of papers, communicated at the event.

3. Manuscripts should be submitted either online at www.editorialmanager.com/lnm to Springer's mathematics editorial in Heidelberg, or electronically to one of the series editors. Authors should be aware that incomplete or insufficiently close-to-final manuscripts almost always result in longer refereeing times and nevertheless unclear referees' recommendations, making further refereeing of a final draft necessary. The strict minimum amount of material that will be considered should include a detailed outline describing the planned contents of each chapter, a bibliography and several sample chapters. Parallel submission of a manuscript to another publisher while under consideration for LNM is not acceptable and can lead to rejection.

4. In general, **monographs** will be sent out to at least 2 external referees for evaluation.

   A final decision to publish can be made only on the basis of the complete manuscript, however a refereeing process leading to a preliminary decision can be based on a pre-final or incomplete manuscript.

   Volume Editors of **multi-author works** are expected to arrange for the refereeing, to the usual scientific standards, of the individual contributions. If the resulting reports can be forwarded to the LNM Editorial Board, this is very helpful. If no reports are forwarded or if other questions remain unclear in respect of homogeneity etc, the series editors may wish to consult external referees for an overall evaluation of the volume.

5. Manuscripts should in general be submitted in English. Final manuscripts should contain at least 100 pages of mathematical text and should always include

   – a table of contents;
   – an informative introduction, with adequate motivation and perhaps some historical remarks: it should be accessible to a reader not intimately familiar with the topic treated;
   – a subject index: as a rule this is genuinely helpful for the reader.
   – For evaluation purposes, manuscripts should be submitted as pdf files.

6. Careful preparation of the manuscripts will help keep production time short besides ensuring satisfactory appearance of the finished book in print and online. After acceptance of the manuscript authors will be asked to prepare the final LaTeX source files (see LaTeX templates online: https://www.springer.com/gb/authors-editors/book-authors-editors/manuscriptpreparation/5636) plus the corresponding pdf- or zipped ps-file. The LaTeX source files are essential for producing the full-text online version of the book, see http://link.springer.com/bookseries/304 for the existing online volumes of LNM). The technical production of a Lecture Notes volume takes approximately 12 weeks. Additional instructions, if necessary, are available on request from lnm@springer.com.

7. Authors receive a total of 30 free copies of their volume and free access to their book on SpringerLink, but no royalties. They are entitled to a discount of 33.3 % on the price of Springer books purchased for their personal use, if ordering directly from Springer.

8. Commitment to publish is made by a *Publishing Agreement*; contributing authors of multiauthor books are requested to sign a *Consent to Publish form*. Springer-Verlag registers the copyright for each volume. Authors are free to reuse material contained in their LNM volumes in later publications: a brief written (or e-mail) request for formal permission is sufficient.

Addresses:
Professor Jean-Michel Morel, CMLA, École Normale Supérieure de Cachan, France
E-mail: moreljeanmichel@gmail.com

Professor Bernard Teissier, Equipe Géométrie et Dynamique,
Institut de Mathématiques de Jussieu – Paris Rive Gauche, Paris, France
E-mail: bernard.teissier@imj-prg.fr

Springer: Ute McCrory, Mathematics, Heidelberg, Germany,
E-mail: lnm@springer.com

Printed in the United States
By Bookmasters